Polymers
Properties and Applications

3

A. Knop · W. Scheib

Chemistry and Application of Phenolic Resins

With 111 Figures

Springer-Verlag Berlin Heidelberg GmbH 1979

Dr. ANDRE KNOP, Bakelite GmbH, Iserlohn-Letmathe, Germany
Dr. WALTER SCHEIB, Prien am Chiemsee, Germany

This volume continues the series *Chemie, Physik und Technologie der Kunststoffe in Einzeldarstellungen*, which is now entitled *Polymers/ Properties and Applications*.

ISBN 978-3-540-09051-9   ISBN 978-3-662-11309-7 (eBook)
DOI 10.1007/978-3-662-11309-7

Library of Congress Cataloging in Publication Data:
Knop, Andre, 1941 – Chemistry and application of phenolic resins. (Polymers; 3). Includes bibliographical references and index. 1. Phenolic resins. I. Scheib, W., 1908 – joint author. II. Title. III. Series.
TP 1180.P 39 K 56   668.4'22   78-10967

Originally published by Springer-Verlag Berlin Heidelberg New York in 1979
Softcover reprint of the hardcover 1st edition 1976

Typesetting and printing: Schwetzinger Verlagsdruckerei GmbH, Schwetzingen.
Bookbinding: Konrad Triltsch, Graphischer Betrieb, Würzburg.
2152/3020-543210

Preface

"The sciences as a whole are slowly
but gradually drifting away from life
and are only returning after a detour".

Goethe

Detours should be avoided. The picture we are presenting here of the current theory in phenolic resin chemistry and the technical application of phenolic resins is based on day-to-day experiences in research, production and marketing, however, with the background of economic relevance.

This book, then, is not to be regarded as a systematic collection and evaluation of the literature, although the literature up to July, 1978 has generally been taken into consideration.

The audience to which this book is directed is wide-ranging: chemists, engineers, marketing professionals and students.

We show where the first fully synthetic polymers, phenolic resins, stand today and what their future is. Taking a look back over their development, one is only more deeply convinced that after a wide variety of adaptions, they still possess the technical and economic strengths which allow for their further market growth and with it, a full appreciation of their value.

We would like to extend our gratitude to all friends and promotors, in particular to those who helped and encouraged us with advice and assistance.

Frankfurt, January 1979

Andre Knop   Walter Scheib

Table of Contents

Abbreviations

ASTM American Society of Testing and Materials
BHMP Bis(hydroxymethyl)phenol, dimethylolphenol
BP Boiling point
BR Butyl rubber
CHP Cumene hydroperoxide
COD Chemical oxygen demand
CR Chloroprene rubber
DIN Deutsche Industrie Normen
DMF Dimethylformamide
DPM Diphenylmethane
DSC Differential scanning calorimetry
DTA Differential thermal analysis
EPA Environmental Protection Agency
F Formaldehyde
FA Furfuryl alcohol
FAO Food and Agriculture Organization
FDA Food and Drug Administration
GC Gas chromatography
GPC Gel permeability chromatography
HPL High-pressure laminate
HPLC High-performance liquid chromatography
HMP Hydroxymethylphenol
HMTA Hexamethylenetetramine
IB Internal bond (= tensile strength vertical to the surface)
IR Polyisobutylene rubber
LC_{50} Medium lethal concentration
MAK Maximum workplace exposure, 8 h
MOE Modulus of elasticity
MP Melting point
MW Molecular weight
MWD Molecular weight distribution
NBR Nitrile butadiene rubber
NMR Nuclear magnetic resonance
OSHA Occupational Health and Safety Administration
P Phenol
PB Particle board
pbw Parts by weight
PF Phenol-formaldehyde resin
PS Patent Specification

PVAc	Polyvinyl acetate
PVB	Polyvinyl butyral
PVF	Polyvinyl formal
Py	Pyridine
RH	Relative humidity
SG	Specific gravity
TGA	Thermogravimetric analysis
THMP	Tris(hydroxymethyl)phenol, Trimethylolphenol
UF	Urea-formaldehyde resin
UMP	Urea-melamine-phenol resin
VC	Volatiles content

Units

Force	1 kp	$= 9{,}80665$ N ≈ 10 N
Mechanical tension	1 kp/cm^2	$= 0{,}0981$ N/mm^2 $\approx 0{,}1$ N/mm^2
	1 N/mm^2	= 145 psi
Pressure	1 at	= 1 kp/cm^2 = 0,980665 bar $\approx$ 1 bar
	1 Pa	= 10^{-5} bar
Temperature	°F	= °C · 1,8 + 32
Dynamic viscosity	1 cP	= 1 mPa. s
Heat quantity	1 kcal	= 4,187 kJ
Thermal conductivity	1 W/Km	= 0,86 kcal/m h °C
	1 W/Km	= 0,579 BTU/ft h °F
	1 W/Km	= 6,95 BTU in/ft^2 h
Length	1 mm	= 0,0394 in
	1 m	= 3,2808 ft
Area	1 mm^2	= 0,0016 sq in
	1 m^2	= 10,764 sq ft
Mass	1 kg	= 2,2046 lb
Density	1 g/cm^3	= 62,41 lb/ft^3

1. Historical and Economic Development of Phenolic Resins

1.1. History

On July 13, 1907, Leo H. Baekeland applied for his famous "heat and pressure" patent[1] for the processing of phenol-formaldehyde resins. This technique made possible the worldwide application of the first wholly synthetic polymer material (only cellulose derivatives were known before). Even from his first patent application[2] of February 18, 1907, it was clear that Baekeland, more than this predecessors, was fully aware of the value of phenolic resins. Before his involvement with phenolic resins Baekeland had worked on photographic problems with the same intensity. His success in developing a fast-copying photographic paper, known throughout the world under the name Velox, gave him the financial independence which allowed him to build his own research laboratory in his home in Yonkers, New York. There, starting in 1905, he devoted his whole time to the investigation of phenolic resins. However, the first patent covering phenolic resins (as substitute for hard rubber) was granted to A. Smith in 1899[3]. A. von Bayer found in 1872 while studying phenolic dyes, that phenol reacting with formaldehyde was converted to a colorless resin[4]. He first noticed that a reddish-brown resinous mass was produced during

Fig. 1.1. Leo Hendrik Baekeland (1863–1944)

Fig. 1.2. L. H. Baekeland's laboratory

the reaction of bitter almond oil with pyrogallic acid. However, nothing was done with this resinous material. Ter Meer[5], A. Claus and E. Trainer[6] continued the experiments. Claus and Trainer obtained a resinous material from 2 mol of phenol and 1 mol of formaldehyde and hydrochloric acid. After the non-converted phenol was distilled off, a soluble resin was obtained with a MP of 100 °C. However, they also could not think of application for this material and reported disappointedly: "It is not possible to crystallize this resinous material."

Claisen[7] and Kleeberg[8] continued the experiments after the company Merklin and Lösekam brought the formaldehyde on to the market in 1889. Kleeberg obtained a cross-linked, insoluble resin using an excess of formaldehyde and hydrochloric acid in a vigorous reaction. There was no interest in the product obtained. After laboratory investigations performed by Manasse[9] and Lederer[10], the Bayer company[11] applied for a patent for a process for the production of *o*- and *p*-hydroxybenzylalcohol, but without mentioning a formation of the resin. Speier[12] obtained an insoluble material from resorcinol, formaldehyde and ammonia as catalyst which could be used as an antiseptic. Speier, Smith and Luft were the first ones to draw the attention to technical applications for curable phenolic resins. Smith in particular pointed out the valuable properties of the new material which did not melt, was a good insulating material and could serve as a substitute for ebonite and wood[3]. Luft[13] tried to flexibilize the brittle material obtained by Smith, by addition of solvents, glycerin and organic acids. He recommended the following applications for his plasticized phenolic resins: water-proof coatings for fabrics, fibers which are carbonized to form filaments for light bulbs, acid- and alkali-resistant vessels, billiard balls, buttons, handles, imitation amber and corals when colorants and fillers are added. Almost at the same time, the Louis Blumer company[14] applied for a patent for the production of synthetic resins as substitutes for shellac. The solid, soluble phenolic resins made with organic acids as catalysts were the first commercial-scale phenolics in the world sold under the trade name Laccain. In February 1903, Henschke[15] continued the experiments of Manasse and, using alkali hydroxide as catalyst for the P/F reaction, obtained an insoluble resin. Further improvements to the preparation of phenolic resins were made by Fayolle[16] and Story[17]. Story worked without catalysts. Laire[18] tried to find a substitute for copal and damar by heating phenol alcohols. He obtained high melting condensation products which were not soluble in low boiling alcohols, but which could be dissolved in turpentine or camphor oil.

So that when Baekeland started with his studies of phenolic resins, the following facts were already known[19]:

- Phenols and formaldehyde are converted to resinous products in the presence of acidic and alkaline catalysts. These may be permanently fusible and soluble in organic solvents or heat-curable depending upon the preparation conditions.
- Phenolic resins were already being sold as substitutes for shellac, ebonite, horn and celluloid. These are colorable, can be mixed with fillers and under the influence of heat shaped in molds into solid parts.

However, economic production of molded parts was not yet possible. The "heat and pressure" patent[1] became the turning point, indicating clearly the importance of economic processing techniques for market acceptance. Phenolic resins mixed with fillers could be hardened in a press or an autoclave, which was called "bakelisator", under pressure at temperatures above 100 °C in a considerably short time and without the formation of blisters. According to the first Baekeland patent[2], phenol, formaldehyde, catalysts and fibrous cellulosic material were reacted (in the cellulose matrix) at elevated temperatures. The impregnation of the fibrous material can be improved by application of vacuum and pressure, infusible products being obtained only if formaldehyde was used in excess. Soon afterwards he recommended[20] the impregnation of the cellulosic fibers with liquid phenolic resins, acid catalyzed resins were being used at this stage. According to a patent application by Lebach in February 1907[21] insoluble and infusible condensation products, useful as plastic materials, could be obtained if phenol is reacted with a surplus formaldehyde using neutral or basic salts as catalysts. In the same year Baekeland also patented a process for the preparation of phenolic resins using alkaline catalysts, preferably ammonia, NaOH and Na_2CO_3. A patent was granted to him in the USA[22] but not in Germany because of the lack of inventive steps considering previous publications by Henschke[15]. It was in this patent, however, that resin manufacture was described for the first time just as it is carried out today:

- The reaction is performed in a closed vessel with a reflux condenser to prevent loss of volatile materials.
- The reaction is interrupted when the desired viscosity is obtained.
- Distillation is performed in a vacuum and can be continued until a solid product which is still soluble in alcohols is obtained.

In 1908, the cure of phenolic resins at ambient temperatures by addition of strong inorganic acids was reported by Lebach[23].

Between 1907 and 1909 Baekeland conducted small-scale trials with a few industrial companies and as result, patented numerous applications for phenolic resins till 1909:
Molding compounds can be made of pulverized, fusible phenolic resins made in alkaline environment and fillers, and molded to a shaped part of high toughness, strength and chemical resistance[24].
Phenolic resins are excellent binders for abrasive materials[25].
Ammonia-catalyzed solid resins in organic solvents can be used for valuable varnishes[26] and coatings for food containers[27].
Temperature and steam-resistant lining materials can be made of phenolic resin impregnated asbestos fibers, paper or cloth[28].
Phenolic resins are useful for coating wood, yielding a hard and abrasive resistant surface of high gloss or can be applied as adhesive for veneer facing[29].
For production of fiber boards an aqueous wood fiber pulp is mixed homogeneously with phenolic resin. After a drying process similar to that used in the manufacturing of paper, the fiber mats obtained are hardened between hot metal plates under pressure[30].

On February 5, 1909, at a meeting of the New York Section of the American Chemical Society, Baekeland reported for the first time the results of his thorough studies of phenolic resins, which he called "Bakelite". His report was received with great enthusiasm by a large audience. He stated his theory that the reaction of phenols with formaldehyde in the presence of catalysts occurs in three phases:

The formation of a soluble initial condensation product, which could be liquid, viscous or solid and which he called A,
the formation of a solid intermediary condensation product, which could still swell in solvents and which he designated as B, and
the formation of an infusible and insoluble product C.

In 1909, Lebach suggested calling the liquid curable resin "resol", the B-phase-material "resitol" and the hardened phenolic resin "resit", while Aylsworth recommended the name "condensite". In the same year Baekeland proposed the designation "novolak" for the fusible, thermoplastic resin, indicating the suggested substitution of shellac.

The production of paper laminates and laminated paper tubes made of liquid or dissolved resins[32)], the manufacture of noiselessly running cog-wheels[33)], phenolic resin putties and glues to bond various materials and impregnating resin for coils and similar electrical devices[34)] were also suggested by Baekeland at this early stage. The low reactivity of *o*- and *p*-cresol was mentioned[35)] and recommended as a means of delaying the hardening reaction and increasing plasticity. The high reactivity of *m*-cresol had already been mentioned[36)]. In another application[37)] the use of phenyl- and cresyl-phosphates was recommended for making PF-resins more flexible. Also, tung oil was recommended as plasticizing additive[38)] for impregnating- and coating resins and the resin preparation method mentioned. A further process for manufacturing phenolic resin bonded fiber boards was patented by him in 1915[39)]. The phenolic resin solution added to the fiber pulp is precipitated on the fibers by the addition of acidic salts according to this disclosure.

After these successful preliminary studies, the time came to put what had been learned into practice. After a visit by Baekeland in June and July 1909, to Germany, the companies Rütgerswerke AG and Knoll & Co, together with Baekeland, founded the "Bakelite Gesellschaft mbH" at Erkner near Berlin on May 25, 1910[40)]. This was the first company in the world to produce synthetic resins. On October 10, 1910, Baekeland founded the General Bakelite Company, in the USA and later other companies in England, France, Japan and Canada. On March 22, 1922 the "Bakelite Corporation" was founded, incorporating Redmanol Chemical Products Company and Condensite Company. This corporation was taken over by the Union Carbide & Carbon Corporation in 1939. The first customers of the German Bakelite Corp. were the big electrical companies. They mainly used shellac for the manufacturing of laminated paper.

In 1910–1911, the production of cresol molding compounds[41)] was also started. At that time, cresol was preferred to phenol because it was cheaper, J. W. Aylsworth, a co-founder of the Condensite Company, also contributed a lot to the development of resins[42)] and the production of molding compounds. In 1910, he found[43)] that novolaks which he obtained from 3 mol of phenol and 2 mol of formaldehyde, could be very favorably cured by the addition of hexamethylenetetramine or trioxymethylene.

The addition of an organic acid anhydride – like stearic acid anhydride – was supposed to bind the ammonia and vapor which are liberated during the curing process. In another patent[44)] he described different methods of producing molding compounds composed of resin, HMTA and fillers on heated rolls, or by an extrusion process and by grinding to appropriate size. Since the cure by application of pressure and

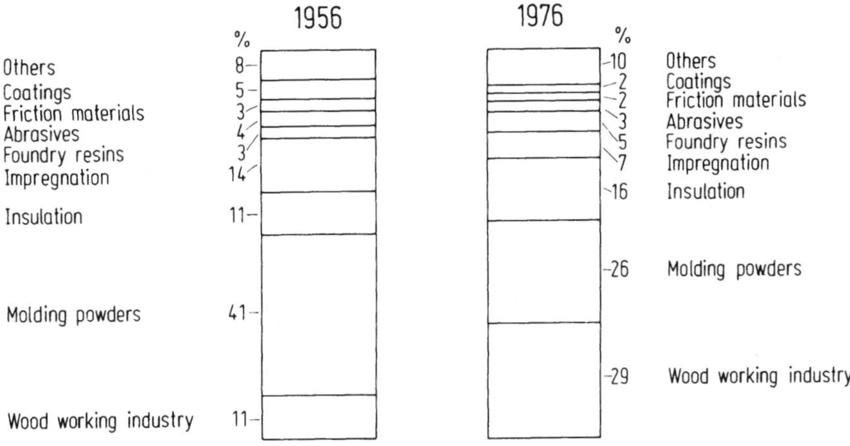

Fig. 1.3. Phenolic resin markets in the USA in 1976 and 1956. Source: Modern Plastics Int.

heat was patented in favor of Baekeland, Aylsworth had to get the preforms cured in the oven without pressure. The temperature was therefore risen only very slowly up to 140 °C. The production of "fast-molding compounds", started in 1924, when the mutual experiences of Baekeland and Aylsworth bore fruit. Very fine particles of HCl-catalyzed novolak were mixed with HMTA, wood flour, colorants and additives and homogenized on hot rolls. This material was than molded in a heated press.

It was to be expected that in this new area of early plastic materials which promised so much success, soon a large number of patents would appear, which did not constitute essential advantages. Baekeland, owner of about 400 patents, had many legal battles with inventors and he expressed his experience quite drastically in the following sentence[45]: "It is an axiom that the test of a valuable invention is that it will be infringed, or attacked by those whose thinking cells are passive until some inventor roused them out of their mental drowsiness".

1.2. Market Position

The historical development of phenolic resins as described in Chapter 1.1 showed clearly that invention and innovation are two different things. L. H. Baekeland was not the inventor of phenolic resins. However, he provided the technical prerequisites which made it possible to introduce them on to the market.

Today, the most important fields of application are the wood working industry, molding- and insulation compounds. More than 2/3 of all phenolic resins are used in these three fields. But also all classic applications established by Baekeland could maintain their position (Fig. 1.3).

A similar breakdown of the markets also applies for West-Germany. However, it does not include those countries like France and Italy where melamine resins are used for the manufacture of particle boards and fiber boards for outdoor applications. There, molding compounds take the first place. Also in Great Britain, the wood

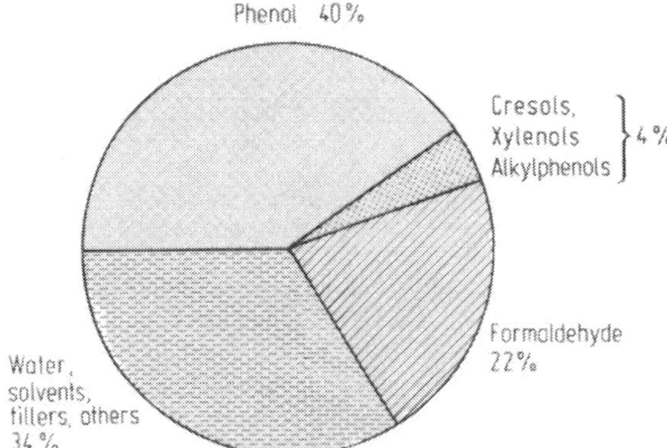

Fig. 1.4. Composition of a "typical" phenolic resin under consideration of all fields of application

working industry plays only a minor role. The determination of the position quantity-wise is more difficult. Phenolic resins have a very heterogeneous composition. Technical resins contain considerable amounts of water, solvents and fillers, as shown for a "typical" phenolic resin in Figure 1.4. Also in figures expressed in dry weight, the high amount of fillers in molding compounds or the volatility of the phenol alcohols and monomers, phenol and formaldehyde, must be considered.

National production and consumption figures given in the literature, only in the rarest cases include the necessary addition if they refer to dry resin content, and are therefore hard to compare. Captive production which is difficult to evaluate, plays a considerable role for the production of mineral wool binder resins and impregnating resins for decorative laminates. The order of magnitude of phenolic resin production in largely closed economic areas can well be estimated on the basis of the phenol production there (Table 2.3, Chapter 2) which is more accurately known[47, 48, 55].

In the phenolic resin production broken down into countries, the USA is leading by a clear margin ahead of the USSR, Japan and West-Germany (Table 1.1).

According to these figures, almost 70 years after the production of technical phenolic resins started, about 4% of the whole world-wide produced plastics today (including cellulosics, excluding elastomers) are phenolic resins. This quantity does not

Table 1.1. Phenolic resin production of the leading countries in comparison to the total plastics production in 1977

	Plastics 1.000 to	Thermosets 1.000 to	Thermoset share %	Phenolics 1.000 to	Phenolics share %	Ref.
West-Germany	6.270	1.924	31	230[a,c]	3.6	49, 50
France	2.648	424	16	104[d]	3.9	51
GB	2.340	810	35	75[d]	3.2	52
Italy	2.850	520	19	88[d]	3.1	53
USA	14.791	2.910	20	638[b]	4.3	46
Japan	5.704	1.377	24	246[a]	4.4	54
USSR[e]	3.582			348[a]	9.7	55

[a] As delivered; [b] Dry weight based; [c] Assumption; [d] Not disclosed; [e] Production in 1976.

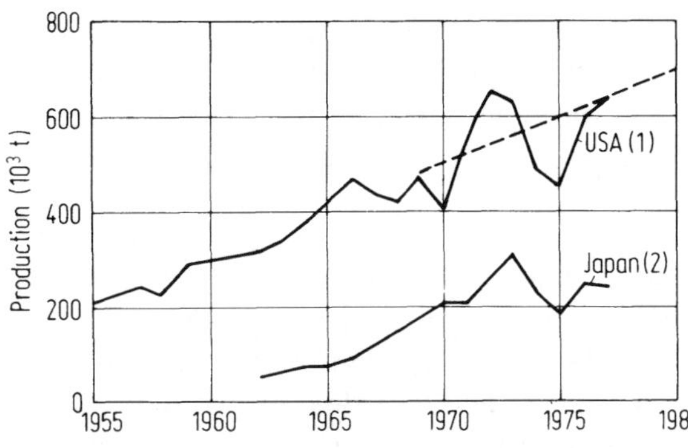

Fig. 1.5. Phenolic resin production in in the USA and Japan. (1) Based on dry weight; (2) as delivered. Source: Mod. Plast. Int.

seem to be high in the first instance. However, one must not compare the phenolic resins, which have to be considered engineering plastics, with the (later developed) commodity plastics. In the plastic scale they range about at the position of polyester resins or polyurethanes. It is typical for their application that they are always used in combination with reinforcing fillers and fibers where they have adhesive functions. The total volume of phenolic resin bonded materials, i. e. particle boards, fiber insulation materials, foundry sands, grinding wheels, is extraordinarily high.

The economic importance of phenolic resins today rather proves that they are irreplaceable in the various engineering fields and distinct areas of daily life. High temperature resistance, infusibility, and flame retardancy are the key properties which will contribute to a further market growth.

Considering the multitude of fields of application, it is not astonishing that the phenolic resin consumption is a very sensitive indicator for the general economic situation (Fig. 1.5). The smoothed curve of the phenolic resin production in the USA from 1957–1977 indicates a high average increase of 5% per annum. In Japan, within the last 10 years, there was an average increase of about 10%. The really high

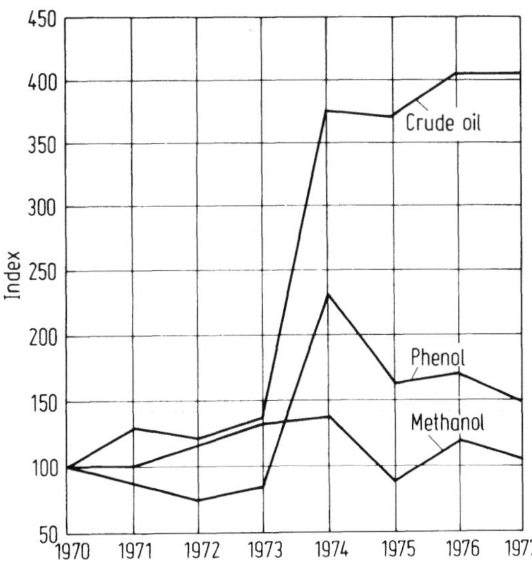

Fig. 1.6. Price index of basic chemicals for phenolic resin production in comparison to crude oil price, West Germany. 1970 = 100. Source: Verband der Chemischen Industrie, Frankfurt

rates of growth are to be found in the field of technical resins (Fig. 1.3). The relative share of the phenolic molding compounds in the total market of the phenolic resins went down since 1956 from 41% to 29% in 1966 and 26% in 1976. The highest increase of phenolic resin consumption can be observed in the field of wood working materials (plywood and particle boards), as well as thermal insulation.

Finally, the question arises, if phenolic resins will still be able to compete in the future. Even under long term consideration, phenolic resins, compared to other plastic materials, benefit by an extraordinarily well secured raw material supply and relative price stability. The point of time when the petroleum reserves will actually be exhausted shall not be discussed herein, but it is without question that the time will come closer when basic chemicals derived from coal will be able to compete. If crude oil is no more available, the leading organic basic chemicals today, ethylene and propylene, must then be produced by relatively expensive processes (e.g. ethanol dehydration, Fischer-Tropsch-Synthesis). However, the basic raw materials of the phenolic resin production, phenol, cresols and methanol (formaldehyde) – as primary products of modern coal gasification processes – will be available under very economic conditions (see also Fig. 2.1, Chapter 2).

References

1. Baekeland, L. H.: US-PS 942699 (13. July 1907); DE-PS 233803
2. Baekeland, L. H.: US-PS 949671 (18. Feb. 1907)
3. Smith, A.: DE-PS 112685 (1899)
4. Bayer, A.: Ber. dtsch. chem. Ges. *5*, 25 (1872); *5*, 1095 (1872)
5. ter Mer, E.: Ber. dtsch. chem. Ges. *7*, 1200 (1874)
6. Claus, A., Trainer, E.: Ber. dtsch. chem. Ges. *19*, 3009 (1886)
7. Claisen: Liebigs Ann. *237*, 261 (1887)
8. Kleeberg: Liebigs Ann., *263*, 283 (1891)
9. Manasse, O.: Ber. dtsch. chem. Ges. *2409*, 27 (1894)
10. Lederer, L.: J. prakt. Chemie *50*, 223 (1894)
11. Farbenfabr. Bayer: DE-PS 85588 (1894), DE-PS 87335 (1895)
12. Speier, A.: DE-PS 99570 (1897)
13. Luft, A.: DE-PS 140552 (1902)
14. Blumer, L.: DE-PS 172877 (1902)
15. Henschke, F.: DE-PS 157553 (1903)
16. Fayolle, E. H.: FR-PS 335584 (1903)
17. Story, W. H.: DE-PS 173990 (1905)
18. de Laire: DE-PS 189262 (1905)
19. Thinius, K.: Studien zur Geschichte der Plaste X, Plaste und Kautschuk *23*, 746 (1976)
20. Baekeland, L. H.: US-PS 942852 (13. July 1907)
21. Lebach, H.: DE-PS 228639 (1907)
22. Baekeland, L. H.: US-PS 942809 (1907)
23. Lebach, H.: GB-PS 27096 (1908) to Knoll & Co.
24. Baekeland, L. H.: US-PS 939966 (1909)
25. Baekeland, L. H.: US-PS 942808 (1909)
26. Baekeland, L. H.: US-PS 954666 (1909)
27. Baekeland, L. H.: US-PS 957137 (1909)
28. Baekeland, L. H.: US-PS 941605 (1909)
29. Baekeland, L. H.: US-PS 1019408 (1909)
30. Baekeland, L. H.: US-PS 1160362 (1909)

31. Baekeland, L. H.: Journal of Industrial and Engineering Chemistry *1*, 149 (1909)
32. Baekeland, L. H.: US-PS 1019406 (1910)
33. Baekeland, L. H.: US-PS 1160364 (1910)
34. Baekeland, L. H.: US-PS 1156452 (1911) and 1213144(1914)
35. Baekeland, L. H.: US-PS 1306681
36. Baekeland, L. H.: US-PS 1088677
37. Baekeland, L. H.: US-PS 1439056 (1918)
38. Baekeland, L. H.: US-PS 1018385
39. Baekeland, L. H.: US-PS 1160365 (1915)
40. N. N. 60 Jahre Bakelite Kunstharze, Kunststoff Rundschau *17*, 231 (1970)
41. Du Bois, J. H.: Plastics World 57 (1968)
42. Aylsworth, J. W.: US-PS 1029737 (1909)
43. Aylsworth, J. W.: US-PS 1020593 (1910)
44. Aylsworth, J. W.: US-PS 1090439 (1911)
45. Baekeland, L. H.: Ind. and Engng. Sci. *6*, 90 (1914)
46. N. N. Modern Plastics Int., Jan. 1978, P. 43
47. N. N. Chem. Ind. *29*, 525 (1977)
48. N. N. Japan Chemical Week, 828, March 11, 1976
49. Verband Kunststofferzeugende Industrie e. V., Frankfurt
50. N. N. Chemische Industrie *29*, 437 (1977)
51. N. N. Informations Chimie *178*, 83 (1978)
52. N. N. Die Kunststoffindustrie in Großbritannien, Kunststoffe 68, 499 (1978)
53. N. N. Caoutchoucs et Plastiques *579*, 49 (1978)
54. Japanese Ministry of International Trade and Industry, Jap. Chem. Week, July *14*, 3, 1977
55. N. N. Chem. Ind. *30*, 26 (1978)

2. Raw Materials

Phenolic resins are produced by the reaction of phenols with aldehydes. The simplest representatives of these types of compounds, phenol and formaldehyde, are by far the most important. As an average, considering all applications, the production of 1 ton of phenolic resin requires approximately 440 kg phenol (containing about 10% cresols and xylenols) and 220 kg formaldehyde as well as solvents, additives and water.

2.1. Phenols

Phenols are a family of aromatic compounds with the hydroxyl group bonded directly to the aromatic nucleus. They differ from alcohols in that they behave like weak acids and dissolve readily in aqueous sodium hydroxide, but are insoluble in aqueous sodium carbonate. Phenols are colorless solids with the exception of some liquid alkylphenols. The most important phenols are listed in Table 2.1 below. Data regarding the molecular structure of phenols and cresols are listed in Section 3.1.

Table 2.1. Physical properties of some phenols[1–6]

Name		MW	MP °C	BP °C	pK_a 25 °C
Phenol	hydroxybenzene	94.1	40.9	181.8	10.00
o-Cresol	1-methyl-2-hydroxybenzene	108.1	30.9	191.0	10.33
m-Cresol	1-methyl-3-hydroxybenzene	108.1	12.2	202.2	10.10
p-Cresol	1-methyl-4-hydroxybenzene	108.1	34.7	201.9	10.28
p-tert.Butylphenol	1-tert.butyl-4-hydroxybenzene	150.2	98.4	239.7	–
p-tert.Octylphenol	1-tert.octyl-4-hydroxybenzene	206.3	81/83	284	–
p-Nonylphenol	1-nonyl-4-hydroxybenzene	220.2	–	295	–
2,3-Xylenol	1,2-dimethyl-3-hydroxybenzene	122.2	75.0	218.0	10.51
2,4-Xylenol	1,3-dimethyl-4-hydroxybenzene	122.2	27.0	211.5	10.60
2,5-Xylenol	1,4-dimethyl-2-hydroxybenzene	122.2	74.5	211.5	10.40
2,6-Xylenol	1,3-dimethyl-2-hydroxybenzene	122.2	49.0	212.0	10.62
3,1-Xylenol	1,2-dimethyl-4-hydroxybenzene	122.2	62.5	226.0	10.36
3,4-Xylenol	1,3-dimethyl-5-hydroxybenzene	122.2	63.2	219.5	10.20
Resorcinol	1,3-dihydroxybenzene	110.1	110.0	281.0	–
Bisphenol-A	2,2-bis(4-hydroxyphenyl)propane	228.3	157.3	–	–

2.1.1. Physical Properties of Phenol

The melting point of pure phenol at 40.9 °C is considerably lowered by traces of water, approximately 0.4 °C per 0.1% of water. A water content of ≤ 6% renders it liquid even at room temperature. To produce phenolic resins, a mixture of 90% of phenol / 10% of water is preferably used. Above 65.3 °C phenol can be mixed with water at any ratio[7]. During the cooling period of those solutions, which may contain 28–92% of water, two phases are being developed, phenol/water and water/phenol.

Phenol is highly soluble in polar organic solvents, but not very soluble in aliphatic hydrocarbons. Phenol crystallizes in the form of colorless prisms. Exposed to air, phenol rapidly develops a reddish color, especially if it contains traces of copper and iron. This happens if phenol is reacted in copper clad or iron reactors, or if phenolic resins are stored in iron barrels. Additional safety and technical data of phenol are listed in Table 2.2 below:

Table 2.2. Technical and safety data of phenol[8]

Flash point	°C	79
Ignition point	°C	605
Explosion limits	vol.-%	2–10
MAK-value	ppm	5
MAK-value	mg/m^3	19
Vapor pressure, 20 °C	mbar	0.2
Threshold odor concentration	ppm	0.5

The toxicology and physiology of phenol is described in Chapter 5.

2.1.2. Supply and Use of Phenol

The largest amount of phenol is required for the production of phenol-formaldehyde resins[9, 10] as shown in the following Table 2.3:

Table 2.3. Breakdown of phenol consumption (1976) in %

%	U.S.A.	West-Europe	Japan
Phenolic resins	46	35	54
ε-Caprolactame	16	21	–
Bisphenol-A	14	10	29
Adipic acid	3	8	–
Others	21	26	17

In 1978, only approximately 3% of the world production of phenol was gained from coal. Among the synthetic processes, the cumene process in the most frequently used. The basic raw materials for the cumene process and thus for the production of phenol are benzene and propylene. To help understand the dependence of the availability and price for phenolic resins on the crude oil situation, the base material supply is given in Figure 2.1.

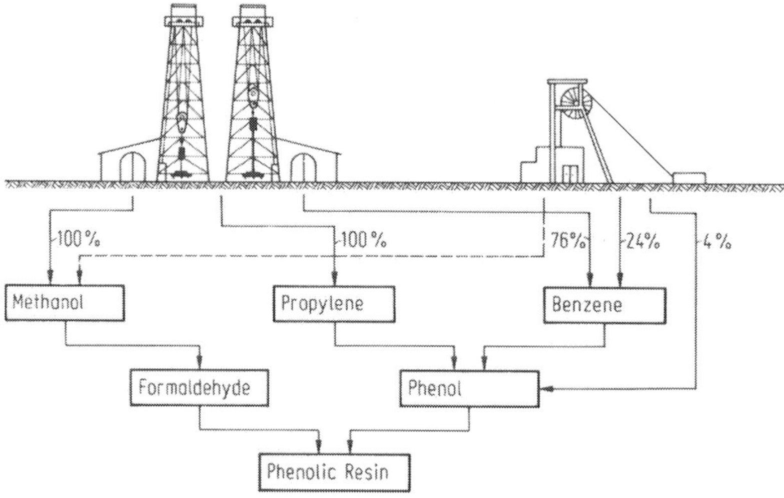

Fig. 2.1. Raw materials origin for the production of phenolic resins. West-Germany 1978

Figure 2.2 shows the percentages of benzene and propylene that go into the production of different derivatives including cumene[11, 12].

In 1978, the phenol capacities amounted to 1.4 million tons in Western Europe and 1.57 million tons in the USA[9]. For the years 1978–1983, an increase in phenol consumption of approximately 5–6% p. a. is expected in Europe. In Europe Phenolchemie, Shell and ICI plan increases of approximately 100.000 tons per year each. In the USA the US Steel Corp., Gulf Oil and General Electric have announced increasing capacities for 1978–1981[9].

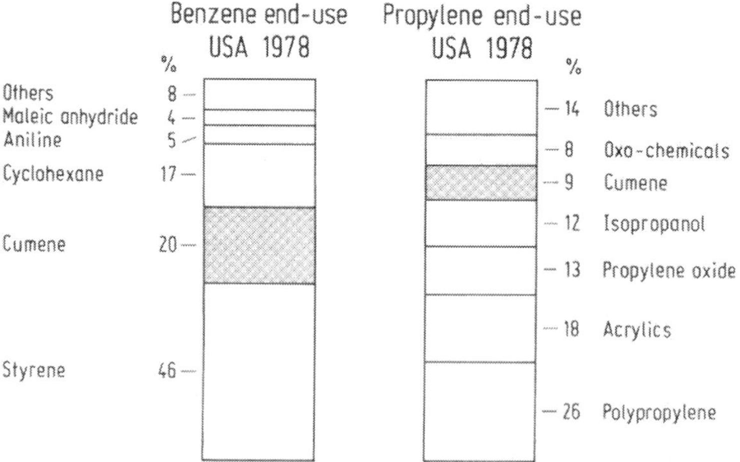

Fig. 2.2. Breakdown of benzene and propylene consumption in the USA[11, 12]

2.1.3. Phenol Production Processes

The cumene process is by far the most important synthetic process[13, 14] for the production of phenol, and today, probably accounts for 90% of the synthetic phenol capacity in the Western world.

A two step oxidation process based on toluene was developed by Dow Chemical. Kalama Chemical in USA and DSM in The Netherlands are working according to this process 2.1.

$$\underset{\text{Toluene}}{C_6H_5CH_3} + 1{,}5\ O_2 \xrightarrow[(Co)]{140\,°C/3\,bar} \underset{\text{Benzoic acid}}{C_6H_5COOH} \ (+H_2O) \longrightarrow C_6H_5OH + CO_2 \qquad (2.1)$$

The hydrolysis of halogenated aromatic compounds, in 1930 developed by Raschig, was later improved to the Raschig-Hooker process. The first step, the oxychlorination of benzene with hydrochloric acid in the presence of a copper-on-alumina catalyst at 275 °C is followed by hydrolysis of the chlorobenzene with water (steam) on a copper-promoted calcium phosphate catalyst at 400–450 °C.

The sulfonation process, the oldest one, is almost of no importance nowadays.

Cumene Process (Hock Process)

In Germany, the phenol synthesis based on cumene was discovered by H. Hock and published by him and Sho Lang[8].

Soon after World War II the first pilot plant was constructed jointly by Rütgerswerke and Bergwerksgesellschaft Hibernia with Hock's assistance. The commercial production was first developed by the Distillers Co. (GB) and Hercules Powder (USA).

The cumene (isopropylbenzene) required for the Hock process is produced[15] by alkylation of benzene with propylene by use of a solid phosphoric acid catalyst (UOP-Process, Eq. 2.2).

$$\underset{\text{Benzene}}{C_6H_6} + \underset{\text{Propylene}}{CH_3-CH=CH_2} \xrightarrow{\text{Phosphoric acid}} \underset{\text{Cumene}}{C_6H_5-CH(CH_3)_2} \qquad (2.2)$$

Cumene is oxidized with oxygen in air in the liquid phase to cumene hydroperoxide (CHP) according to reaction (2.3), yielding small amounts of dimethylbenzyl alcohol and acetophenone as by-products. The mechanism mentioned in parenthesis for the acid-catalyzed peroxide decomposition (Eq. 2.4) was postulated by Seubold and Vaugham[16].

CHP, a liquid compound with relatively low vapor pressure, is stable at normal temperature and conditions, but decomposes very rapidly under acidic conditions

$$C_6H_5{-}CH(CH_3)_2 + O_2 \text{ (Air)} \xrightarrow{\text{Catalyst}} C_6H_5{-}C(OOH)(CH_3)_2 \qquad \Delta H_1 = -116 \text{ kJ/mol} \tag{2.3}$$

Cumene hydroperoxide CHP

$$\text{CHP} \xrightarrow[-H_2O]{H^+} \left[C_6H_5{-}C(\bar{O}|^+)(CH_3)_2 \xrightarrow{\text{Rearragement}} C_6H_5{-}O{-}\overset{+}{C}(CH_3)_2 \xrightarrow{+H_2O} C_6H_5{-}O{-}C(\overset{+}{O}H_2)(CH_3)_2 \right] \longrightarrow \tag{2.4}$$

$$\longrightarrow C_6H_5OH + CH_3{-}\overset{\overset{O}{\|}}{C}{-}CH_3 + H^+ \qquad \Delta H_2 = -253 \text{ kJ/mol} \tag{2.5}$$

Phenol Acetone

and higher temperatures. During the second stage of the process, the concentration of CHP and separation of unreacted cumene occur. The concentrated reaction product is then converted in a splitting plant by use of sulfuric acid as catalyst to a crude mixture of phenol and acetone, which also contains α-methylstyrene as a by-product[14]. Then, various purification and distillation steps follow. α-Methylstyrene can also be hydrogenated and returned to the process.

CHP is a potentially hazardous material. Therefore, many safety regulations have to be observed and safety equipment must be installed in technical plants[17].

The world's first phenol-from-cumene plant was started up in Montreal, Canada, in 1952.

2.1.4. Cresols and Xylenols – Synthesis Methods

Cresols[7,18], hydroxy derivatives of toluene, commonly designated as methyl phenols, exist in three isomers depending on the relative position of the methyl towards the hydroxyl group. The molecular configuration is described in Section 3.1. The main source of cresols was originally coal tar. Today, however, synthetic processes dominate, mainly based on toluene and phenol[19]. The importance of the petroleum industry as

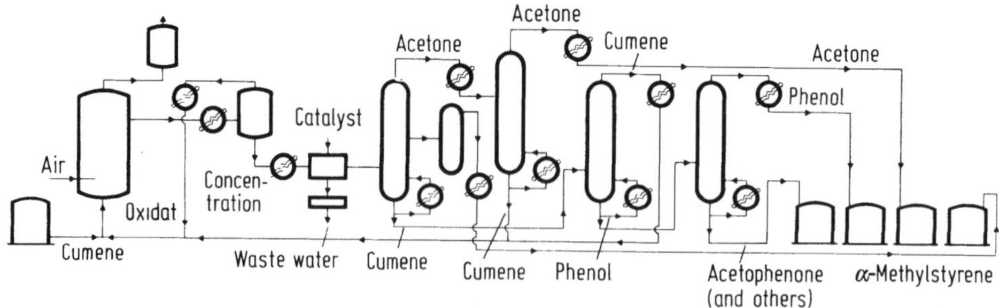

Fig. 2.3. Flow diagram of the cumene process, (Drawing: Phenolchemie, D-4390 Gladbeck)

Fig. 2.4. Phenol production according to the cumene process, oxidation plant (Photo: Phenolchemie, D-4390 Gladbeck)

a source of cresols and xylenols is relatively insignificant. Starting with toluene, the cresols are obtained either by sulfonation, by alkylation with propylene, or by chlorination. In the sulfonation process, the main product is the *para* derivative together with some *ortho* derivative. In the chlorination process the *meta* isomer prevails (about 50%) with an approximately equal *o*/*p* ratio. This route is advantageous for resin-grade cresols. The chemistry of the toluene alkylation is very similar to the cumene process with differences in the oxidation step. Toluene is reacted first with propylene in the presence of $AlCl_3$ or other catalysts to obtain a mixture of cymenes. In this process the *m*/*p* ratio of approximately 2 : 1 by less than 5% *o*-cymene is reported[20].

Mitsui and Sumitomo constructed industrial plants in Iwakuni and Ohita respectively, each with a capacity of 10,000 tons per year which have been operating satisfactorily since 1970.

$$\text{Toluene } (C_6H_5CH_3) + \underset{\text{Propylene}}{CH_2{=}CH{-}CH_3} \xrightarrow{\text{Catalyst}} \underset{\text{Cymene}}{CH_3C_6H_4CH(CH_3)_2} \xrightarrow{\text{Oxidation}} \tag{2.6}$$

$$\underset{\text{Cymene hydroperoxide}}{CH_3C_6H_4C(CH_3)_2OOH} \xrightarrow{H^+} \underset{\text{m-Cresol}}{CH_3C_6H_4OH} + \underset{\text{Acetone}}{CH_3{-}\overset{O}{\overset{\|}{C}}{-}CH_3} \tag{2.7}$$

More important among the synthetic processes is the production of cresols and xylenols based on alkylation of phenol with methanol[18]. In the gas phase process (Koppers and Pitt-Consol, USA; CRM, Great Britain), the methanol and phenol vapors pass an aluminium oxide catalyst at approximately 350 °C under moderate pressure. Mainly *o*-cresol and 2,6-xylenol are obtained. If 2,6-xylenol is desired as main product which is used for PPO production, magnesium oxide is employed as catalyst. The following purification and separation is accomplished by vacuum distillation for *o*-cresol (99% purity) and crystallisation for 2,6-xylenol (98%). The further purification of 2,6-xylenol as needed for PPO is made by counter-current extraction with aqueous sodium hydroxide solution (Pitt-Consol). The yield of the gas phase process is more than 90% with regard to the phenol used and more than 85% with regard to methanol.

Other processes operated by Chemische Werke Lowi and UK-Wesseling, are performed in the liquid phase. The Lowi process is carried out at 300–350 °C at a pressure of 40–70 bar, Al-methylate is used as catalyst. Thus, mainly *o*-cresol is obtained. Higher methanol ratio favors the formation of 2,4 and 2,6-xylenol. By transalkylation of xylenols in the presence of phenol the yield of cresols can be increased.

UK-Wesseling produces *o*- and *p*-cresol of 99% purity, 2,6-xylenol of 98% and 2,4-xylenol of 92% purity by use of zinc bromide as catalyst. The synthesis of *p*-cresol, used mainly for BHT or similar antioxidants, is also performed by sulfonation of toluene (Sherwin Williams, USA, 20.000 tons per year capacity).

2.1.5. Alkylphenols

Phenolic compounds with a saturated carbon side chain containing a minimum of three carbon atoms should be termed alkylphenols. These alkylphenols are produced from phenols or cresols by Friedel-Craft alkylation with olefins, mainly isobutene, diisobutene or propylene[7]. At low temperatures, e. g. below 50 °C, *o*-substitution predominates. *o*-Alkylphenols can be rearranged to *p*-isomers by heating up to 150°C with acid catalysts. High yield of *o*-derivates is achieved by use of Ca-, Mg-, Zn- or Al-phenoxides as catalysts at a temperature of about 150 °C.

In the resin area, alkylphenols are used for the production of coating resins because of their good compatibility with natural oils and increased flexibility or as cross-linking agents in the rubber industry. Other uses include antioxidants, surfactants and phosphoric acid esters.

2.1.6. Phenols from Coal and Petroleum

An average of approximately 1.5% crude phenols, mainly phenol (~ 0.5%) as well as *o*-, *m*- and *p*-cresols, 2.3-, 2.4-, 2.5-, 2.6- and 3.5-dimethylphenol, is found in coal tar[21]. Phenols are further obtained from condensates of coke oven gases and waste waters of coal gasification plants. The extraction is performed either according to the older Pott-Hilgenstock process using benzene/sodium hydroxide[22] or according to Lurgi's Phenosolvan process with diisopropyl ether as solvent[23]. The extraction of phenols from coal tar is performed with diluted sodium hydroxide (8–12%), followed by

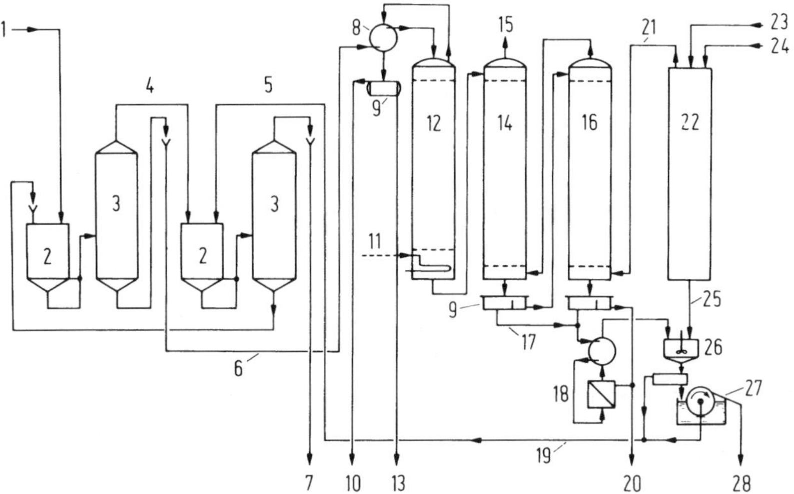

Fig. 2.5. The recovery of phenols from coal tar by alkali-extraction according to Franck/Collin[21].
1 Carbolic oil; 2 Mixing equipment; 3 Separating column; 4 1st Extraction; 5 2nd Extraction; 6 Raw phenoxide oil; 7 Phenol-free carbolic oil; 8 Condenser; 9 Separator; 10 Steam distilled oil; 11 Steam; 12 Steam distillation column; 13 Waste water; 14 Precipitation column; 15 Exhaust gas; 16 2nd Precipitation column; 17 Calcium hydroxide; 18 Cooling equipment; 19 Sodium hydroxide; 20 Raw phenols; 21 Carbon dioxide; 22 Lime kiln; 23 Limestone; 24 Coke; 25 Calcium oxide; 26 Causticizing; 27 Lime sludge filter; 28 Calcium carbonate

precipitation of the crude phenols with carbon dioxide. The flow diagram of an extraction plant is shown in Figure 2.5. The crude phenoxide solution contains approximately 0.5% non-phenolic components – neutral oils (hydrocarbons) and pyridine bases – which have to be removed prior to precipitation, mostly by steam distillation[21].

Other processes recommend selective solvents to extract phenol, e. g. aqueous methanol (Metasolvan process). However, only Lurgi's Phenoraffin process which uses an aqueous sodium phenoxide solution as selective solvent has achieved technical importance. The selectivity of NaOH is superior to all other recommended solvents.

A further source of cresols is the petroleum industry, particularly in the USA. During the catalytic cracking process various phenolic compounds are formed. Similar to the recovery of phenolic compounds from coal tar, the extraction is carried out with diluted sodium hydroxide solution. After the separation of the phenols by precipitation with carbon dioxide, further treatment follows by distillation. Thereby, phenol (BP 181.8 °C) and *o*-cresol (BP 191.0 °C) can be recovered technically pure immediately.

The separation of *m*- and *p*-cresol is only possible with special chemical or physicochemical methods due to the similar boiling points. This is also applicable to xylenols. In the urea process, the mixture of *m*- and *p*-cresol is heated with urea. Upon cooling a crystalline addition compound of *m*-cresol and urea is formed. Further, *p*-cresol when gently heated to 90 °C forms a crystalline addition compound with anhydrous oxalic acid. *m*-Cresol with sodium acetate results in adducts with low solubility. The chemical processes use the differences in the reaction rate of sulfonation (sulfuric acid process) followed by crystallization and hydrolysis or the different behaviour in the alkylation with isobutylene (isobutylene process) and following dealkylation with sulfuric acid.

2.1.7. Other Phenolic Compounds

Cashew Nut Shell Liquid (CNSL)

An important phenolic compound from natural sources is cashew nut shell liquid (CNSL). This liquid from the shells of cashew nuts, which grow mainly in Southern India, has become a useful raw material in the manufacture of special phenolic resins to be used for coating, laminating, and brake lining resin formulations[24]. Those resins possess outstanding resistance to the softening action of mineral oils and high resistance to acids and alkalies.

The CNSL, when obtained by a special heat treatment which includes decarboxylation, contains a mixture of mono- and diphenols (2.8) with an unsaturated C_{15} side chain in the meta position, thereby exhibiting high reactivity towards formaldehyde.

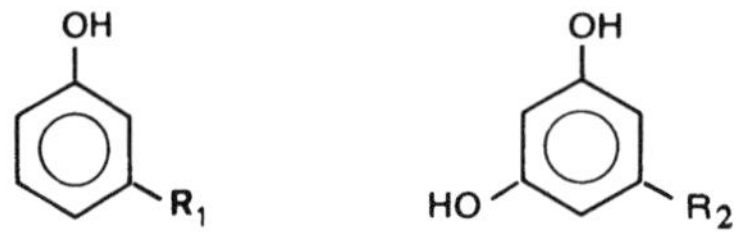

OH, R_1: Cardanol (90%)   OH, HO, R_2: Cardol (10%)   (2.8)

$R_1 = -(CH_2)_7-CH{=}CH-(CH_2)_5-CH_3$
$R_1,R_2 = -(CH_2)_7-CH{=}CH-CH_2-CH{=}CH-(CH_2)_2CH_3$
$R_1,R_2 = -(CH_2)_7-CH{=}CH-CH_2-CH{=}CH-(CH_2)_2CH_3$

The reacted CNSL-resin, the hardening reaction includes polymerization and polyaddition, yields a solid infusible product which in powdered form ("friction dust") retains high binding power at raised temperatures and is used in brake lining formulations[25].

Resorcinol

Resorcinol, as a dihydric phenol (1,3-dihydroxybenzene) is a very interesting intermediate material for the production of thermosetting resins. However, it is only used for special applications due to its relatively high price. The reaction rate with formaldehyde is considerably higher compared with that of phenol[26]. This is of great technical importance for the preparation of cold setting adhesives[27]. Resorcinol or resorcinol-formaldehyde prepolymers can be used as accelerating compounds for curing phenolic resins. The addition of 3–10% of such compounds permits shorter cure cycles in particle board and grinding wheel production. Furthermore, the adhesion of textile materials, e. g. tire cord to rubber, can be greatly improved by pretreating them with resorcinol-formaldehyde resin.

Resorcinol can otherwise be used as intermediate material for azo- and triphenylmethane and other dyes, pharmaceuticals, cosmetics, tanning agents, textile treating agents and antioxidants[28].

The only commercial process[29] for the production of resorcinol is the alkali fusion of *m*-benzenedisulfonic acid according to the Eq. (2.9).

$$\text{Benzene} + 2\,SO_3 \xrightarrow{H_2SO_4} \text{Benzene disulfonic acid} \xrightarrow[-H_2O]{NaOH} \text{Resorcinol} \tag{2.9}$$

As an intermediate stage, the disodium salt of the acid is formed. It is then fused with sodium hydroxide in a nickel alloy tank. The melt is dissolved in water and the resulting slurry acidified with sulfuric acid. The resorcinol is then recovered by counter-current extraction and purified by distillation. Plants with a capacity of 10,000 tons per year each are operated by Koppers, USA, and Hoechst, West-Germany.

Another route, not used commercially at the present time, is the Hock process starting with benzene by alkylation with propene (2.10) (Hibernia AG, West-Germany).

$$C_6H_4(CH(CH_3)_2)_2 \xrightarrow{O_2} C_6H_4(C(CH_3)_2OOH)_2 \xrightarrow{H^+} C_6H_4(OH)_2 + 2\,CH_3-\overset{\overset{\large O}{\|}}{C}-CH_3 \tag{2.10}$$

The oxidation is carried out with air at about 90 °C, the decomposition with diluted sulfuric acid in acetone or other solvent.

Bisphenol-A

Bisphenol-A is the common name for 2,2-bis(4-hydroxyphenyl)propane[30]. In 1923, the commericial production of bisphenol-A was introduced by Chemische Werke Albert in Germany, by the addition of acetone to phenol using hydrochloric acid as a catalyst. The importance of the colorless BPA/F-resins for the coating industry will be discussed in Chapter 13.

Today, the main use for BPA is the production of epoxide resins (~ 65%) and polycarbonate. Sulfuric acid is used in the newer process because of problems associated with the volatility and corrosiveness of hydrochloric acid. Sulfur compounds like thioglycolic acid or mercaptans further increase the reaction rate of the acid-catalyzed addition of carbonyl compounds to phenols[31].

While the purity of BPA made by the sulfuric acid process is satisfactory for the use in formaldehyde resins, high purity BPA is needed for the production of epoxy resins and especially for polycarbonate. This is normally accomplished by recrystallization from toluene or by crystallization as phenol-BPA adduct.

2.2. Aldehydes

Formaldehyde[32] is the almost exclusively used carbonyl component for the synthesis of technically relevant phenolic resins. Special resins can also be produced with other aldehydes, for example acetaldehyde, furfural or glyoxal, but have not achieved greater technical importance. Ketones are very seldom used.

The physical properties of aldehydes are compiled in Table 2.4.

2.2.1. Formaldehyde, Properties and Processing

Formaldehyde is produced by dehydrogenation of methanol, over either an iron oxide/molybdenum oxide catalyst or over a silver catalyst. Because of hazards in handling mixtures of pure oxygen and methanol, air is used as oxidizing gas[35]. Oxygen is used to burn the developing hydrogen.

$$\underset{\text{Methanol}}{CH_3{-}OH} + 1/2\,O_2 \xrightarrow{\text{Catalyst}} \underset{\text{Formaldehyde}}{HC{\overset{\displaystyle O}{\lessgtr}}_{H}} + H_2O \qquad \Delta H = -159\ \text{kJ/mol} \tag{2.11}$$

When the silver catalyst is used, the reaction mixture of methanol and air is prepared so as to be over the upper flammability limit; this is reversed when the oxide catalyst is used[35]. Reactor effluent passes to the absorption train where formaldehyde and other condensables are recovered by condensation and absorption in recirculating formalin streams. The raw formaldehyde solution is then purified by stripping out the unconverted methanol. Formaldehyde assay may be adjusted by regulating amounts of water added to the absorber column or by subsequent diluting in storage tanks. Inhibitors are added to retard the formation of paraformaldehyde in storage.

The Formox process works with a mixture of iron oxide and molybdenum oxide as catalyst. The reaction proceeds at relatively low temperatures, between 250–400 °C, almost up to completion (95–98%). As a side reaction, the formaldehyde produced is oxidized to yield carbon monoxide and water (2.12).

Table 2.4. Physical properties of some aldehydes[1, 33]

Type	Formula	MP °C	BP °C
Formaldehyde	$CH_2{=}O$	−92	−21
Acetaldehyde	$CH_3CH{=}O$	−123	20.8
Propionaldehyde	$CH_3{-}CH_2{-}CH{=}O$	−81	48.8
n-Butyraldehyde	$CH_3(CH_2)_2{-}CH{=}O$	−97	74.7
Isobutyraldehyde	$(CH_3)_2CH{-}CH{=}O$	−66	61
Glyoxal	$O{=}CH{-}CH{=}O$	15	50.4
Furfural	CH — CH / ‖ ‖ H / CH C–C=O / \ O / (furan ring, 2-CHO)	−31	162

$$CH_2{=}O + 1/2\ O_2 \longrightarrow CO + H_2O \qquad \Delta H = -215\ kJ/mol \qquad (2.12)$$

The Perstorp/Reichhold, Montecatini, Nissui-Topsoe, CdF, Lummus and Hiag/Lurgi processes function in accordance with this method[36, 37, 38].

In the BASF and Monsanto processes a silver catalyst is used[34, 39]. Here, in general, methanol is partially oxidized and dehydrogenized at 330–450 °C on silver crystals or silver nets. The BASF process uses a vapor/methanol/air mixture. The conversion is considerably high at approximately 90%. Silver catalyst processes with an incomplete methanol conversion performed at 330–380 °C have been developed by Degussa and ICI.

A further process is based on the direct oxidation of methane with oxygen from air at approximately 450 °C and 10–20 bar on an aluminum phosphate contact. This process, however, has not yet achieved any technical importance.

Table 2.5. Properties of formaldehyde[34]

Molecular weight	30.03	
Boiling point	−19.2	°C
Melting point	−118	°C
Enthalpy of solution in H_2O	62	kJ/mol
Enthalpy of formation (25 °C)	−116 ± 6	kJ/mol
Free energy	−110	kJ/mol
Self-ignition temperature	300	°C
Explosion limits in air	7–72	vol % CH_2O
Dissociation constant in H_2O 0 °C	$1.4 \cdot 10^{-14}$	
Dissociation constant in H_2O 50 °C	$3.3 \cdot 10^{-13}$	

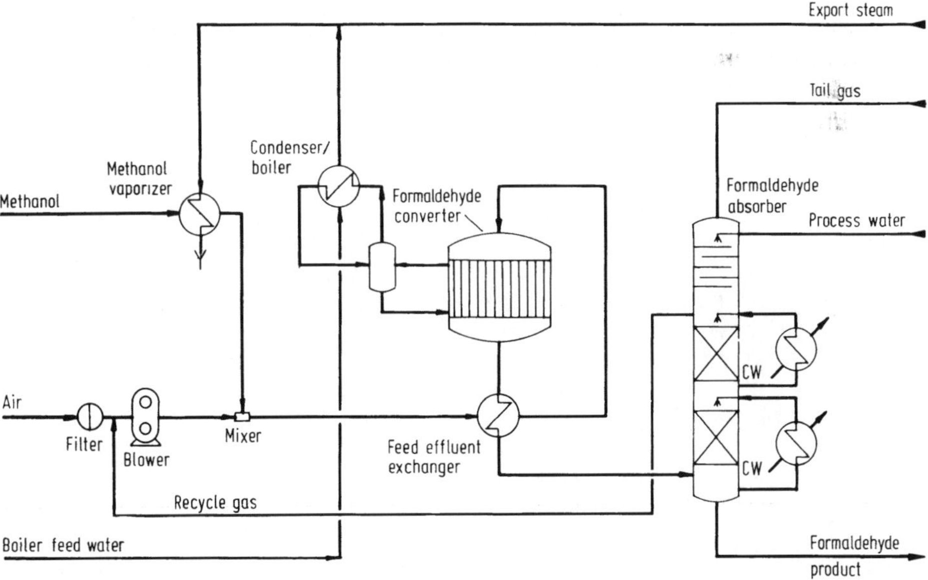

Fig. 2.6. Flow diagram of the iron oxide-molibdenum oxide catalyzed formaldehyde production process[36] (Drawing: Haldor Topsoe A/S DK-2800 Lingby)

Formaldehyde, a colorless, pungent, irritating gas, is found in aqueous solution almost exclusively as polymethylene glycol (2.14).

$$HO{-}CH_2{-}OH \quad (2.13) \qquad HO{-}[{-}CH_2{-}O{-}]_n{-}CH_2{-}OH \quad (2.14)$$

The portion of formaldehyde $CH_2{=}O$ in aqueous solution[32] is very low (<0.01%). The MWD is indicated in Section 3.2. Further equilibria exist in the presence of methanol[33], which adds to the stabilization by endcaping forming a hemiformal (2.15).

$$CH_3OH + HO{-}CH_2{-}OH \rightleftharpoons CH_3{-}O{-}CH_2{-}OH + H_2O \qquad (2.15)$$

Under the influence of acids, hemiacetals react to form acetals by eliminating water. A very important reaction is the formation of HMTA from ammonia and formaldehyde. The overall reaction is indicated in Eq. (2.21); the reaction mechanism is discussed in Section 3.3.2. By the catalytic action of strong bases, e. g. sodium hydroxide, formaldehyde undergoes a disproportionation reaction, known as Cannizzaro reaction, yielding methanol and formic acid according to Eq. (2.16).

$$2\,CH_2{=}O + NaOH \rightleftharpoons CH_3OH + CH_3{-}C\begin{matrix}{=}O\\ {-}ONa\end{matrix} \qquad (2.16)$$

Formaldehyde solutions always contain minor quantities of formic acid due to the Cannizzaro reaction, generally around 0.05%. The formic acid content can easily be determined by titration with sodium hydroxide.

The Kriewitz-Prins reaction[40] may have some importance in modification reactions of phenolic resins with unsaturated compounds. Olefins can react with carbonyl compounds under non-free radical conditions to result predominantly in unsaturated alcohols (2.17), *m*-dioxanes (2.18) and 1.3 glycols (2.19).

$$(CH_3)_2C{=}CH_2 + CH_2O \xrightleftharpoons{+H^+} (CH_3)_2C^+{-}CH_2{-}CH_2OH \xrightleftharpoons{-H^+} (CH_3)(CH_2{=})C{-}CH_2{-}CH_2OH \qquad (2.17)$$

$$(CH_3)_2C^+{-}CH_2{-}CH_2OH \xrightarrow[-H^+]{+CH_2O} \text{m-Dioxane: } (CH_3)_2C\,{-}CH_2{-}CH_2{-}O{-}CH_2{-}O{-} \text{ (ring)} \qquad (2.18)$$

m - Dioxane

$$\text{m-Dioxane} \xrightleftharpoons[-H^+]{+H_2O} (CH_3)_2C(OH){-}CH_2{-}CH_2{-}OH \qquad (2.19)$$

1,3-Glycol

As in the hydroxymethylation of phenols, the hydroxymethylene carbonium ion $^{\oplus}CH_2-OH$ is the alkylating agent.

When storing formaldehyde solutions, attention should be given to the fact that at lower temperatures or higher concentrations paraformaldehyde may separate. The stabilization of aqueous formaldehyde solutions can be achieved with alcohols, preferably methanol. Urea, melamine, methylcellulose and guanidine derivatives are also recommended for stabilization among other compounds. Proper storage containers should be of stainless steel; iron containers are not suitable. Containers with plastic lining or RP containers may also be used.

The major use of formaldehyde is in the production of thermosetting resins based on phenol, urea and melamine.

Table 2.6. Breakdown of formaldehyde consumption in the USA, Europe and Japan (1976)

%	USA	W. Europe	Japan
Urea resins	28	37	47
Phenolic resins	21	12	9
Melamine resins	6	5	7
Polyacetal	9	4	5
Pentaerythrol	5	4	7
Hexamethylenetetramine	7	4	7
Others	24	34	16

As shown in Table 2.6, approximately 55% of the total formaldehyde consumption occurs in the production of thermosetting resins. Since 30–55% aqueous solutions are being used for resin production, it is often the case that the consumers install their own small formaldehyde plants which are supplied with methanol by a large central methanol plant. In spite of the high transportation costs for formaldehyde solutions the relatively small capital investment will be soon repayed. For this reason, there are about 53 formaldehyde plants in Western Europe at the present time with a total capacity of 4.3 million tons per year (USA 4 million tons per year) in 1976.

Formaldehyde is the more reasonably priced component in phenolic resins. As an average, throughout all fields of application, about 1.6 mol formaldehyde including HMTA per mol phenol are used.

Instead of formaldehyde solutions of 30–55%, higher concentrated aqueous or alcoholic ones may be used. These solutions are produced by dissolving paraformaldehyde in water at 80–100 °C by addition of a small quantity (1%) of NaOH or ter-

Table 2.7. Specification of a formaldehyde solution

Formaldehyde		37.0 ± 0.1%
Methanol	max.	0.5%
Formic acid	max.	0.02 %
Chloride	max.	0.5 ppm
Fe	max.	0.12 ppm
Al	max.	0.25 ppm

tiary amines as depolymerization catalyst. It is also possible to concentrate formaldehyde solutions by adding paraformaldehyde.

A specification of resin grade formaldehyde solutions is shown in Table 2.7.

2.2.2. Paraformaldehyde

Paraformaldehyde[34, 41] is a white, solid, low molecular polycondensation product of methylene glycol with the characteristic odor of formaldehyde. The degree of polymerisation ranges between 10 and 100 . Types of paraformaldehyde common in the trade contain approximately 1–6.5% of water. The preparation of paraformaldehyde is performed by distillation of 30–37% aqueous formaldehyde solutions. According to the conditions (temperature, time, pressure) different types of paraformaldehyde are obtained. The values in Table 2.8 show that paraformaldehyde is not a defined compound.

Table 2.8. Properties of paraformaldehyde[41]

Content of formaldehyde	90–97%
Content of free water	0.2–4%
Specific weight	1.2–1.3 g/cm^3
Flash point	71 °C
Melting range	120–170 °C

Paraformaldehyde is only very seldom used for resin production because of its high price compared with aqueous formaldehyde solutions and because of problems associated with the exothermal heat evolution (Chapter 4). Paraformaldehyde and an acid catalyst may be used to cure novolak resins. However, the odor and high formaldehyde loss make it unattractive. Products obtained are of poorer quality than when HMTA is used.

On the other hand, paraformaldehyde is used almost exclusively to crosslink resorcinol prepolymers, e. g. in cold setting structural wood adhesives. Lower curing temperatures are adequate because of the higher reactivity of resorcinol, thus formaldehyde evolution is greatly reduced. The reactivity of paraformaldehyde depends on the degree of polymerization. A fairly accurate reactivity test method is the resorcinol test. This test indicates the period of time in minutes in which an alkaline resorcinol/paraformaldehyde mixture heats up to 60 °C due to the "condensation" reaction.

2.2.3. Trioxane and Cyclic Formals

Trioxane, a cyclic low molecular weight derivative of formaldehyde or methylene glycol, is a colorless solid (MP 62–64 °C, BP 115 °C) and can be prepared by the heating

$$\text{Trioxane} \xrightarrow[H_2O]{BF_3} HO-CH_2\left[O-CH_2\right]_n OH \longleftrightarrow CH_2O + H_2O \qquad (2.20)$$

Trioxane Poly-oximethylene Formaldehyde

of paraformaldehyde or a formaldehyde solution (60–65%) in the presence of 2% of sulfuric acid[34]. Trioxane can be used as formaldehyde spending and curing agent for phenolic resins.

Cyclic formals: 1,3-dioxolane, 4-phenyl-1,3-dioxolane and 4-methyl-1,3-dioxolane have been recommended to cure novolaks and because of their solvent action, for low-pressure laminating resins with reduced viscosity[42, 43].

2.2.4. Hexamethylenetetramine, HMTA

HMTA, used almost exclusively for cross-linking of novolak resins, is prepared from formaldehyde and ammonia according to Eq. (2.21).

$$6\,CH_2O + 4\,NH_3 \rightleftharpoons (CH_2)_6N_3 + 6\,H_2O \qquad (2.21)$$

The reaction is reversible. HMTA is split at elevated temperatures, generally above 115 °C, depending on the medium[44]. In aqueous solutions, HMTA is easily hydrolyzed; the corresponding reaction mechanism is indicated in Section 3.3.2. HMTA is often used as catalyst in the resole formation reaction instead of ammonia, yielding equivalent results.

Table 2.9. Physical properties of hexamethylenetetramine

Molecular weight	140.2
Specific weight g/cm^3	1.39
Behaviour when heated	sublimation at 270–280 °C
Solubility in 100 g H_2O/20 °C	87.4 g
Solubility in 100 g H_2O/60 °C	84.4 g

HMTA is highly soluble in water, relatively easy to dissolve in chloroform, and less soluble in methanol or ethanol. The aqueous solution shows a weak alkaline action with a pH range between 7–10. Finely ground HMTA tends to cause dust explosions.

A further application for HMTA is in the explosives industry: cyclotrimethylenetrinitramine is obtained by nitration (Hexogen, Cyclonite, RDX). In the pharmaceutical field, it is used as disinfectant (Urotropin). In the rubber industry HMTA is used as an activator for the vulcanization reaction[28].

2.2.5. Furfural

Furfural, sometimes called furfurol, is a colorless liquid with chemical properties similar to benzaldehyde. Commercial production starts with residues of annual plants like maize cobs, bagasse or rice hulls. These naturally occurring pentosans are hydrolized by diluted sulfuric acid to furfural which is then isolated by steam distillation. With alkaline catalysts the first step in the reaction with phenol is similar to that of formaldehyde, yielding α-(*o*- or *p*-hydroxyphenyl)-furfuryl alcohol[45] (Eq. 2.22).

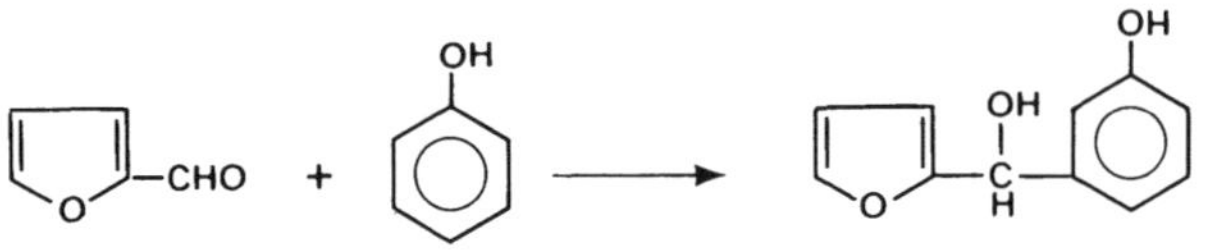

(2.22)

The further reaction mechanism is not established. Furane ring scission and reaction with another phenol nucleus may occur leading to relative wide internuclear distances.

Also the double bond may be involved in the polymerization reaction and leads to instability under acidic conditions. The phenol-furfural resins show enhanced flexibility, low melt viscosity and a low viscosity index. Furfural is used in combination with formaldehyde for the preparation of resins for grinding and friction materials. Phenol-furfural resins may be prepared by continuous addition (30 min) of furfural to a phenol melt at 135 °C and refluxing for 3.5 hr[46]. Similarly mixed formaldehyde-furfural resins can be prepared[47].

An important derivative of furfural is furfuryl alcohol. It is obtained from furfural by hydrogenation (2.2.3). Furfuryl alcohol/PF resin blends and acidic catalysts are used in the foundry industry for the no-bake and hot-box core-making process and for the preparation of acid resistant cements. Furfuryl alcohol is also used for the production of furane resins.

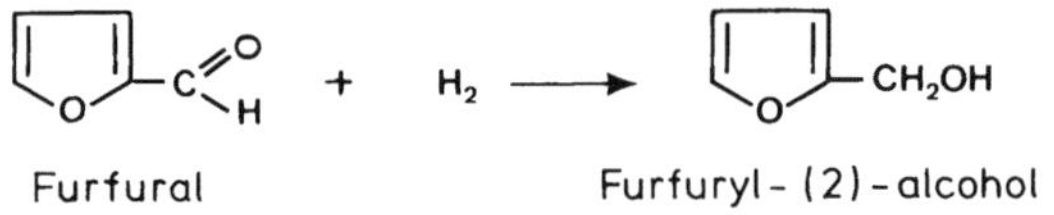

$$+ \; H_2 \longrightarrow \; -CH_2OH \qquad (2.23)$$

Furfural — Furfuryl-(2)-alcohol

2.2.6. Other Aldehydes

The higher aldehydes react with phenol at considerably lower rates. Acidic catalysts are preferred[48], e. g. for the preparation of certain binuclear antioxidants. A base-catalyzed reaction is not practical with acetaldehyde or higher aldehydes since they undergo rapid aldol condensation and self-resinification reactions. Novolak resins are prepared under strong acidic conditions, generally in a water-free system, by continuous aldehyde addition to the phenol melt. The preferred mol ratio phenol/aldehyde is between 1 : 0.8 to 1 : 1.3[49]. Only acetaldehyde and butyraldehydes are, however, of limited commercial importance, i. e. for rubber modification and antioxidants. The chemical structure of acetaldehyde novolak resins corresponds to those obtained by the reaction of acetylene and phenol with cyclohexylamine as catalyst[50].

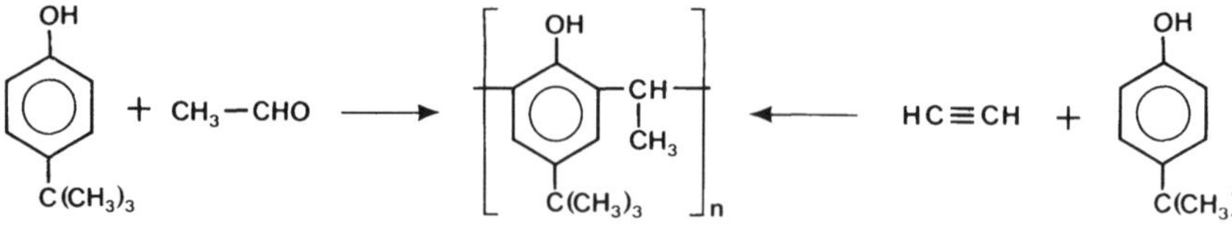

$$+ \; CH_3-CHO \longrightarrow \left[-CH(CH_3)- \right]_n \longleftarrow HC{\equiv}CH \; + \qquad (2.24)$$

References

1. Weast, R. C. (ed.): Handbook of Chemistry and Physics, 48th ed. Cleveland: The Chemical Rubber Co. 1967
2. Conrad, F. M.: Phenols. In: Encyclopedia of Polymer Science and Technology, Vol. 10, P. 73. New York: Interscience Publishers 1969

3. Bolton, P. D., Hall, F. M., Reece, J. H.: Spectrochim. Acta *22*, 1149 (1966)
4. Chen, D. T. J., Laidler, K. J.: Trans. Faraday Soc. *58*, 480 (1962)
5. Ko, H. C., O Hara, W. F., Hu, T., Hepler, L. G.: J. Chem. Eng. Data *9*, 122 (1964)
6. Kortüm, G., Vogel, W., Andrusson, K.: Dissociation Constants of Organic Acids in Aqueous Solution, London: Butterworth 1961
7. Böcker, E.: Phenole and Alkylphenole. In: Ullmanns Encyclopädie d. Techn. Chem.; Vol. 13, 3. Ed. München: Urban und Schwarzenberg: 1962
8. Hock, H., Lang, S.: Chem. Ber. *77*, 257 (1944)
9. N. N.: Chem. Ind. 29. Sept. 1977, 525
10. N. N.: Japan Chemical Week, 628, March 11 (1976)
11. Hayman, H.: CEP, Sept. 1977, 29
12. Ponder, T. C.: Hydrocarbon Proc. July 1978 P. 189
13. Jordan, W., Cornils, B.: Phenole. In: Methodicum Chimicum, Vol. 5, P. 105, Stuttgart: Thieme 1975
14. Kropf, H.: Chem.-Ing.-Techn., *36*, 759 (1964)
15. Pujado, P. R., Salazar, J. R., Berger, C. V.: Hydrocarbon Proc., March 1976, P. 91
16. Seubold, F. H., Vaugham, W. E.: J. Amer. Chem. Soc. *75*, 3790 (1953)
17. Flemming, J. B., Lambrix, J. R. Nixon, J. R.: Hydrocarbon Proc., Jan. 1976 (185)
18. McNeil, D.: Cresols. In: Kirk-Othmer: Encyclopedia of Chemical Technology, 2. Ed., Vol. 6, London: Interscience 1966
19. N. N.: Les cresols, Informations Chimie *94*, 127 (1971)
20. K. Ito: Hydrocarbon Proc., August 1973, P. 89
21. Franck, H.-G., Collin, G.: Steinkohlenteer. Berlin, Heidelberg, New York: Springer 1968
22. Wurm, H.-J.: Chem. Ing.-Techn., 48, 840 (1976)
23. Lurgi GmbH: Dephenolization of Effluents by the Phenosolvan Process, Technical Bulletin
24. Nylen, P., Sunderland, E.: Modern Surface Coatings: New York: Interscience Publishers 1965
25. BP Chemicals Int. Ltd.: Cellobond Products for Friction Materials, Technical Bulletin
26. Van Gils, G. E.: I & EC Product Research and Development 7/2, June 1968, P. 151
27. Rhodes, P. H.: Mod. Plastics, August 1947, 145
28. Koppers Co. Inc.: Resorcinol, Technical Bulletin
29. Allen, F. L.: Chem. Engng. 25, 118 (1967)
30. Conrad, F. M.: Bisphenol-A. In: E. G. Hancock (ed.): Benzene and its Industrial Derivatives. London: Ernst Benn Ltd. 1975
31. Barclay, R., Sulzberg, Th.: Bisphenols and their Bis (Chloroformates). In: J. K. Stille, T. W. Campbell (ed.): Condensation Monomers. New York: Willey 1973
32. Walker, J. F.: Formaldehyde. ACS Monograph No. 159, 3. Ed. (1964)
33. Zabicka, J. (ed.): The Chemistry of the Carbonyl Group, Interscience, 1970
34. Diehm, H., Hilt, A.: Formaldehyde. In: Ullmanns Encyclopädie der techn. Chem., Vol. 11, 4. Ed., Weinheim: Verlag Chemie 1976. Diehm, H.: Chemical Enging. *27*, 83 (1978)
35. Maux, R.: Air Best for Formaldehyde and Maleic, Hydrocarbon Proc., March 1976, P. 90
36. Haldor Topsoe A/S: Nissui Topsoe Formaldehyde Process, Technical Bulletin
37. Chauvel, A. R.: Hydrocarbon Proc., Sept. 1973, 179
38. N. N.: Hydrocarbon Proc., Nov. 1965, 216; Nov. 1969, 183; Nov. 1975, 150
39. N. N.: Hydrocarbon Proc., Nov. 1973, 135
40. Krauch, A., Kuntz, W.: Reaktionen der Organischen Chemie, Heidelberg: Hüthig 1976
41. Degussa: Paraformaldehyde, Technical Bulletin
42. Heslinga, A., Schors, A.: J. Appl. Polymer Sci. *8*, 1921 (1964)
43. Keutgen, W. A.: Phenolic Resins. In Kirk-Othmer: Encyclopedia of Chemical Technology, Vol. 15, 2. Ed., New York: Interscience 1968
44. Degussa: Hexamethylentetramin, Technical Bulletin
45. Porai, A. E., Koshitz et al.: Kunststoffe *23*, 27 (1933)
46. Brown, L. H.: Ind. Engng. Chem. *44*, 2673 (1952)
47. Mikes, J. A.: J. Polymer Sci. *53*, 1 (1961)
48. Martini, J. C.: J. Org. Chem. *35*, 2904 (1970)
49. BASF AG: DE-OS 2605 482 (1976)
50. I. G. Farb.: DE-PS 645 112 (1932)

3. Reaction Mechanisms

The conditions, mainly pH and temperature, under which reactions of phenols with formaldehyde are carried out, have a profound influence on the character of the products obtained. Three reaction phases have to be considered: formaldehyde addition to phenol, chain growth or prepolymer formation at temperatures < 100 °C and finally the cross-linking or hardening reaction at temperatures above 100 °C. The rate of the phenol-formaldehyde reaction at pH 1 to 4 is proportional to the hydrogen ion concentration, above pH 5 it is proportional to the hydroxyl ion concentration, indicating the change in reaction mechanism. Two prepolymer types are obtained depending on pH.

Novolaks are obtained by the reaction of phenol and formaldehyde in the acidic pH region. In general, the reaction is carried out at a molar ratio of 1 mol phenol to 0.75–0.85 mol of formaldehyde. Novolaks are mostly linear condensation products linked with methylene bridges of a relatively low MW up to approximately 2,000. These resins are soluble and permanently fusible, i.e. thermoplastic, and are cured only by addition of a hardener, almost exclusively formaldehyde applied as HMTA, to insoluble and infusible products.

Resols are obtained by alkaline reaction of phenols and aldehydes, whereby the aldehyde is used in excess. P/F molar ratios between 1:1.0 to 1:3.0 are customary in technical resols. These are mono- or polynuclear hydroxymethylphenols (HMP) which are stable at room temperature and, by application of heat, seldom of acids, are transformed into three-dimensionally cross-linked, insoluble and infusible polymers (resits) over different intermediate stages (resitols). However, the limited storage stability of resols at ambient temperature must be taken into consideration.

The methylene bridge is thermodynamically the most stable cross-linkage. It is prevalent in completely cured phenolic resins. Theoretically, 1.5 mol of formaldehyde is needed for the complete three-dimensional cross-linking of 1 mol of phenol (3.1). A higher proportion of formaldehyde is used in technical resins. On an average, considering all fields of application, approximately 1.6 mol formaldehyde are used. This excess is necessary to meet distinct technical requirements, for instance resin efficiency or free phenol content.

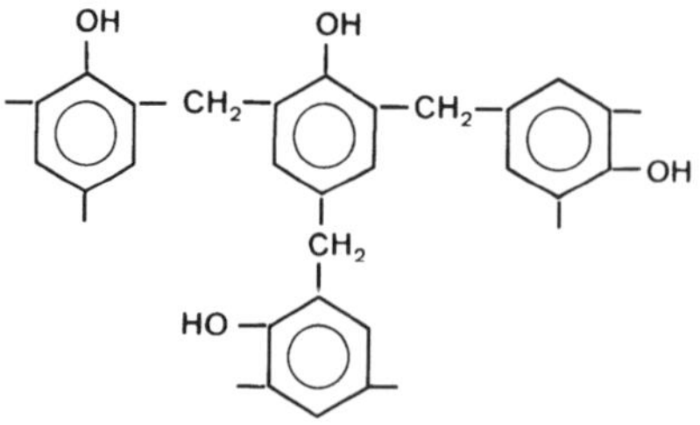

(3.1)

Phenols as condensation monomers have a functionality of 1 to 3 according to the substitution (Table 3.1). The functionality of the aldehydes always has the value 2.

Table 3.1. Functionality of different phenols in the reaction with aldehydes

Functionality		
1	2	3
1,2,6-Xylenol	*o*-Cresol	Phenol
1,2,4-Xylenol	*p*-Cresol	*m*-Cresol
	1,3,4-Xylenol	1,3,5-Xylenol
	1,2,5-Xylenol	Resorcinol
	p-tert.-Butylcresol	
	p-Nonylphenol	

3.1. Molecular Structure and Reactivity of Phenols

The molecular configuration in solution and crystal structure[1] of phenol is determined by a strong inclination to form hydrogen bonds[2]. In the solid state phenol forms H-bonded chains in the form of a threefold spiral[2]. In solution e.g. in benzene containing small amounts of water, trimolecular species (3.2), Ph_3, $Ph_2 \cdot H_2O$, $Ph \cdot 2\,H_2O$ are formed[3].

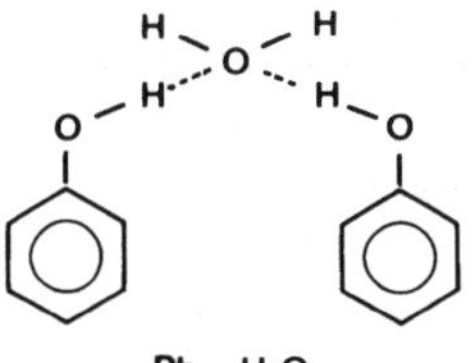

$Ph_2 \cdot H_2O$ (3.2)

For Ph_3 a cyclic structure was proposed. 2-Hydroxymethylphenol forms a strong intramolecular hydrogen bond[4]. The tendency and extent of H-bonding of phenols can be easily detected by NMR-chemical shift or by infrared frequency change. A linear relation between the thermodynamic functions for H-bond formation and pK_a-values exists. In Table 3.2 it is shown that alkyl substituted phenols have only a little lower acidity compared to phenol. This is also confirmed by calculated electron densities listed in Table 3.5. The differences between phenol and cresols are very small. More pronounced is the effect of bulky substituents in ortho-position because of steric factors (Table 2.1, Chapter 2). Hydroxymethylphenols are stronger acids than phenol. Phenols in their electronically excited states are more acidic than the ground state molecules as deduced from spectroscopic data[5,6].

The hydroxyl group is in the benzene ring plane even for 2,6-di-tert-butylphenol[11]. *Ortho*- and *meta*-substituted phenols exist as *cis* (α) and *trans* (β) isomers with refer-

Table 3.2. Acidity of phenols at 25 °C

Compound	pK_a(25 °C)	Ref.
Phenol	10.00	7,8
o-Cresol	10.33	7,8
m-Cresol	10.10	7,8
p-Cresol	10.28	7,8
2-Hydroxymethylphenol	9,84	9,10
4-Hydroxymethylphenol	9,73	9,10
2,4-Dihydroxymethylphenol	9,69	9
2,4,6-Trihydroxymethylphenol	9,45	9

ence of the orientation of the hydroxyl group. If the group in the *o*-position can act as a H-bond acceptor, the *cis* form is energetically preferred. For instance, this is also found for *ortho*-cresol and confirmed by the calculated energies indicated in Table 3.3[12]. For bulky substituents the *trans*-isomer is favoured.

The hydroxyl group is an inductive electron withdrawing (−I) and conjugatively electron releasing (+R) group. Both effects favour *para*-substitution. Steric reasons also decrease the accessability of the *ortho* positions. In comparison with other activating groups the following order with decreasing activating power may be given:

$NMe_2 > NHMe > NH_2 > OH > OMe > Me$

The oxide ion group in the phenoxide ion is a very strong activating substituent, stronger than NR_2, and more *ortho* directing than the hydroxyl group. The calculations (Table 3.4) indicate clearly the higher electron density of the *ortho* position compared to the *para* position in the phenoxide ion. This differentiation is not so marked in the neutral molecules. The direct experimental comparison of phenol substitution rates with benzene, to indicate the relative activating power of the hydroxyl group on the benzene nucleus, is impracticable because of the extremely large difference[11] in the reaction rates in the order of 10^{10}).

The formaldehyde-phenol reaction corresponds to an electrophilic aromatic substitution in acidic as well as in alkaline environments. It is generally assumed that this reaction type involves the rate determining formation of a π-complex followed by rapid loss of a proton. The actual pathway of reaction, however, is much more complicated with phenols because of solvent interactions and inter- and intramolecular hydrogen bond formation, the abnormal and wide variation in *ortho*/*para* ratio supports this.

The attack on the *para* position is favoured by polar solvents and acidic conditions, while the attack on the *ortho* position is favored by nonpolar solvents, alkaline conditions and group II metal oxide-, hydroxide-, or acetate catalysts. The results of kinetic examinations are described in Section 3.3.3.

In the last few years, because of the fast growth of the electronic computers, it has been possible to employ more sophisticated quantum theoretical calculations which not only include the π-electrons, but all valence electrons of large molecules.

These theoretical calculations have proved to be a powerful tool in structural chemistry. One of the most applied semi-empirical calculations which includes the 1 s electron of hydrogen, the 2 s, 2 p_x, 2 p_y, 2 p_z electrons of carbon is the CNDO/2(Complete Neglect of Differential Overlap)-method introduced by Pople et al.[13]. This method has been used by J. V. Knop[12] for the calculation of the electronic structure for the ground and the excited singlet and triplet states of the following molecules (Table 3.3):

Table 3.3. CNDO/2 Calculations. Ground state energy and dipole moment of phenols[12]

Compound	Ground state energy [eV]	Dipole moment [Debye]
Phenol	–1512.32185	1,52
Phenoxide ion	–1486.42812	7,89
ortho-Cresol (α)	–1720.89644	1,75
ortho-Cresol (β)	–1720.68707	1,46
meta-Cresol (α)	–1720.64869	1,92
meta-Cresol (β)	–1720.65463	1,60
para-Cresol	–1720.64234	1,79
ortho-Hydroxymethylphenol	–2111.80143	1,55
meta-Hydroxymethylphenol	–2111.70331	0,23
para-Hydroxymethylphenol	–2123.93774	1,92

Table 3.4. CNDO/2 and CNDO/S calculations. Electron densities of the ground and first excited states of phenols[12]

Compound	Position	Electron density		
		Ground state	1st-Singlet	1st-Triplet
Phenol	1	3,836	3,834	3,910
	2	4,005	4,037	4,018
	3	3,995	4,079	4,032
	4	4,011	3,971	4,041
	5	3,997	4,069	4,041
	6	4,015	4,063	4,020
	7	6,378	6,282	6,274
ortho-Cresol (α)	1	3,851	3,846	3,913
	2	3,964	4,015	3,982
	3	4,001	4,078	4,051
	4	4,016	3,991	4,052
	5	4,005	4,076	4,038
	6	4,011	4,051	4,037
	7	6,374	6,278	6,279

Table 3.4. (continued)

Compound	Position	Electron density		
		Ground state	1st-Singlet	1st-Triplet
meta-Cresol (β)	1	3,843	3,864	3,920
	2	4,015	4,082	4,027
	3	3,944	3,982	3,995
	4	4,018	3,975	4,042
	5	3,999	4,120	4,058
	6	4,023	4,024	4,028
	7	6,367	6,278	6,273
para-Cresol	1	3,847	3,813	3,924
	2	4,007	4,094	4,038
	3	4,005	4,076	4,029
	4	3,959	3,890	4,007
	5	4,005	4,103	4,048
	6	4,017	4,080	4,030
	7	6,368	6,280	6,273
ortho-Hydroxymethylphenol	1	3,842	3,893	3,905
	2	3,962	3,922	3,981
	3	4,001	4,104	4,054
	4	4,010	4,042	4,050
	5	4,001	3,981	4,044
	6	4,004	4,100	4,045
	7	6,380	6,298	6,283
para-Hydroxymethylphenol	1	3,861	3,746	3,937
	2	4,003	4,188	4,060
	3	4,020	4,107	4,024
	4	3,894	3,803	4,024
	5	4,026	4,173	4,056
	6	4,011	4,157	4,042
	7	6,381	6,294	6,292

The electron density distribution of the neutral molecules is not a sufficient basis for the interpretation of the reported kinetic data. This applies especially to the excited states (1st singlet and triplet state). An increased differentiation of the electron density distribution is found for the corresponding ions (Table 3.5). The electron density of the *para*-position in the phenoxide ion is remarkably higher than that

Table 3.5. CNDO/2 and CNDO/S calculations. Electron densities of the ground and first excited states of phenoxide ions[12)]

Compound	Position	Electron density		
		Ground state	1st-Singlet	1st-Triplet
Phenoxide ion	1	3,901	3,815	3,991
	2	4,048	4,167	4,007
	3	4,022	4,254	4,144
	4	4,063	3,946	4,206
	5	4,022	4,254	4,144
	6	4,048	4,167	4,007
	7	6,765	6,387	6,493
ortho-Hydroxymethyl-phenoxide ion	1	3.848	4,056	3,904
	2	3,970	3,989	3,979
	3	4,030	4,175	4,175
	4	4,054	4,382	4,166
	5	4,029	4,238	4,158
	6	4.050	4,127	4,102
	7	6.594	6,224	6,453
para-Hydroxymethyl-phenoxide ion	1	3,910	3,812	3,957
	2	4,047	4,198	4,041
	3	4,031	4,244	4,211
	4	3,944	3,839	4,089
	5	4,036	4,291	4,172
	6	4.042	4,203	4,029
	7	6.761	6,464	6,512

of the *ortho*-position and is therefore an adequate explanation of the *ortho*/*para* ratio stated by the experiment. The different effect of the hydroxymethyl group on the electron density of the remaining *ortho* and *para*-positions compared to the phenoxide ion is also very interesting. The electron density of the remaining *ortho*-position is increased by the introduction of the hydroxymethyl group into the *ortho*-position, whereas that of the *para*-position is decreased.

The electron density of both *ortho*-positions is decreased by the introduction of the methylol group into the *para*-position.

3.2. Formaldehyde-Water and Formaldehyde-Alcohol Equilibria

Formaldehyde is by far the most reactive carbonyl compound. In aqueous medium a very fast acid and base catalyzed hydration reaction of formaldehyde to methylene glycol occurs[14, 15]. The equilibrium indicated in Eq. (3.3) far on the side of methylene glycol, can be estimated by UV-spectroscopy ($n-\pi^*$ transition of the carbonyl group), by NMR or by polarographic methods[16, 17]

$$CH_2=O + H_2O \underset{k_2}{\overset{k_1}{\rightleftharpoons}} HO-CH_2-OH \tag{3.3}$$

$$K_d = \frac{[CH_2O]}{[HOCH_2OH]} = 1.4 \cdot 10^{-14}$$

Methylene glycol is found in aqueous solutions as a low molecular condensation polymer. It is also obtained by dissolving paraformaldehyde. The concentration of monomeric non-hydrated formaldehyde is very low, generally less than 0.01%[14]. The MWD of polymethylene glycol in a 40% aqueous solution is indicated in Table 3.6.

Table 3.6. Molecular weight distribution of the polymethylene glycol in aqueous solution of 40% at 35 °C[15]

n	%	n	%
1	26.81	6	5.88
2	19.36	7	3.87
3	16.38	8	2.50
4	12.32	9	1.58
5	8.96	10	0.99

The methylene glycol is present as monomer only in very dilute aqueous formaldehyde solutions (1–2%). The depolymerization of aqueous polyoxymethylene

$$HO+CH_2-O]_{\overline{n}}H + H_2O \rightleftharpoons HO+CH_2O]_{\overline{n-1}}H + HO-CH_2-OH \tag{3.4}$$

glycol (3.4) in the presence of acidic and basic catalysts is of importance for the overall reaction rates for resol and novolak formation.

Alcohols are often present in the P/F reaction. Methanol is present at least in small amounts (~1%) because the formaldehyde production process starts from methanol. In addition it can be formed from formaldehyde during storage by disproportionation (Cannizzaro reaction). Furthermore, methanol may be added because it is very efficient in stabilization of concentrated aqueous formaldehyde solutions. The chain termination prevents the formation of low soluble polymers so that precipitation or turbidity will be omitted. Alcohols can react with aqueous formaldehyde in a neutral pH to form hemiformals (3.5). Diformals are not formed under these conditions.

$$R{-}OH + HO{-}CH_2{-}OH \rightleftharpoons R{-}O{-}CH_2{-}OH + H_2O \tag{3.5}$$

Also the reaction between hydroxymethylphenols and methylene glycol must be considered. The extent of this reaction with the hydroxymethyl group (3.6) as well as with the phenolic hydroxyl group (3.7) has been studied by high-resolution NMR[18].

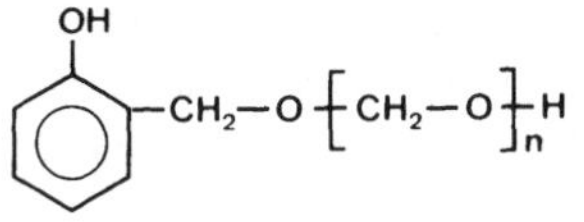

OH
$C_6H_4(OH){-}CH_2{-}O{-}[CH_2{-}O]_n{-}H$ (3.6)

$C_6H_5{-}O{-}[CH_2{-}O]_n{-}H$ (3.7)

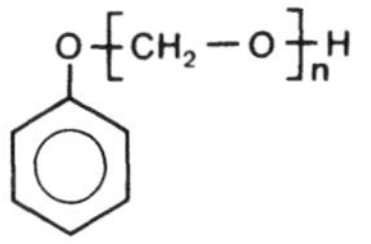

Peaks for n = 0, 1, 2, 3 have been identified. In a mixture of 70 parts of 40% formalin and 100 parts of phenol approximately 10% of the phenol has reacted with methylene glycol to form phenol hemiformal.

Formaldehyde is therefore consumed in the P/F-reaction also to form hemiformals which constitute a potential source of formaldehyde and may be detected by usual titrimetric methods, but is otherwise not more directly involved in the hydroxymethylation reaction. This should be a sufficient reason for the apparent reduction of the reaction rate at rising conversion as stated by Zsavitsas[19].

3.3. Phenol-Formaldehyde Reaction under Alkaline Conditions

The reaction between formaldehyde and phenol in the alkaline pH-range was first mentioned in 1894 by L. Lederer[20] and O. Manasse[21]. It is therefore occasionally also designated as the Lederer-Manasse reaction. At a pH above 5, *bis*- and *tris* alcohols are formed as well as *mono* alcohols and other compounds. The simplest product of this reaction, 2-hydroxybenzylalcohol (saligenin), was already isolated from the glucoside salicyn by hydrolysis with diluted acids in 1845[22].

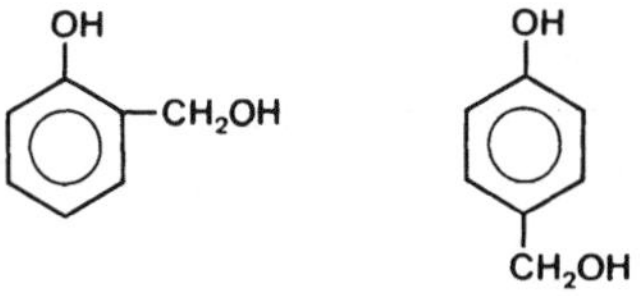

(3.8)

o-Hydroxymethylphenol MP 86°C

p-Hydroxymethylphenol MP 124...126°C

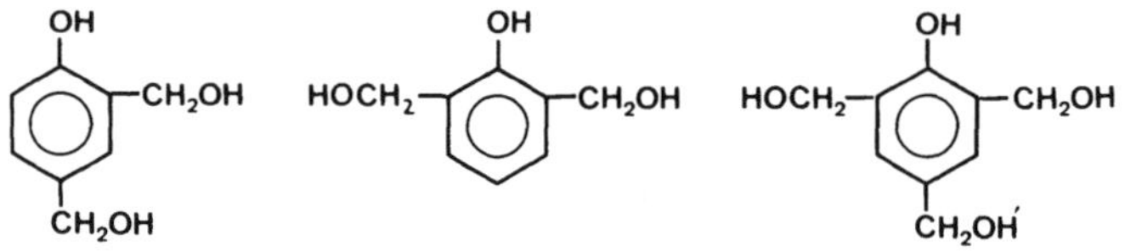

(3.9)

o,p-Dihydroxymethylphenol MP 93 °C   *o,o'*-Dihydroxymethylphenol MP 101 °C   Trihydroxymethylphenol MP 79. . .82 °C

3.3.1. Inorganic Catalysts and Tertiary Amines

Sodium hydroxide, ammonia and HMTA, sodium carbonate, calcium-, magnesium- and barium hydroxide and tertiary amines are used as catalysts in the alkaline hydroxymethylation reaction. In aqueous solutions, as used in all technical processes, formaldehyde is present as methylene glycol. Phenol reacts quickly with alkali hydroxide to form the phenoxide ion which is stabilized by resonance according to Eq. (3.10)

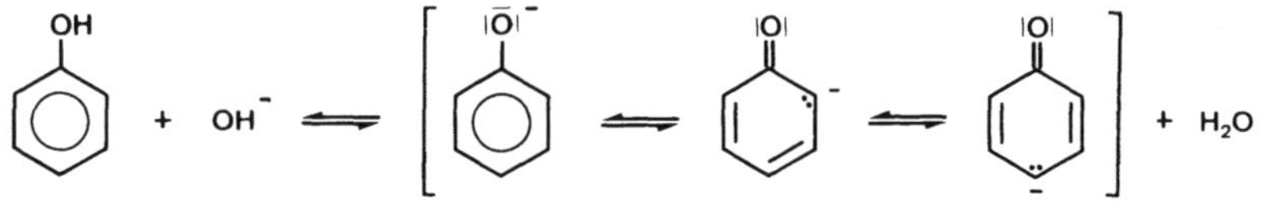

(3.10)

In the following reaction, catalyzed by alkalis, C-alkylation in *ortho* and *para* position occurs almost exclusively. *Meta* substitution is hardly not evident.

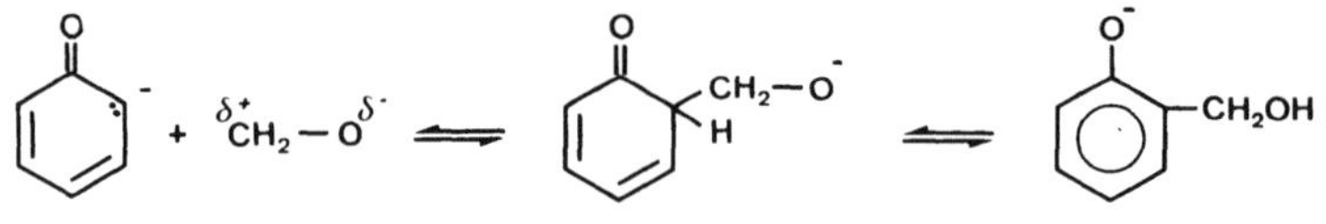

(3.11)

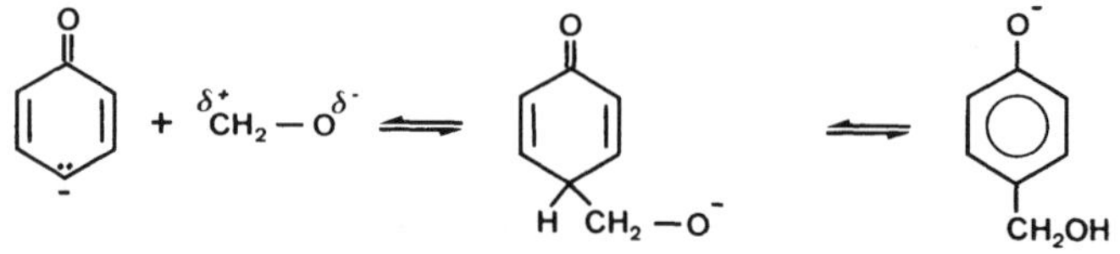

(3.12)

The quinoide transition state is stabilized by proton shifting as indicated in the Eqs. (3.11) and (3.12). This reaction mechanism was recommended for dilute solutions[24]. The monomethylol derivative continues to react with formaldehyde, forming two dimethylol- and one trimethylol compounds.

The kinetics of the base-catalyzed phenol-formaldehyde reaction has been thoroughly researched[23–27] and is relatively well understood. In general, a second order reaction type was found, with the exception of the ammonia catalyzed reaction, which rather corresponds to one of the first order. The general expression for the overall reaction rate is:

$$\text{reaction rate} = k[Ph^-][\text{methylene glycol}]$$

It must be pointed out, however, that the actual constitution of the hydroxyalkylating agent in the alkaline catalyzed reaction is not yet fully understood. It is not clear how methylene glycol would react with the phenoxide ion[19]. The concentration of non-hydrated formaldehyde is too low to explain the reaction rates.

$$HO{-}CH_2{-}OH \rightleftharpoons {}^{+}CH_2{-}OH + OH^- \qquad (3.13)$$

$$HO{-}CH_2{-}OH \rightleftharpoons \overset{\delta+}{CH_2}{-}\overset{\delta-}{O} + H_2O \qquad (3.14)$$

A deviating reaction mechanism was proposed very early by Claisen, later by Walker[14] and others. The presence of hemiformals 3.7 in aqueous phenol-formaldehyde solutions has been proven by the means of NMR[18].

O–CH₂OH ⇌ O⁻ CH₂OH (3.15)

As evidence for the formation of hemiformals as intermediates, followed by a shift of the hydroxymethyl group according to Eq. (3.15), the absence of the reaction, if the phenolic group is etherified, was pointed out. This statement, however, is not correct since the nucleophily of the phenoxide anion is the decisive parameter in the alkaline hydroxymethylation.

A series of experimental results indicates that the constitution of the transition state is considerably more complex than indicated in the Eq. (3.11). The strongest evidence is the dependence of the *ortho/para* substitution ratio on the type of catalyst[26, 28]. The *ortho/para* ratio decreases from 1.1 at pH 8.7 to 0.38 at pH 13.0. It has been recognized that the *ortho* substitution is considerably enhanced if metal hydroxides of the first and second main group along the series K < Na < Li < Ba < Sr < Ca < Mg are used as catalysts (Fig. 3.1). Even more distinct is the effect of the hydroxides of transition metals. The *ortho* substitution is the more favored, the higher the chelating strength of the cation. The directing effect of the Fe, Cu, Cr, Ni, Co, Mn and Zn ions is explained by Peer as formation of chelates as transient compounds according to formula (3.16). Boric acid also has a strong *ortho*-directing effect (3.17).

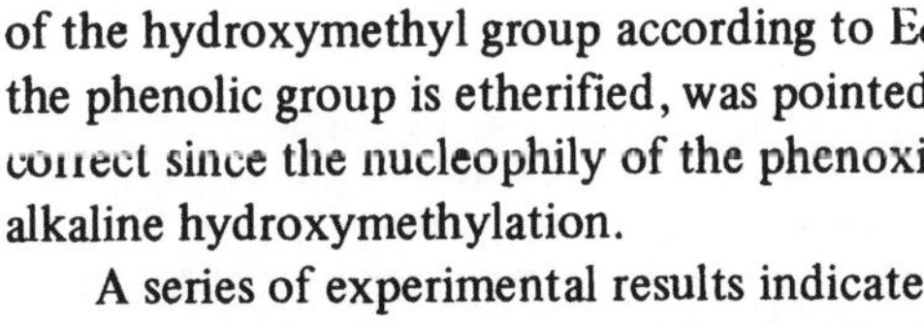

Ph, OH, Me, O, O, CH₂, H ⇌ Ph, OH, Me, O, O, H, CH₂ (3.16)

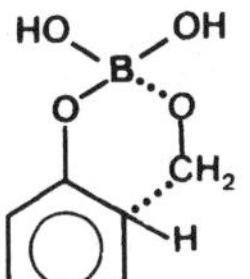

HO, OH, B, O, O, CH₂, H (3.17)

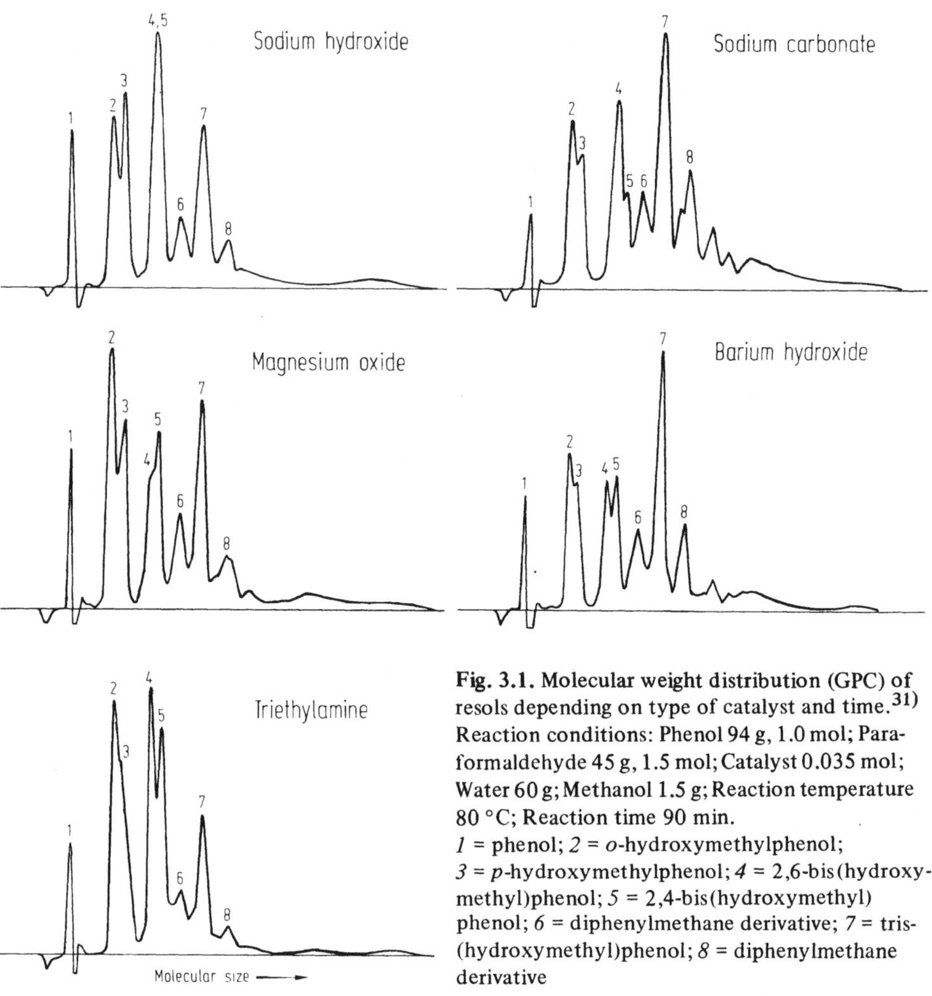

Fig. 3.1. Molecular weight distribution (GPC) of resols depending on type of catalyst and time.[31] Reaction conditions: Phenol 94 g, 1.0 mol; Paraformaldehyde 45 g, 1.5 mol; Catalyst 0.035 mol; Water 60 g; Methanol 1.5 g; Reaction temperature 80 °C; Reaction time 90 min.
1 = phenol; *2* = *o*-hydroxymethylphenol; *3* = *p*-hydroxymethylphenol; *4* = 2,6-bis(hydroxymethyl)phenol; *5* = 2,4-bis(hydroxymethyl) phenol; *6* = diphenylmethane derivative; *7* = tris-(hydroxymethyl)phenol; *8* = diphenylmethane derivative

The formation of "high ortho" novolaks if magnesium oxide or zinc oxide are used as catalysts has already been stated by Bender[29, 30]. The different activity of some of the more frequently used catalysts and their effect on MWD examined by GPC is shown in Figure 3.1[31]. Identical reaction conditions have been used for the production of the resols. The catalyst was neutralized with hydrochloric acid and the resin analyzed without previous distillation. The highest degree of *ortho* orientation is observed when zinc acetate is used (Fig. 3.5), followed by magnesium oxide and triethylamine as catalysts (Fig. 3.1).

The interpretation of the orientation and directing effects depending upon the type of catalyst must be performed with care. The concentration of any particular alcohol in the reaction mixture does not only depend upon the rate of formation, but also upon the rate of disappearance due to further reaction, i.e. it also depends upon the mol ratio of phenol to formaldehyde and the reaction time. The order of appearance of the individual methylol phenols in the GPC chromatogram depends despite of molecular size also on the number of hydroxyl groups because of solvent interaction.

3.3.2. Ammonia, HMTA and Amine-Catalyzed Reactions

Ammonia-catalyzed resols differ clearly from all other resols by their characteristic yellow color, which is to be attributed to the azomethine group – CH = N –, and also by their higher average MW. This color is also typical for novolaks cured with HMTA. It was shown[32, 33] that these prepolymers contain dibenzylamine (3.18) and tribenzylamine structures (3.19).

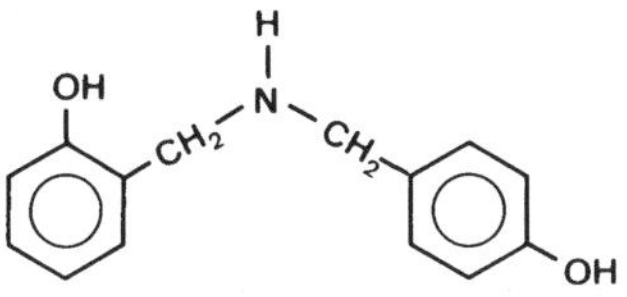

(3.18)

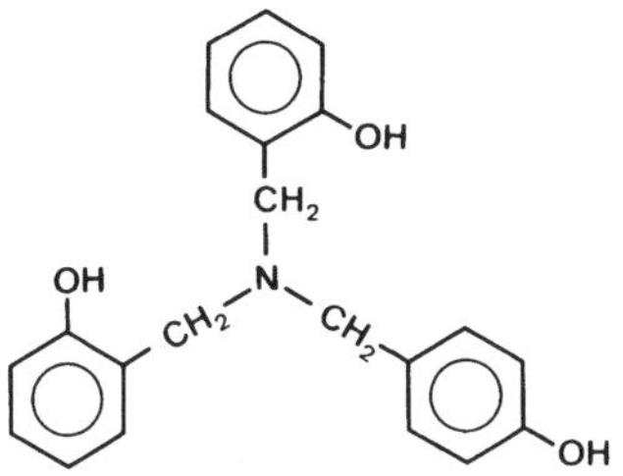

(3.19)

The reaction between phenols, aldehydes and amines can also be formally treated as an aminoalkylation according to the reaction sequence shown in equation (3.20), commonly designated as the Mannich-reaction[34].

$$\mathrm{Y{-}H} + \mathrm{O{=}C}\begin{matrix}\nearrow R_1 \\ \searrow R_2\end{matrix} + \mathrm{H{-}N}\begin{matrix}\nearrow R_3 \\ \searrow R_4\end{matrix} \xrightarrow{-H_2O} \mathrm{Y{-}}\overset{R_1}{\underset{R_2}{\mathrm{C}}}\mathrm{{-}N}\begin{matrix}\nearrow R_3 \\ \searrow R_4\end{matrix} \qquad (3.20)$$

H-acidic compound (Phenol)	Carbonyl compound (Formaldehyde)	Amine (Ammonia)	α-Aminoalkylated compound

Phenols are O–H acidic as well as C–H acidic compounds. However, in general, the substitution always takes place in the aromatic nucleus.

If amines are present at the same time, the carbonyl component is confronted with 2 nucleophiles. For this reason it must be determined whether the formaldehyde reacts with the phenol or with the amine first. The reaction with ammonia or aliphatic amines leads to crystalline hexahydrotriazines. The formation of HMTA from formaldehyde and ammonia is a very fast reaction in which the formation of the bis-hydroxymethylamine (reaction 3.21) is the slowest and, therefore, rate determinating step[35]. In diluted aqueous systems, the reaction stops at the stage of the α-aminoalcohols. Aldimines are only formed to a very small extent. All reaction steps are reversible.

The hydrolysis of HMTA also leads to aminomethylated products (reaction 3.22) so that HMTA or an equivalent amount of ammonia may be used interchangeably as catalysts giving equivalent results. HMTA[36] is used as hardener for curing novolaks

$$NH_3 + CH_2O \underset{fast}{\rightleftharpoons} NH_2CH_2OH \underset{slow}{\overset{CH_2O}{\rightleftharpoons}} NH(CH_2OH)_2$$

$$NH_2CH_2OH \xrightarrow[fast]{+NH_3\ -H_2O} NH_2CH_2NH_2 \tag{3.21}$$

$$NH_2CH_2NH_2 + NH(CH_2OH)_2 \xrightarrow{-2\,H_2O} \text{cyclo-}(NH-CH_2-NH-CH_2-NH-CH_2) \tag{3.22}$$

$$\text{cyclo-}(NH-CH_2-NH-CH_2-NH-CH_2) + NH(CH_2OH)_2 \rightleftharpoons \text{bicyclic } (CH_2-N-CH_2)(NH)(CH_2)(NH)(CH_2-N-CH_2) \overset{CH_2O}{\rightleftharpoons} \text{hexamethylenetetramine } (CH_2)_6N_4 \tag{3.23}$$

(Section 3.4.4). A simple reaction which is also of technical importance, is the formation of 2,4,6-tris-dimethylaminomethylphenol from phenol, formaldehyde and dimethylamine (reaction 3.24). The solubility of phenols in sodium hydroxide is frequently lost at the aminomethylation.

$$C_6H_5OH + 3\,CH_2O + 3\,HN(CH_3)_2 \longrightarrow 2,4,6\text{-}[(CH_3)_2N-CH_2]_3C_6H_2OH \tag{3.24}$$

The MWD of resols catalyzed by high amounts of ammonia deviate considerably from other resols. Tertiary amines lead to prepolymers which are similar to sodium hydroxide-catalyzed resols in their structure (Fig. 3.1). The structure of ammonia-catalyzed resols allows the technical production of solid resols with a melting range between 40 and 60 °C. Solid, pulverized resols are required for many applications, for instance for the production of molding compounds, brake lining mixtures, abrasive materials and coating resins. Resins with even higher melting points would be desirable for storage stable and free-flowing powders. However, this cannot be realized technically, since the curing reaction which starts between 90–100 °C prohibits a higher residue temperature while the discharging of the melt and its cooling requires considerable time. Ammonia-catalyzed resol solutions are of considerable importance for the production of impregnating resins for electrical laminates and coatings. In these cases, the removal of the catalyst (by precipitation or washing) is not required to obtain favorable dielectrical properties or high corrosion resistance.

3.3.3. Reaction Kinetics of the Base-Catalyzed Hydroxymethylation

It is the aim of kinetic research work to determine how a reaction proceeds. However, the determination of reliable kinetic data for phenolic resin formation is complicated because the reaction conditions, including temperature, type and amount of catalyst, and mol ratios and polarity of the solvents will exert a profound influence on the results obtained. The purity of the raw materials, and even the nature of the reaction vessel, will play significant roles. The identification of the reaction products is also relatively difficult[37]. Therefore, it is not surprising that the reported kinetic data differ considerably. For the interpretation of kinetic data, it must be considered that the *ortho*-position in phenol and *p*-hydroxymethylphenol is present twice. It is more reasonable, for the above mentioned reasons, to compare the relative reaction rates, which take in account the availability of the nuclear positions ("concentration"). The NaOH catalyzed reaction was the most thoroughly examined one. The relative reaction rates found are compiled in Table 3.7. The rate constant of the formation of *o*-HMP was set equal to one.

Table 3.7. Relative positional reaction rates in the phenol-formaldehyde reaction, *ortho*-substitution set as unity

		Relative reaction rates		
		Freeman and Lewis[25] 1954	Zsavitsas and Beaulieu[15] 1967	Eapen and Yeddanapalli[27] 1968
Phenol	→ *o*-methylolphenol	1.00	1.00	1.00
Phenol	→ *p*-methylolphenol	1.18	1.09	1.46
o-Methylolphenol	→ *o-o'*-dimethylolphenol	1.66	1.98	1.75
o-Methylolphenol	→ *o-p*-dimethylolphenol	1.39	1.80	3.05
p-Methylolphenol	→ *o-p*-dimethylolphenol	0.71	0.79	0.85
o-p-Dimethylolphenol	→ trimethylolphenol	1.73	1.67	2.04
o-o'-Dimethylolphenol	→ trimethylolphenol	7.94	3.33	4.36

One of the most complete studies was published by Freeman and Lewis[25]. The reaction was performed at 30 °C and a molar ratio of phenol/formaldehyde/NaOH of 1:3:1. One mol of formaldehyde was used per reactive nuclear position during the separate examination of the secondary reactions by use of model compounds such as 2- and 4-HMP and 2,4- and 2,6-BHMP.

Thus, it was possible to determine the rate constants of all competitive secondary reactions. The appearance and disappearance of each individual methylol compound was followed by quantitative paper chromatography.

The *para*-position in phenol shows a slightly higher relative reactivity towards formaldehyde than the *ortho*-position. However, *o*-HMP is produced at a higher rate due to the fact that two *ortho*-positions are available. Since the methylol group was regarded as electron attracting, a decrease in the reactivity of the remaining positions towards further formaldehyde addition was expected. This has also been observed in

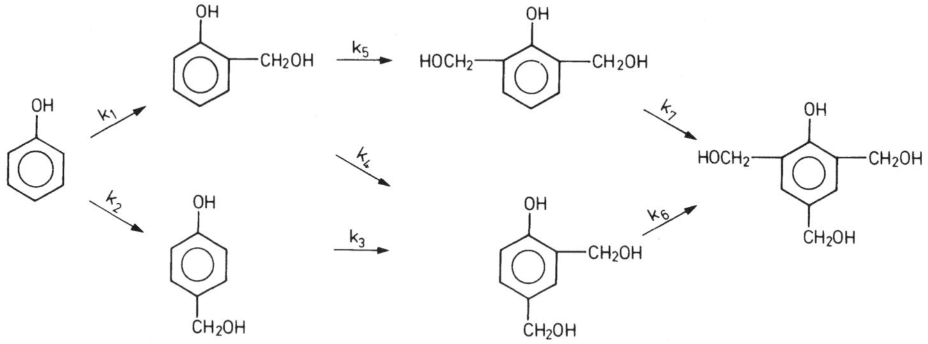

$k_1 = 6{,}2 \cdot 10^{-6}\ l\ mol^{-1}\ s^{-1}$
$k_2 = 10{,}5 \cdot 10^{-6}\ l\ mol^{-1}\ s^{-1}$
$k_3 = 7{,}5 \cdot 10^{-6}\ l\ mol^{-1}\ s^{-1}$
$k_4 = 7{,}3 \cdot 10^{-6}\ l\ mol^{-1}\ s^{-1}$
$k_5 = 8{,}7 \cdot 10^{-6}\ l\ mol^{-1}\ s^{-1}$
$k_6 = 9{,}1 \cdot 10^{-6}\ l\ mol^{-1}\ s^{-1}$
$k_7 = 41{,}7 \cdot 10^{-6}\ l\ mol^{-1}\ s^{-1}$

Fig. 3.2. Reaction steps and individual rate constants of the phenol-formaldehyde reaction according to Freeman and Lewis[25)]

the case of the *p*-HMP. However, a considerable increase in reactivity has been observed at the incorporation of the methylol group into the *ortho*-position. The latter is also found with the higher methylolated phenols, especially remarkable with 2,6-BHMP. The different effect of the hydroxymethyl group led to uncertainity in interpreting the kinetic data. The CNDO calculation of the electron densities of the *ortho*- and *para*-HMP ions (Table 3.5) confirms clearly the reduced electron density of the *ortho*-position of *p*-HMP and the increased electron density of the remaining *ortho*-position in the *o*-HMP ion, compared to phenoxide ion. The *para*-position in the *o*-HMP ion shows a reduced electron density in this calculation compared to phenoxide ion. However, the substitution rate reported does not correspond to this result.

The results of Freeman and Lewis were confirmed later by examinations performed by Zsavitsas[19, 24)] who improved the identification procedure by gas-liquid chromatographic separation of phenol alcohols[38)]. A theoretical calculation based upon pK_a-values, resulted in excellent agreement with the experimental values. The increasing positional reactivity of the system as the methylol substitution proceeds aiso has considerable practical importance. Due to the marked tendency towards the formation of polyalcohols, technical resols contain considerable amounts of unreacted phenol, even if a relatively high formaldehyde ratio is chosen. This is a disadvantage not only for reasons of resin efficiency in the technical processes but because it may also cause environmental problems.

The influence of methyl substitution on the reactivity was studied by Sprung[39)]. This study covered the reaction of phenol, cresols and xylenols with paraformaldehyde without addition of water using triethanolamine as catalyst. The results which are indicated in Table 3.8 must be considered as relative rates of formaldehyde consumption.

m-Cresol adds formaldehyde at nearly three times the rate of phenol. The methyl group introduced in *ortho* or *para* position renders the nucleus less reactive. How-

Table 3.8. Relative reaction rates of various phenols with formaldehyde[39]

Compound	Relative reactivity
2,6 Xylenol	0.16
o-Cresol	0.26
p-Cresol	0.35
2,5-Xylenol	0.71
3,4-Xylenol	0.83
Phenol	1.00
2,3,3-Trimethylphenol	1.49
m-Cresol	2.88
3,5-Xylenol	7.75

ever, an increase of the relative rates was also observed at methyl substitution in *ortho*- and *para*-position in a later study[27]. Here, further research work taking into account common reaction conditions is desirable. Halogens, though *ortho*- and *para*-directing, deactivate all nuclear positions in the phenol molecule. All *meta*-directing groups, for instance the nitro and sulfonic group, reduce the reactivity to a great extent.

The rate of methylolphenol formation is a function of pH above pH 5. Tertiary amines and quaternary ammonium hydroxides are comparable to sodium hydroxide in their activity, while mono- and diethylamine or ammonia (Fig. 3.1) are essentially weaker catalysts.

The kinetics of the alkali-catalyzed addition of formaldehyde to dihydroxy-diphenyl methanes (DPM) has been studied by quantitative paper chromatography[40]. The order of reactivity of the three DPM's is found to be *o,p'*>*p,p'*>*o,o'*-DPM. The low reactivity of *o,o'*-DPM is explained on the basis of the formation of a chelate ring in its singly ionized form. The rate constants for the addition of formaldehyde to DPM's are of about the same order of magnitude as those of the condensation of methylolphenols at low alkali concentrations. But at higher alkali concentrations, the rate of formaldehyde addition is increased while that of the condensation reactions in general is reduced (Fig. 3.4).

3.3.4. Prepolymer Formation

Apart from formaldehyde addition, selfcondensation reactions of the methylphenols or condensation with phenol take place at the usually applied conditions and temperatures between 60 and 100 °C leading to prepolymer formation. Below 60 °C and at high pH, the condensation reaction is negligible. The condensation reactions of the methylphenols have been investigated in particular by three research groups led by Zinke, von Euler and Hultzsch. The opinions of these teams only partially agree. The methodical procedures were comparable. Model compounds (mono- and di-methylolphenols) were synthetized, and, as they separate upon heating, the formaldehyde and water were estimated quantitatively. The evidence of composition of the intermediate compounds was given. A comprehensive description of these studies can be found in the book by Megson[37] and Martin[41]. Two reactions 3.25 and 3.26 are discussed above all.

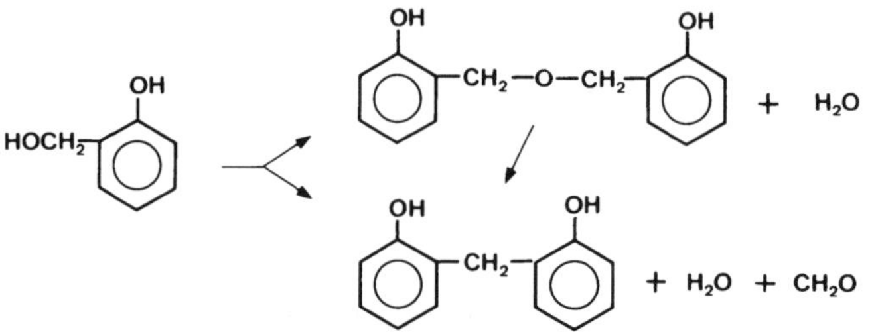

(3.25)

(3.26)

It seems at first easy to decide on the actual reaction pathway using the amount and ratio of water to formaldehyde being separated off. In fact, however, water and formaldehyde are found in changing proportions according to the composition of the methylol compound and temperature. In particular, lower amounts of formaldehyde are found as it was expected – only 0.6 mol or less – and in most cases higher amounts of water. The conditions of the thermal treatment of model compounds can, however, hardly be compared with actual reaction conditions as they are used for resin production and cure.

It was estimated later[42] that by far the most important reaction under strong alkaline conditions is the formation of diphenylmethanes according to Eq. (3.26).

The formation of dihydroxydibenzyl ether according to Eq. (3.25) is very unlikely under strong alkaline conditions. This reaction, however, is considered the prevalent one under neutral or weak acidic conditions and temperatures up to 130 °C as they normally exist at the curing of resols[43]. Above 130–150 °C methylene bridge formation becomes predominant[37] and, at still higher temperatures, a number of ill-defined reactions become important[44].

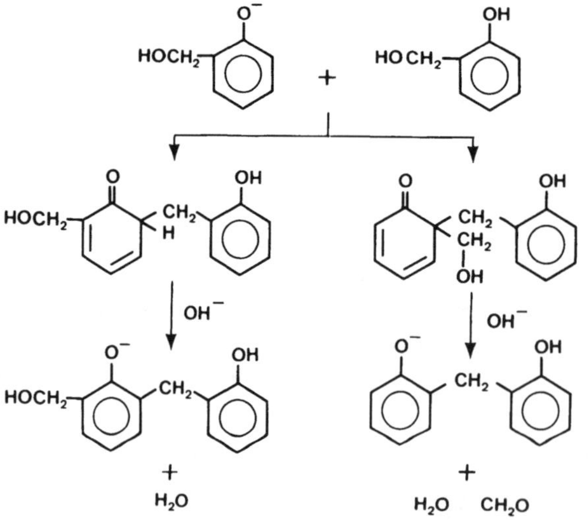

(3.27) (3.28)

It was shown that the condensation reaction between two HMP's proceeds substantially faster than the reaction between HMP and phenol[42]. The methylol group has a strong activating effect on the condensation reaction. The reaction between HMP's may occur by separation of a proton (3.27) or formaldehyde (3.28). Both reactions are believed to be S_N2 mechanisms, involving attack of a methylol substituent with displacement of a hydroxyl group[45].

The ratio of the reactions (3.27) and (3.28) depends upon the structure of the methylolphenols involved and the reaction conditions.

Under neutral or weakly acidic conditions and at lower temperatures linear dimethylene ether linkages containing polymers have been prepared from 2,6-bis-(hydroxymethyl)-4-methylphenol[46].

$$HOCH_2-C_6H_2(OH)(CH_3)-CH_2OH \xrightarrow[130\,^\circ C]{} \left[-CH_2-C_6H_2(OH)(CH_3)-CH_2-O-\right]_n + n\,H_2O \quad (3.29)$$

Dibenzyl ether formation is probably a S_N2 reaction involving displacement of a hydroxyl group catalyzed by protons supplied either by the phenolic group or by the neutralizing acid.

3.3.5. Resole Cross-linking Reactions. Quinone Methides

Heat curing, by far the most important hardening process for resols, is performed at temperatures between 130 and 200 °C. Since this is a polycondensation reaction, the molecular weight increase depending on conversion shows a different course in comparison to a polymerization reaction[47].

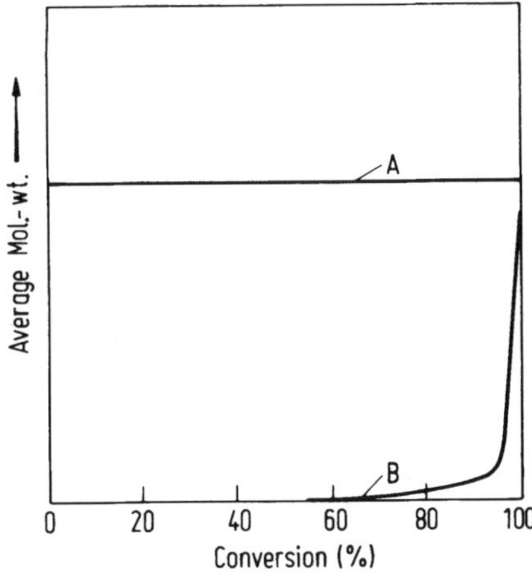

Fig. 3.3. Dependence of the average molecular weight on conversion, A = radical polymerization, B = polycondensation reaction

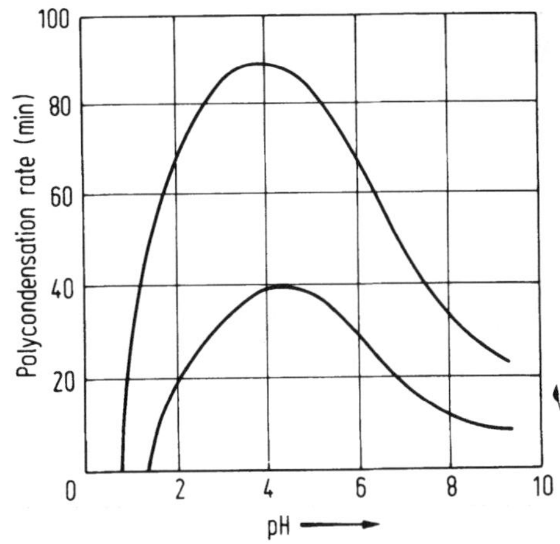

Fig. 3.4. Effect of pH and P/F-ratio on polycondensation rate of resol resins at 120 °C. According to Keutgen, Lit.[43], Chapt. 2

Since the curing reaction occurs under different conditions than prepolymer formation in aqueous solutions, deviating reaction mechanisms are possible. This is confirmed by the constitution of the bridge linkages and functional groups present in cured resols. The methylene bridge prevails as the most thermodynamically stable linkage.

Acid curing

The acid curing of phenol resols is only of limited technical importance. Resols can be cured by the addition of a variety of strong organic and inorganic acids at ambient temperature. *p*-Toluene sulfonic acid, hydrochloric acid and phosphoric acid are mostly used. Phenol sulfonic acid is preferred in newer techniques because of its excellent compatibility and its ability to be incorporated into the polymer backbone, thereby minimizing acid migration and subsequent corrosion problems. The products resulting from acid curing are well established[48]. Hultzsch[49] found that numerous hydroxymethylphenols under acidic conditions result in dihydroxydiarylmethane derivatives. The reaction mechanism corresponds to the second step in the hydroxymethylation in acidic medium, e.g. novolak formation, and is mentioned in detail in the following chapter. Applications for acid cured phenolics are mentioned in Chapter 12.2, 14 and 19.2.

Heat Curing

Hultzsch[50, 51] and von Euler[52, 53] considered quinone methides to be intermediates in the heat curing of resols (and prepolymer formation). In chemical structure[54], quinone methides are similar to quinones, but differ in their π-electron charge distribution[55]. Thus, they show a high reactivity towards electrophiles as well as towards nucleophiles. The structure can be written as resonance between the quinoid and benzoid structure (3.30).

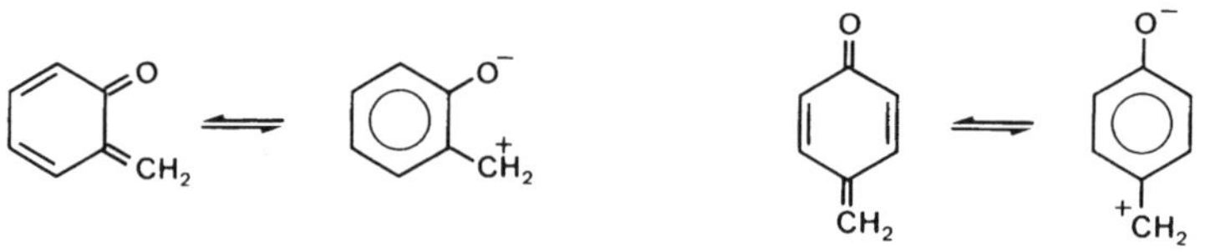

O O⁻ CH₂ ⁺CH₂ (3.30)

o-Benzoquinone methide (6-methylene-2,4-cyclohexadiene-1-one)

p-Benzoquinone methide (4-methylene-2,5-cyclohexadiene-1-one)

The reactivity of quinone methides is therefore generally so high that they cannot be isolated under normal conditions. In the absence of reactive compounds, they react among one another forming colorless dimers (3.31), trimers (3.32) or polymers.

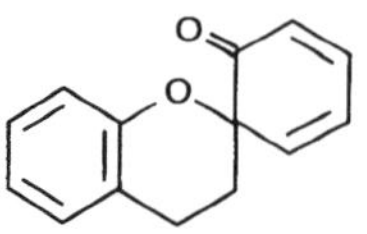

(3.31)

o-Quinone methide dimer

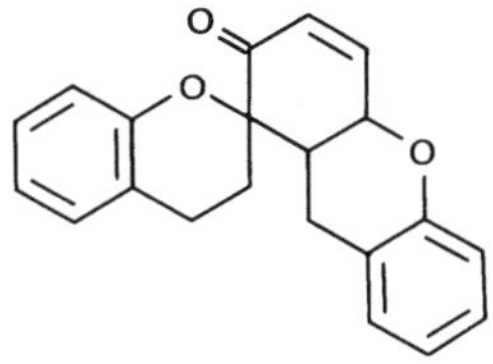

(3.32)

o- Quinone methide trimer

The preferred reaction is the Michael type addition of CH-acidic compounds (3.33).

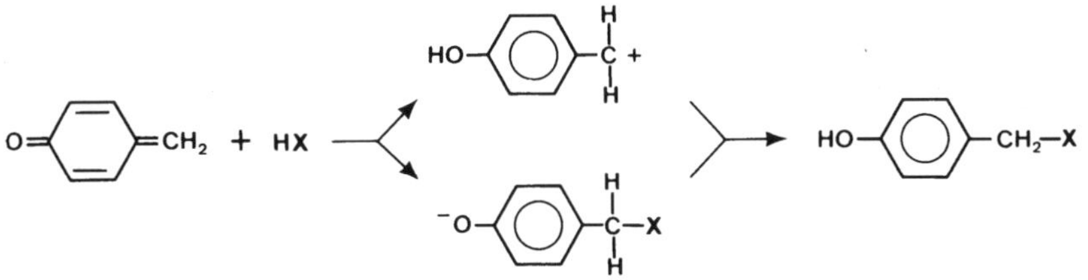

(3.33)

With hydroxyl compounds the reaction rate, which is greatly accelerated by traces of acids, is reduced in the following sequence[55]:

$$PhOH > H_2O > CH_3OH > C_2H_5OH \qquad (3.34)$$

In the presence of olefins, addition yielding chroman derivatives, e.g. flavan (3.35) in the case of styrene, occurs.

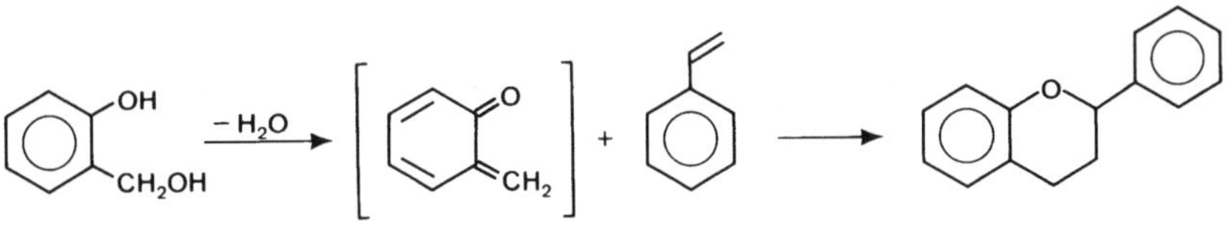

(3.35)

In the past this reaction was regarded as very important for the vulcanization of rubber and modification of phenolic resins with olefinic compounds, for instance rosin, tung oil and others. The reaction mechanism proposed by Hultzsch and separately by von Euler is stressed in Section 16.1.

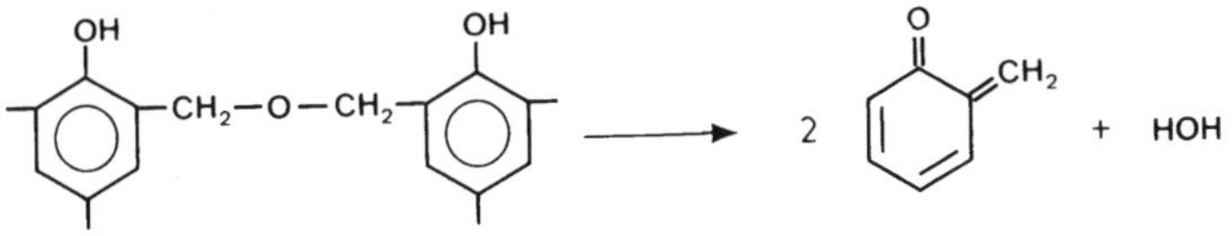

(3.36)

Hultzsch assumed that the quinone methides are derived from the dibenzylethers in accordance with Eq. (3.36). Von Euler pointed out that they can also be derived directly from the hydroxymethylphenols as shown in the reaction (3.35). However, it is very doubtful that quinone methides play an essential role at temperatures below 180 °C. This applies especially in the case of prepolymer formation in aqueous medium, considering the competitive reactions (3.33). At higher temperatures and appropriate conditions (solvent- and water-free etc.) the formation of quinone methides may result, and this could explain the presence of some linking groups identified in cured resins according to the following formulae:

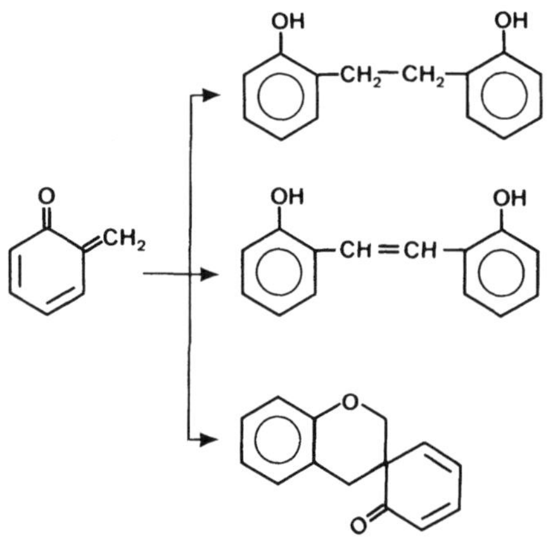

(3.37)

(3.38)

(3.39)

It was suggested that aldehydes also found in cured resins, were formed as a result of a redox reaction (3.40).

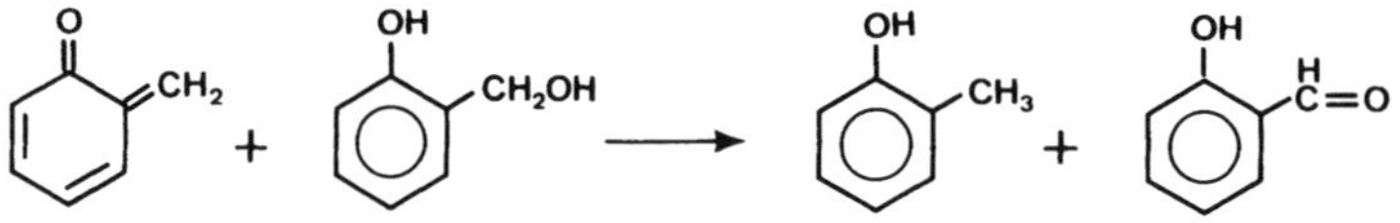

(3.40)

Aldehydes, on the other hand, can develop, according to Zinke, by splitting of dimethylene ether linkages in a disproportionation reaction (3.41).

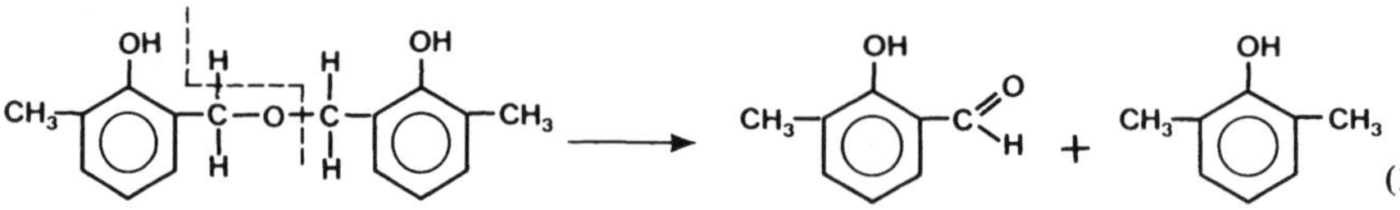

(3.41)

The content of aromatic aldehydes found in cured resols amounts up to 2%. However, it can be considerably higher for some model compounds.

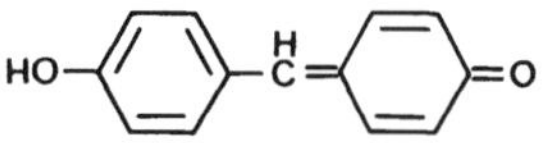

(3.42)

The yellowish-red color of cured resols could be attributed to a relatively stable quinone methide (3.42) which may form by reduction of dihydroxydiphenylmethane.

However, it is not necessary to presume the formation of quinone methides as a reaction step in the heat curing of resols. The most important cross-linking reactions at temperatures between 130–180 °C occur over carbonium ions as indicated in the Eqs. (3.27), (3.28), (3.29). The generation of carbonium ions followed by stabilization under chroman ring formation also takes place in the reaction of resols with olefinic compounds.

3.4. Phenol-Formaldehyde Reactions under Acidic Conditions

3.4.1. Strong Acids

The reaction between phenol and formaldehyde in the acidic pH range occurs as an electrophilic substitution. The catalysts most frequently used are oxalic acid, hydrochloric acid, sulfuric acid, *p*-toluene sulfonic acid or phosphoric acid. In the first step according to Eq. (3.43) the formation of a hydroxymethylene carbonium ion from methylene glycol occurs. This ion is the hydroxyalkylating agent.

$$HO{-}CH_2{-}OH \xrightleftharpoons{H^+} {}^+CH_2{-}OH + H_2O \tag{3.43}$$

$$C_6H_5OH + {}^+CH_2{-}OH \underset{}{\xrightleftharpoons{\text{slow}}} [\text{HO-}C_6H_5(H)(CH_2OH)]^+ \underset{}{\xrightleftharpoons{\text{fast}}} HO{-}C_6H_4{-}CH_2OH + H^+ \tag{3.44}$$

The following addition of the hydroxymethylene carbonium ion to phenol (3.44) occurs relatively slowly and is therefore rate determining. *Meta*-substitution is also not observed in the acidic medium.

However, the methylol group is very unstable under acidic conditions. Benzylic carbonium ions result (3.45) under these conditions, which then react very fast with phenol, yielding dihydroxydiphenylmethane[37, 41, 43] according to Eq. (3.46).

$$HO{-}C_6H_4{-}CH_2OH + H^+ \rightleftharpoons HO{-}C_6H_4{-}CH_2^+ + H_2O \tag{3.45}$$

$$HO{-}C_6H_4{-}CH_2^+ + C_6H_5OH \rightleftharpoons HO{-}C_6H_4{-}CH_2{-}C_6H_4{-}OH + H^+ \tag{3.46}$$

Thus, phenol alcohols cannot be isolated as intermediates in contrary to the alkaline hydroxymethylation. However, their existence as transient species[56] can be detected for instance by NMR-spectroscopy[18]. Under acidic conditions, methylol substitution and methylene bridge formation both occur preferably at the *para* position.

The part of *ortho*-substitution can be further reduced by high acidity, by the use of higher aldehydes, for instance acetaldehyde, or reaction performance in aqueous/alcoholic solution. In the presence of alcohols the space consuming carbonium ion is formed from the hemiacetal which will be the reacting species (3.47).

$$R{-}O{-}CH_2OH \overset{H^+}{\rightleftharpoons} R{-}O{-}CH_2^+ + H_2O \tag{3.47}$$

The selective effectiveness of thioalcohols or thioglycolic acid in the formation of 2,2-bis–(4-hydroxyphenyl)propane (bisphenol-A) from phenol and acetone[57] is to be similarly understood (3.48).

$$C_6H_5OH + CH_3{-}CO{-}CH_3 \xrightarrow{R-SH/H^+} HO{-}C_6H_4{-}C(CH_3)_2{-}C_6H_4{-}OH \tag{3.48}$$

95% p-p′ Isomer

The hemimercaptols formed from ketones and thioalcohols are considered to be more reactive than the hydrated ketones. Therefore, the reaction can be conducted at lower temperature.

In the hydrogen chloride-catalyzed reaction, the intermediate formation of chloromethanol (3.49) has been frequently suggested and the chloromethylation (3.50) has been proposed as the main reaction.

$$CH_2O + HCl \rightleftharpoons HO{-}CH_2{-}Cl \tag{3.49}$$

$$C_6H_5OH + HO{-}CH_2{-}Cl \rightleftharpoons C_6H_4(OH)(CH_2{-}Cl) + H_2O \tag{3.50}$$

But there is no evidence for the formation of chloromethanol under these conditions. The C–Cl bond would be immediately hydrolized in aqueous medium and consequently, it is improbable that chloromethanol is even intermediately formed[58].

3.4.2. Reaction Kinetics in Acidic Medium

In order to obtain reliable kinetic data, the condensation reaction rate of individual phenol alcohols made in separate reactions under acidic conditions must be studied as well as the overall substitution/condensation reaction. In most earlier studies[23, 59–61] the overall reaction rate, which was in most cases found to obey a second-order rate

Table 3.9. Arrhenius energy of activation and entropy of activation of the overall phenol-formaldehyde reaction in acidic medium according to Malhotra and Avinash[62]

pH	k at 80 °C (1 $mol^{-1} sec^{-1}$)	E_A(kJ)	S_A(J/K · mol)
1.14	$0.52.10^{-2}$	250	–581
1.32	$1.53.10^{-3}$	331	–390
2.20	$2.59.10^{-4}$	681	+637
3.00	$7.05.10^{-6}$	794	+737

law, was determined. The rate of formation of dihydroxydiphenylmethane (condensation reaction) was found to be more than five times[41] and 10–13 times, as reported in a second study[62], as fast as the formation of HMP (substitution reaction). The reaction rate is proportional to the H-ion concentration. The activation energy and activation entropy for the overall reaction[62] increase with increasing pH, indicating the change in mechanism at higher pH. Hydrochloric acid was used as catalyst in this study (Table 3.9).

In general, a second formaldehyde addition does not occur at usually applied P/F molar ratios of 1:(0.70–0.85). Mainly linear chain molecules are formed with MW generally below 2,000, that means, at maximum 20 phenol nuclei are connected over methylene bridges. The MWD dependence on the P/F molar ratio and catalyst type is shown in Figure 3.5.

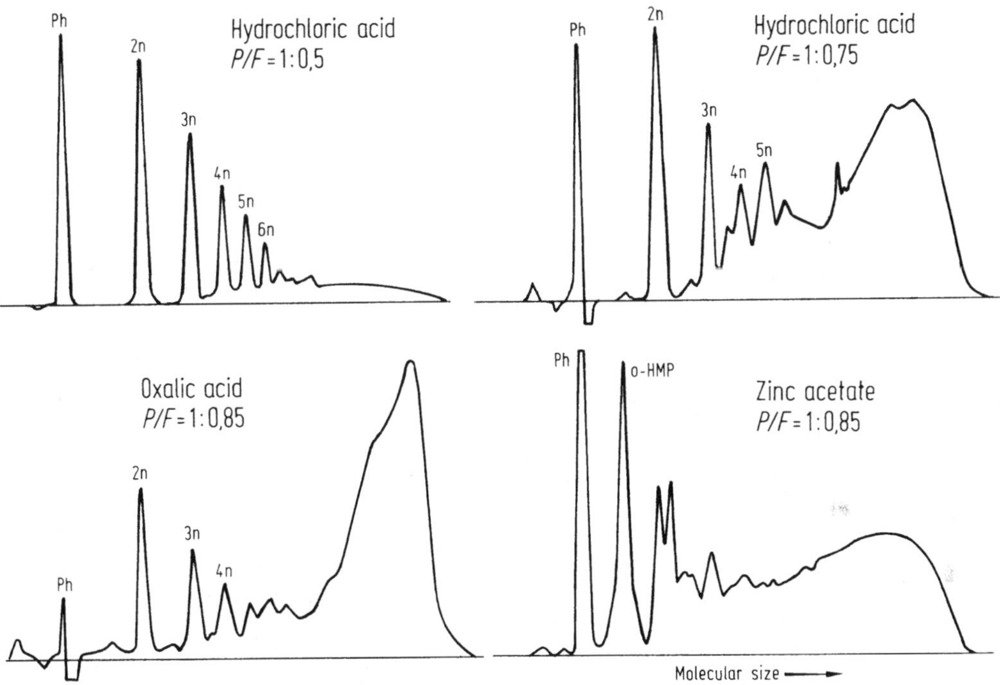

Fig. 3.5. Effect of phenol/formaldehyde ratio and catalyst type on molecular weight distribution[31] estimated by GPC (non distilled)

Table 3.10. Reaction conditions for the preparation of novolaks shown in Figure 3.5

Designation		Catalyst	Time min.	Temp. °C	Phenol/ Formald. mol	Water g	Methanol g
1	HCl	0.020 Mol	90	100	1:0.50	35	0.8
2	HCl	0.020 Mol	90	100	1:0.75	35	0.8
3	Oxalic acid	0.020 Mol	90	100	1:0.85	35	0.8
4	Zinc acetate	0.035 Mol	210	100	1:0.85	35	0.8

The reaction conditions are noted in Table 3.10. The acidic catalyst was neutralized with NaOH and the resin dissolved in DMF (ca. 3%) without prior distillation.

If oxalic acid is used as catalyst instead of hydrochloric acid, no deviations in MWD worth mentioning are to be found. The individual DPM-isomers are not separated in the GPC spectrum.

3.4.3. Reaction under Weak Acidic Conditions. "High-*Ortho*"-Novolak Resins

The novolaks produced in the weak acidic range at pH 4 to 6 and by use of specific catalysts, mainly salts of 2-valent metals[28, 63], have some characteristic properties. They are also designated "high-*ortho*"-novolaks because of the dominating *ortho*-*ortho*-linkage. It can be shown by IR- or NMR-spectroscopy that they also contain dibenzylether structures[64]. They may also contain methylol groups, if the reaction is performed at relatively low temperatures.

$$HO-CH_2-\left[C_6H_3(OH)-CH_2-C_6H_3(OH)\right]_n\left[CH_2-O-CH_2-C_6H_3(OH)\right]_m-CH_2-C_6H_4(OH) \qquad (3.51)$$

It has been observed[29] that these novolaks have high curing rates with HMTA because of the high degree of *ortho*-linkage and consequently free *para* positions which show higher reactivity towards formaldehyde.

Table 3.11. Reactivity of dihydroxydiphenylmethanes with hexamethylenetetramine according to Bender et al.[29]

Nuclear position	MP [°C]	Gel time with 15% HMTA [sec]
o-o'	118,5–119,5	60
o-p'	119–120	240
p-p'	162–163	175

These fast curing novolaks may be of considerable technical importance[65–67], for instance for the production of fast curing molding compounds, foundry granulates and polyol components for the Ashland cold-box process (see Chapter 14). The reaction conditions for the production of "high-*ortho*"-linked novolak resins are compiled as follows:

Table 3.12. Typical reaction conditions for the preparation of "high-ortho"-novolaks

pH	preferably between 4–6
Catalysts:	compounds of 2-valent metals: Ca, Mg, Zn, Cd, Pb, Cu, Co, Ni; preferably acetates
Molar ratio:	large phenol excess
Reaction conditions:	post-reaction at 150–160 °C necessary to convert the hydroxymethyl groups before distillation
Distillation:	120–140 °C maximum distillation temperature

The mechanism of this selective *ortho*-hydroxymethylation which can be attributed to the formation of chelate-like complexes as intermediates, was already mentioned in Section 3.3.1. The preferred catalyst is zinc acetate. The MWD, which differs sharply from other novolaks, is shown in Figure 3.5. The high amount of *o*-HMP (saligenin) and the presence of DPM as well is clearly indicated. The resin was prepared at 100 °C without post-reaction (Table 3.10) and analyzed without prior distillation. The lower conversion in comparison to strong acidic catalysts is also demonstrated.

A further peculiarity of *ortho*-linked novolaks is their high acidity and the remarkable tendency to form complex compounds with 2- and 3-valent metals and nonmetals compared to phenol. Sprengling[68] studied the acidity of various two nuclear phenols by means of titration in anhydrous ethylenediamine and benzene/isopropanol. *o,o'*-DPM showed an extremely high acidity compared to other isomers, being between phenol and oleic acid. This effect designated as "hyperacidity" was attributed to the formation of a strong intramolecular hydrogen bond of one hydrogen, increasing the dissociation tendency of the other (3.52).

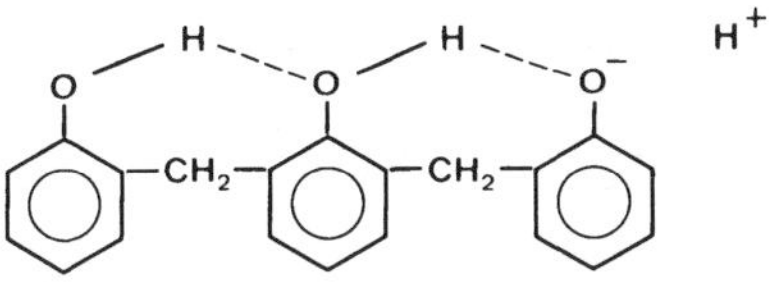

(3.52)

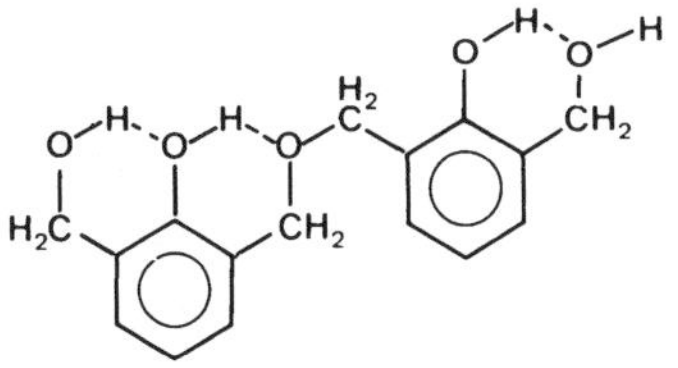

(3.53)

Hultzsch[69] showed that for *p*-alkylphenol resins which have only *o*-*o*-links, anomalies also develop, even when these resins have dimethylene ether bridges ($-CH_2-O-CH_2-$) and methylol groups. It is obvious that here other nuclear distances exist and, therefore, the formation of a "hyper-acidic hydrogen" is not possible. These resins do not show hyperacidity; however, they distinguish themselves by a remarkable affinity towards 2-valent metals. This "base reactivity" is technically utilized for the modification of polychloroprene adhesives with alkylphenolic resins and magnesium- and/or zinc oxide to increase adhesion and bond strength at higher temperature. If alkylphenol resins, dissolved in aromatic hydrocarbons, are treated with finely distributed magnesium oxide, a certain portion of oxide will dissolve. These resins are able to bind magnesium oxide chemically up to 10%. On the other hand, it can be shown that novolaks exclusively bonded with methylene bridges cannot take up magnesium inspite of increased acidity. The reaction with magnesium leads to complexes with the coordination number 4 according to formula (3.54). Similarly, other ions can also be incorporated in phenolic resins[70] under the formation of metal containing resins, for instance the metals Ca, Ba, Cu, Ni, Co, Pb, Mn, Cr and Fe. Some special metal incorporating reactions are mentioned in Section 8.3.

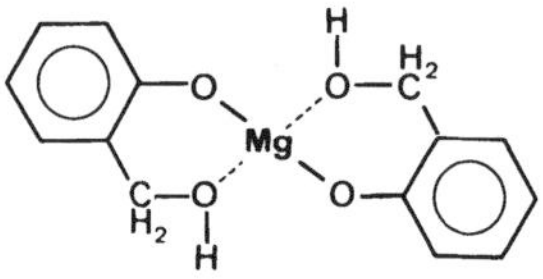

(3.54)

3.4.4. Novolak Cross-linking Reaction with HMTA

Novolak resins, which are thermoplastics in nature, must generally be cured by addition of an auxiliary chemical cross-linking agent. The most widely used curing agent is HMTA. Paraformaldehyde or trioxane are of only limited importance. Novolaks, generally made from phenol and formaldehyde at a P/F ratio of 1:0.8 may be cured by addition of 8–15% of HMTA. However, 9–10% seems to be the most commonly employed HMTA level, yielding the best overall performance. The properties of cured parts are determined to a great extent by the ratio of the two reactants. Physical properties of HMTA are to be found in Section 2.2, the mechanism of HMTA formation from ammonia and formaldehyde in Section 3.3.2. By heating with 2,4-xylenol[71] and 2,6-xylenol[72] as model substances in an autoclave, dibenzylamines were obtained, which at higher temperatures (> 160 °C) resulted in DPM-derivatives. Benzoxazine (3.55) was identified by means of NMR[18] if 2,4-xylenol had been heated with HMTA to 135 °C in air. If further heated with an excess of 2, 4-xylenol, conversion to di- and/or tribenzylamines was noted.

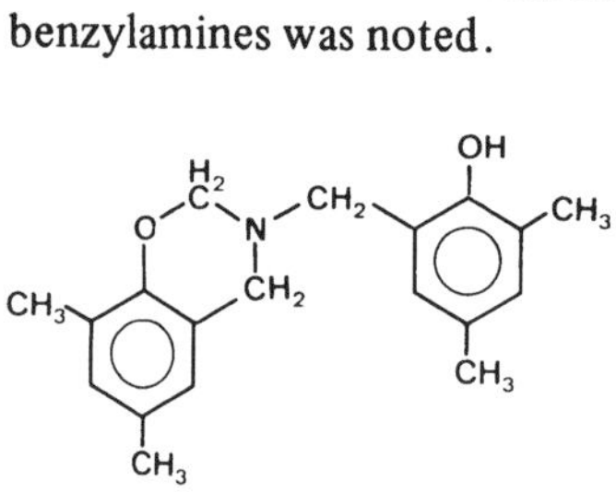

(3.55)

Phenol and HMTA form definite molecular addition complexes (1 Ph : 1 HMTA, orthorombic and 3 Ph: 1 HMTA, monoclinic) which were identified by DTA-traces and X-ray diffraction patterns[74].

However, such addition compounds did not seem to be involved as intermediate compounds in the hardening reaction of novolaks. The small amount of water which is always present (min. 0.1–0.5%) in novolak resins leads to hydrolysis of HMTA in reverse of Eqs. (3.22), (3.23) and formation of amino-methylol compounds[35]. Due to the acidic nature of phenol, carbonium ions are generated from these α-amino alcohols, which then interact with phenol to form secondary and tertiary benzylamine containing chain molecules in a Mannich type reaction. The opinion that this is a homogeneous acid catalysis is supported by the fact that the reaction rate increases considerably at decreasing pH[73]. Water and free phenol enhance the reaction rate to a great extent.

$$HN(CH_2OH)_2 \tag{3.56}$$

$$N(CH_2OH)_3 \tag{3.57}$$

The reaction of phenols with formaldehyde and formation of methylol compounds occurs in parallel, now catalyzed by hydroxybenzylamines. The reaction of *m*-cresol with HMTA was found to be of first order. The reaction rate constant was evaluated to be $1.9 \times 10^{-4}\ l\ mol^{-1}\ s^{-1}$, HMTA hydrolysis and formation of the aminomethylene ions was the slowest and rate determining step[73]. The activation energy was found to be 64.9 kJ/mol. The formation of oligomers is accompanied by the liberation of a considerable amount of gas, which consists of at least 95% of ammonia. The hardened resin may contain up to 6% chemically bound nitrogen. At higher temperatures the benzylamine groups undergo decomposition reactions not as yet completely defined. Products containing the azomethine group $-CH=N-$ among others, which would account for the yellow color of the oligomers, are formed.

$$HO\text{-}C_6H_2(-)(-)\text{-}CH_2\text{-}NH\text{-}CH_2\text{-}C_6H_2(-)(-)\text{-}OH \tag{3.58}$$

$$HO\text{-}C_6H_2(-)(-)\text{-}CH{=}N\text{-}CH_2\text{-}C_6H_2(-)(-)\text{-}OH \tag{3.59}$$

3.4.5. Reaction with Epoxide resins

Epoxide resins[75,76] may be used instead of HMTA to cross-link phenolic resins if the splitting-off of volatile compounds must be avoided. Epoxidized novolaks, which may be obtained as solid, grindable resins by the reaction of novolak resins with epichlorohydrine, depending on the MW of novolak resin applied, are mainly used.

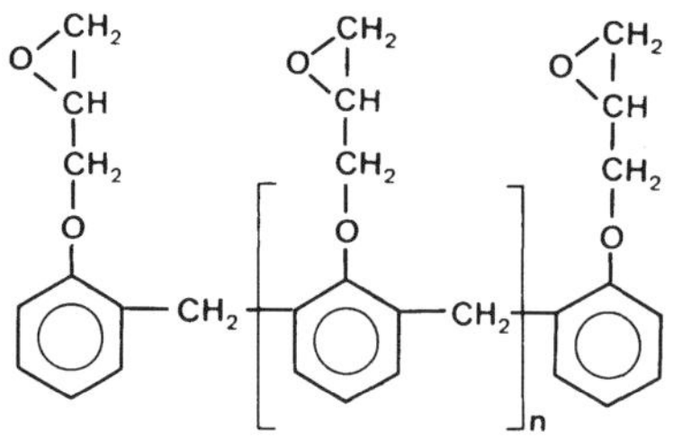

(3.60)

Both novolaks and resols may be employed. The products obtained exhibit high strength, strong adhesion, excellent dielectric properties and improved oxidation resistance. However, they also show a somewhat reduced thermal resistance. Additional cross-linking, thus increasing thermal stability, may be effected by addition of small amounts of HMTA. The ring-opening reaction of epoxides with H-active compounds, e.g. phenols, is a S_N2-type reaction which is catalyzed by inorganic and organic bases, preferably tertiary amines[77]. Similar products may be obtained if the phenolic resin is treated with epichlorohydrine and sodium hydroxyde. The chlorohydrine group obtained in the first step is converted to a second epoxide group which then further reacts with a phenolic hydroxyl group.

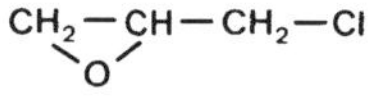

$CH_2-CH-CH_2-Cl$ (epoxide O bridging) (3.61)

Epichlorohydrine

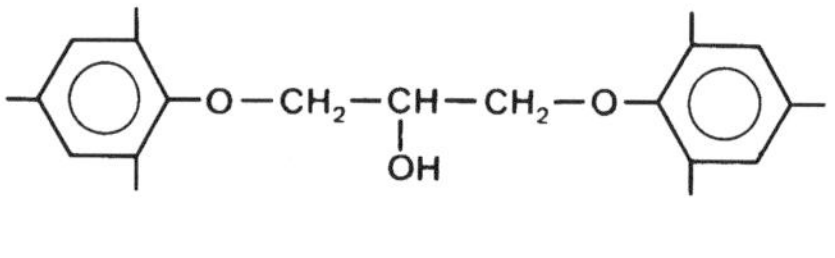

$-C_6H_3-O-CH_2-CH(OH)-CH_2-O-C_6H_3-$ (3.62)

3.4.6. Reactions with Diisocyanates

The reaction of isocyanates with phenols, leading to phenylurethanes (Eq. 3.63), is essentially a nucleophilic attack of the phenol on the carbon atom of the isocyanate group followed by a 1,3-shift of the hydrogen atom. Therefore, if steric factors are neglected, the reactivity of the hydroxyl compound increases as its nucleophilic character increases.

For simplicity the reactions are represented by means of monofunctional compounds.

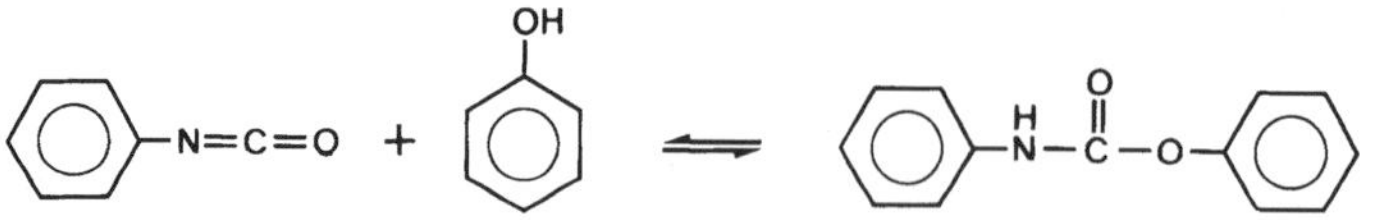

$$C_6H_5-N=C=O + C_6H_5OH \rightleftharpoons C_6H_5-NH-C(=O)-O-C_6H_5 \quad (3.63)$$

The most important competitive reactions in the preparation of urethane-modified phenolic resins are the reactions with the hydroxymethyl group and water. The first step in the reaction with water is the formation of the unstable carbamic acid which decomposes to form an amine and carbon dioxide (Eq. 3.64). The amine immediately reacts with additional isocyanate to form a substituted urea.

$$R{-}N{=}C{=}O + H_2O \longrightarrow \left[R{-}\overset{H}{N}{-}\overset{O}{\overset{\|}{C}}{-}OH \right] \longrightarrow R{-}NH_2 + CO_2 \qquad (3.64)$$

The relative rates of the individual reactions are dependent on the type of catalyst used. The catalysts most commonly employed are tertiary amines (triethylenediamine, N-alkyl morpholines) and metal catalysts, especially stannous octoate[78].

The phenolic resin-diisocyanate reaction is used for fast curing "cold set" foundry resins[79]. 4,4′-Diphenylmethane diisocyanate is the most used cross-linking compound.

However, urethanes tend to thermal dissociation. The ability to dissociate depends upon chemical nature as indicated below[80]:

Aryl–NH–CO–O–aryl	thermostable to about 120 °C
Alkyl–NH–CO–O–aryl	thermostable to about 180 °C
Aryl–NH–CO–O–alkyl	thermostable to about 200 °C
Alkyl–NH–CO–O–alkyl	thermostable to about 250 °C

Phenol-"blocked" isocyanates are used for single component polyurethane coatings or as cross-linkers in polyester powder coatings[81].

References

1. Wyckoff, R. W. G.: Crystal Structures, 2. Ed., Vol. 6. New York: Interscience 1969
2. Rochester, C. H.: Acidity and Inter- and Intramolecular H-Bonds. In: S. Patai (ed.): The Chemistry of the Hydroxyl Group, London: Interscience 1971
3. Saunders, M., Hyne, J. B.: J. Chem. Phys. *29*, 1319 (1958)
4. Richards, R. E., Thompson, H. W.: J. Chem. Soc. (1947) 1260
5. Weller, A.: Progr. Reaction Kinetics *1*, 187 (1961)
6. Wehry, E. L., Rogers, L. B.: J. Amer. Chem. Soc. *88*, 351 (1966)
7. Bolton, P. D., Hall, F. M., Reece, J. H.: Spectrochim. Acta *22*, 1149 (1966)
8. Chen, D. T. Y., Laidler, K. J.: Trans. Faraday, Soc. *58*, 440 (1962)
9. Zsavitsas, A. A.: J. Chem. Engng. Data *12*, 94 (1967)
10. Kortüm, G., Vogel, W., Andrusson, K.: Dissociation Constants of Organic Acids in Aqueous Solution. London: Butterworths (1961)
11. Harper, D. A., Vaughan, J.: Directing and Activating Effects. In S. Patai (ed.): The Chemistry of the Hydroxyl Group, London: Interscience 1971
12. Knop, J. V., Knop, A.: Unpublished
13. Pople, J. A., Santry, D. D., Segal, G. A.: J. Chem. Phys. *43*, 129 (1965)
14. Walker, J. F.: Formaldehyde, ACS Monograph No. 159, 3 Ed. New York: Reinhold Publ. Co. 1964
15. Diehm, H., Hit, A.: Formaldehyd. In: Ullmanns Encyclopädie der techn. Chem. Vol. 11, 4. Ed. Weinheim: Verlag Chemie 1976
16. Zabicky, J. (ed.): The Chemistry of the Carbonyl Group. London: Interscience 1970
17. Moedritzer, K., Wazer, J. V.: J. Phys. Chem. *70*, 2025 (1966)
18. Kopf, P. W., Wagner, E. R.: Polymer Sci., Polymer Chem. Ed. *11*, 939 (1973)

19. Zsavitsas, A. A., Beaulieu, R. D.: Am. Chem. Soc. Div. Org. Coatings and Plastic Preprints *27,* 100 (1967)
20. Lederer, L.: J. prakt. Chemie *50,* 223 (1894)
21. Manasse, O.: Ber. dtsch. chem. Ges. *27,* 2409 (1894)
22. Piria, R.: Liebigs Ann. Chem. *56,* 37 (1845)
23. DeJong, J. I., DeJonge, J.: Rec. Trav. Chim. *72,* 497 (1953)
24. Zsavitsas, A. A.: Am. Chem. Soc. Div. Org. Coatings and Plastic Preprints *26,* 93 (1966)
25. Freemann, J. H., Lewis, C. W.: J. Amer. Chem. Soc. *76,* 2080 (1954)
26. Peer, H. G.: Rec. Trav. Chim. *78,* 851 (1959)
27. Eapen, K. C., Yeddanapalli, L. M.: Makromol. Chem. *119,* 4 (1968)
28. Peer, H. G.: Rec. Trav. Chim. *79,* 825 (1960)
29. Bender, H. L., Farnham, A. G., Guyer, J. W., Apel, F. N., Gibb, T. B.: Ind. Engng. Chem. *44,* 1619 (1952)
30. Union Carbide Corp.: US-PS 2464207 (1949)
31. Knop, A., Trapper, W.: Unpublished
32. Zinke, A., Hanus, F.: Mh. Chemie *78,* 311 (1948)
33. Hultzsch, K.: Chem. Ber. *82,* 16 (1949)
34. Hellmann, H., Opitz, G.: α-Amino-alkylierung. Weinheim: Verlag Chemie 1960
35. Ogata, Y., Kawasaki, A.: Equilibrium Additions to Carbonyl Compounds. In: J. Zabicky (ed.): The Chemistry of the Carbonyl Group, Vol. 2. London: Interscience 1970
36. Degussa: Hexamethylentetramin, Technical Bulletin
37. Megson, N. J. L.: Phenolic Resin Chemistry. London: Butterworth (1958)
38. Higginbottom, H. P., Culbertson, H. M., Woodbrey, J. C.: Analyt. Chem. *37,* 1021 (1965)
39. Sprung, M. M.: J. Amer. Chem. Soc. *63,* 334 (1941)
40. Francis, D. J., Yeddanapalli, L. M.: Makromol. Chem. *119,* 17 (1968)
41. Martin, R. W.: The Chemistry of Phenolic Resins. New York: J. Wiley 1956
42. Yeddanapalli, L. M., Francis, D. J.: Makromol. Chem. *55,* 74 (1962)
43. Kornblum, N., Smiley, R. A., Blackwood, R. K., Iffland, D. C.: J. Amer. Chem. Soc. *77,* 7269 (1955)
44. Hultzsch, K.: Chemie der Phenolharze. Berlin, Göttingen, Heidelberg: Springer 1950
45. Lenz, R. W.: Organic Chemistry of Synthetic High Polymers. London: Interscience 1967
46. Kammerer, H., Kern, W., Henser, G.: J. Polymer Sci. *28,* 331 (1958)
47. Vollmert, B.: Polymer Chemistry. Berlin, Heidelberg, New York: Springer 1973
48. Müller, H. F., Müller, J.: Kunststoffe *37,* 75 (1947)
49. Hultzsch, K.: Kunststoffe 37, 205 (1947)
50. Hultzsch, K.: Ber. dtsch. chem. Ges. *74,* 898 (1941)
51. Hultzsch, K.: Angew. Chem. *A 60,* 179 (1948)
52. Von Euler, H., Adler, E., Cedwall, J. O.: Ark. Kemi. Min. Geol. *14A,* No. 14 (1941)
53. Von Euler, H., Adler, E., Cedwall, J. O., Törngren, O.: Ark. Kemi. Min. Geol. *15A,* No. 11 (1942)
54. Cologne, J., Descotes, G.: α,β-Unsaturated Carbonyl Compounds. In: J. Hammer (ed.): 1,4-Cycloaddition Reactions. New York: Academic Press 1967
55. Wagner, H. U., Gompper, R.: Quinone Methides. In: S. Patai (ed.): The Chemistry of the Quinoid Compounds, Chap. 18, Vol. 2. New York: Wiley 1974
56. Baekeland, L. H.: Ind. Engng. Chem. *4,* 739 (1912)
57. Barclay, R., Sulzberg, Th.: Bisphenols and Their Bis-(Chloroformates). In J. K. Stille, T. W. Champbell: Condensation Monomers. New York: Wiley-Intersciene 1973
58. Olah, G. (ed.): Friedel Crafts and Related Reactions. Interscience (1964), Vol. II, Part II, P. 762
59. Von Euler, H., Dekispezy, S. V.: Z. Phys. Chem. *A 189,* 109 (1941)
60. Jones, T. T.: J. Soc. Chem. Ind. *69,* 102 (1964)
61. Bassow, N. J. et al.: Plaste und Katschuk *6,* 417 (1974)
62. Malhotra, H. C., Avinash: J. Appl. Polymer Sci. *20,* 2461 (1976)
63. Bender, H. L.: Mod. Plastics *30,* 136 (1953)
64. Higginbottom, H. P., Culbertson, H. M., Woodbrey, J. C.: J. Polymer Sci. Part. *A, 3,* 1079 (1965)

65. Reichold Chemicals (Canada) Ltd.: DE-OS-2034 136 (1970)
66. Ashland Oil Inc.: DE-AS 19 20 759 (1969)
67. Cor Tech Research Ltd.: DE-OS 25 16 898 (1975)
68. Sprengling, G. R.: J. Amer. Chem. Soc. *76*,, 1190 (1954)
69. Hultzsch, K., Hesse, W.: Kunststoffe *53*, 166 (1963)
70. Nord-Aviation Société Nationale de Constructions Aeronautiques: DE-AS 1816 241 (1968)
71. Zinke, A., Purcher, S.: Mh. Chem. *79*, 26 (1948)
72. Zinke, A., Ziegler, E.: Ber. dtsch. chem. Ges. *77*, 271 (1944)
73. Kamenskii, I. V., Kuznetsov, L. N., Moisenko, A. P.: Vysokomol. soyed. *A 18/8*, 1787 (1976)
74. Orrell, E. W., Burns, R.: Study of the Novolak-Hexamine Reaction. Plastics and Polymers *36*, 125 (1968)
75. Lee, H., Neville, K.: Handbook of Epoxy Resins. London: McGraw-Hill 1967
76. Potter, W. G.: Epoxide Resins, London: Iliffe Books 1970
77. Ishii, Y., Sakai, S.: 1,2-Epoxides. In Frish, K. C., Reegen, S. L.; Ring Opening Polymerizations, Vol. 2. New York: Marcel Dekker Inc. 1969
78. Frisch, K. C., Rumao, L. P.: Catalysis in Isocyanate Reactions J. Macromol. Sci.-Revs. Macromol. Chem., C *5* (1) 103 (1970)
79. Ashland Oil: DE-PS 1583521 (1967)
80. Wicks, Z. W.: Progr. Organic Coatings *3*, 73 (1975)
81. Myers, R. R., Long, J. S.: Treatise on Coatings. New York: Marcel Dekker Inc. 1967

4. Resin Production

Aqueous, solid or dissolved resins, which will meet a number of special requirements, depending on their future use, can be produced from the basic materials phenol and formaldehyde by variation of the reaction conditions:

- molar ratio of phenol to formaldehyde
- kind and amount of catalyst
- reaction time and reaction temperature
- content of water and free phenol (distillation conditions)
- modification with other aldehydes and phenols
- etherification and/or dissolving in organic solvents
- modification with other compounds.

The production of resins is performed by a discontinuous process. Many continuous processes were described in the patent linterature[1–5] but obviously they were not applied, with the exception of the continuous process for standard novolaks[3] and perhaps for particle board-resols. The multitude of resin specifications demanded by the market and the therefore non-uniform production would render the continuous process uneconomical. The formation of cured resin coats on hot surfaces also is an important reason for the failure of continuous resol processes. Krahl[1] described a process during which the reaction mixture flows through a

Fig. 4.1. First phenolic resin kettle. (Photo: Bakelite GmbH, D-5860 Iserlohn-Letmathe)

Fig 4.2. Production plant of phenolic resins. (Photo: Bakelite GmbH, D-5860 Iserlohn-Letmathe)

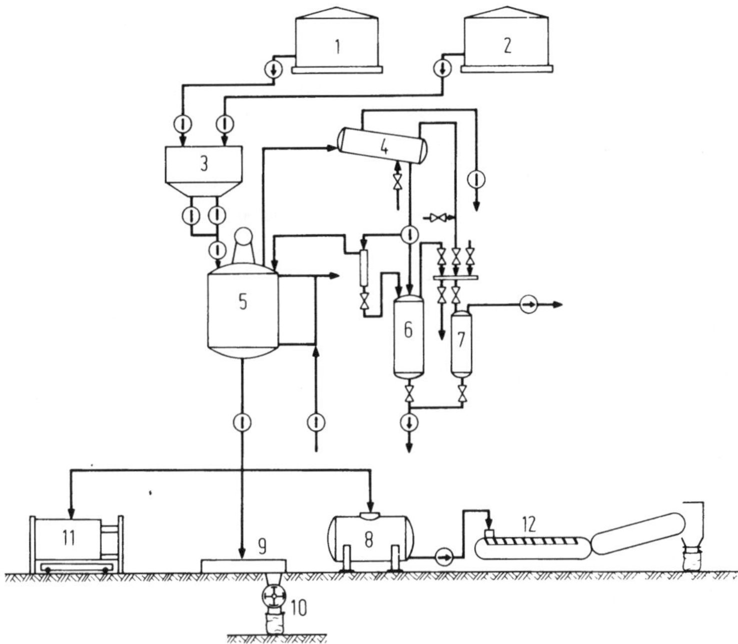

Fig. 4.3. Phenolic resin production plant, batch process. (Drawing: Bakelite GmbH, D-5860 Iserlohn-Letmathe).
1 Phenol; *2* Formaldehyde; *3* Scale; *4* Condenser; *5* Reactor; *6* Condensate receiver; *7* Vacuum; *8* Resin receiver; *9* Resin trough; *10* Mill; *11* Cooling carriage; *12* Cooling belt

number of small reaction chambers, which were arranged in a cascade-like formation. The distillation was also performed continuously.

An essential problem of the batch production of phenolics is the high exothermic heat which also limits the batch volume. Therefore, resins were formerly produced in vessels having a capacity of 3–5m^3, later of 10 m^3, and today reactors with a capacity up to 60 m^3 are used, especially for the production of novolak resins.

The total reaction enthalpy of the phenol-formaldehyde substitution and the following condensation reaction in acidic medium at a molar ratio of 1 : 1 was determined experimentally by Mangold and Petzold[6] and Jones[7] among others to be $\Delta H_0{}^{tot.}$ = 81.1 kJ/mol and 82.3 kJ/mol respectively. Individual values of 20.1 kJ/mol for the substitution reaction, 78.1 kJ/mol for the condensation reaction and 98.2 kJ/mol for the total reaction have been calculated out of the combustion enthalpies[8]. The heat generation per unit of time and peak temperature depend upon the production conditions and the molar ratio.

The use of aqueous formaldehyde, which is generally the custom, has the advantage that the amount of heat formed by the exothermal reaction is absorbed by the water and can be utilized to heat the contents of the vessel and finally be exhausted as evaporation heat. This method prevents an uncontrolled and sometimes explosion-like reaction. The danger of such a reaction is especially serious, if the aqueous formaldehyde solution is even partly substituted by paraformaldehyde in order to either increase the output or to reduce the consumption of energy and time for the distillation.

The reactors are double-walled, closed vessels, which are divided into different heating- and cooling sections (Fig. 4.2). Vessels and auxiliary installations of alloyed stainless steel are used, sometimes nickel clad vessels are also in operation. These materials do not cause the discoloration of the resins. Copper and copper alloys also exhibit a very good resistance to phenol, but they lead to discoloration.

There is a distinct difference between pure and impure phenol as far as the corrosiveness is concerned. Very pure phenol does not attack high-alloy ferritic and austenitic stainless steels even at boiling temperature[9]. Structural parts under stress made of unalloyed steel are liable to stress corrosion cracking, especially at the welding seams[10]. The corrosion of the vessels is higher within the liquid phase than in the

Table 4.1. Corrosion resistance of some materials to pure and used phenol at 240 °C[9]

Material	Corrosion Rate mm/year at 240 °C	
	Pure phenol	Used phenol
Unalloyed steel	0.76	1.09
18/8 Cr-Ni-steel	0.00	0.00
Nickel 99.4	0.00	0.79
Aluminium	56.00	–

vapor phase. The corrosion rate increases very fast at lower pH. Even minor additions of carboxylic acids increase the corrosion. Water also has a considerable influence on the corrosion rate.

A series of examples for the production of various phenolic resins is given in the literature[5,11,12].

4.1. Novolak Resins

Oxalic acid, MP 101 °C, as dihydrate, is the catalyst which is mostly used. Oxalic acid sublimes in a vacuum at about 100 °C and at normal pressure at 157 °C without decomposition. At a still higher temperature (180 °C) it decomposes to carbon monoxide, carbon dioxide and water, so that the washing process can be avoided. Due to its reducing action very bright resins are obtained. Hydrochloric acid, sulfuric acid and phosphoric acid are seldom used. Hydrochloric acid which formerly was very important due to the low costs, is being used less and less because of its high corrosiveness. It has been pointed out that the use of hydrochloric acid, when formaldehyde and hydrochloric acid are present in the gas phase in a concentration of more than 100 ppm, may result in the formation of 1.1'-dichlorodimethyl ether which has a strong carcinogenic effect. The use of maleic acid is recommended for the production of high melting novolaks.

The molar ratio of phenol to formaldehyde is normally within the range of 1 : (0.75–0.85). The influence of the P/F molar ratio on MWD and therefore also on the MP and melt viscosity is shown in Figure 3.5, Chapter 3. While novolaks with a MP of 70–75 °C are used for the production of foundry resins for shell cores, those with MP's between 80–100 °C are used for all other fields of application. These novolaks are reduced to appropriate size, mixed with HMTA and ground to powder in special mills and processed in this form.

In the batch process the phenol, which is stored in tanks of alloyed steel at approximately 60 °C, is transferred to the reaction vessel via a scale and heated to 95 °C. After catalyst addition the formaldehyde solution is fed in under stirring at a rate so that the mixture is softly simmering. When all the formaldehyde solution is added, the temperature is maintained until the formaldehyde is reacted up to a small rest. Then, the water is distilled first at normal pressure and, by further heating up to 160 °C, in the vacuum the unreacted phenol. The separation can be helped along by blowing in overheated steam. As soon as the desired melting point is reached, the resin is drawn off into a heated container and then pelletized and cooled down on a continuous cooling belt (Fig. 4.3).

In the continuous process[3, 13] shown in Figure 4.4 formaldehyde, phenol and catalyst are conveyed from the relevant storage tanks to the first stage reactor, where their quantity is automatically measured and controlled. In this stirring reactor, provided with an external heating jacket, the reaction between phenol and formaldehyde is started. The reaction is then continued and completed in a second stage reactor. The reaction is carried out under pressure up to 7 bar at a temperature level ranging between 120 °C and 180 °C to enhance the reaction rate. The reacted mixture leaving the second stage reactor is conveyed into a flash drum, acting also

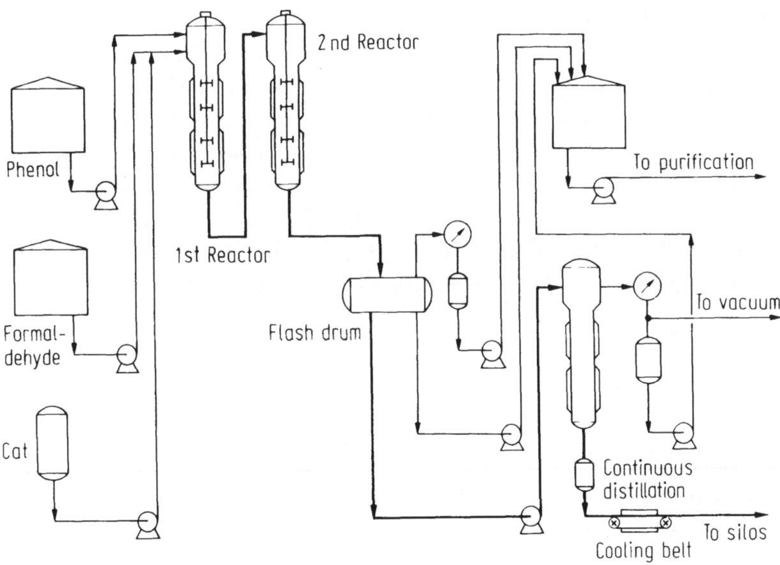

Fig. 4.4. Continuous novolak resin production process[13] (Drawing: Euteco SPA, I-20161 Milano)

as a vapour-liquid separator. The vapour flashed is condensed in a condenser, collected in a receiver and sent to the purification section while the liquid phase separating on the bottom of the flash drum settles in two layers. The upper, aqueous layer containing small amounts of phenol is withdrawn and sent to the purification section, whilst the lower layer, i.e. the resin layer, is pumped on for further treatment.

Since the resin still contains small portions of water, it must be dewatered. This operation is carried out in a vacuum evaporator of special design from which two streams are obtained: the overhead vapor, essentially consisting of water with very small amounts of phenol, is condensed and joined to the other streams to be disposed of in the purification section, while the bottom stream, i. e. dewatered resin, is fed to a flake machine essentially consisting of a water-cooled metal band.

A phenol novolak may be prepared on a laboratory scale according to the following directions[14]:

"A 1,000 ml resin kettle is charged with 130 g of phenol (1.38 moles), 13 ml of water, 92.4 g of 37% aqueous formaldehyde (1.14 moles) and 1 g of oxalic acid dihydrate. The mixture is stirred and refluxed for 30 min. An additional 1 g of oxalic acid dihydrate is then added, and refluxing is continued for another hour. At this point, 400 ml of water are added and the mixture cooled. The resin is permitted to settle for 30 min and the upper layer of water decanted or withdrawn through a siphon. Heating is then begun with the condenser set for vacuum distillation. Water is distilled at 50–100 mm pressure until the pot temperature reaches 120 °C or until a sample of the resin is brittle at room temperature. About 140 g of resin are obtained."

In general, the yield does not exceed 105% based on phenol weight at novolak resin production.

Water has a remarkable influence on the plasticity of the novolaks[15]. Only 1% water content is necessary to reduce the MP of novolaks with a medium MW between 450 and 700 in average of 3.4 °C. This function was evaluated within the range of 0.1 to 4% water[16]. The effect on the melt viscosity is even more remarkable. It was reduced by 90% by increasing the water content from 0.1 to 3.1% at 120 °C (MW 450). As expected, the effect of water as internal lubricant at small concentrations is especially high. An addition of 0.5% water (from 0.2 to 0.7%) reduces the melt viscosity by 50%. The reactivity of novolaks to HMTA increases with increasing water (and free phenol) content, so that the flow distance (a measure for the melt viscosity and reactivity) is reduced inspite of decreasing melt viscosity[16]. The influence of free phenol is less drastic. The MP is lowered according to MW between 2.1–2.9 °C/1% of free phenol. The influence on the melt viscosity is more drastic at higher MW[17].

4.2. Resols

The alkaline P/F reaction is more versatile regarding the molar ratios, catalysts and reaction conditions applied than the production of novolaks. P/F molar ratios between 1 : 1 and 1 : 3 may be applied. The type of catalyst has much higher influence on the molecular structure and MWD than in the acidic range (Section 3.3). Sodium

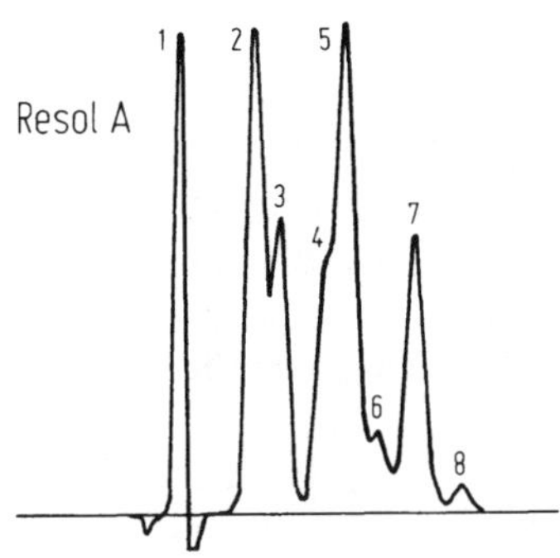

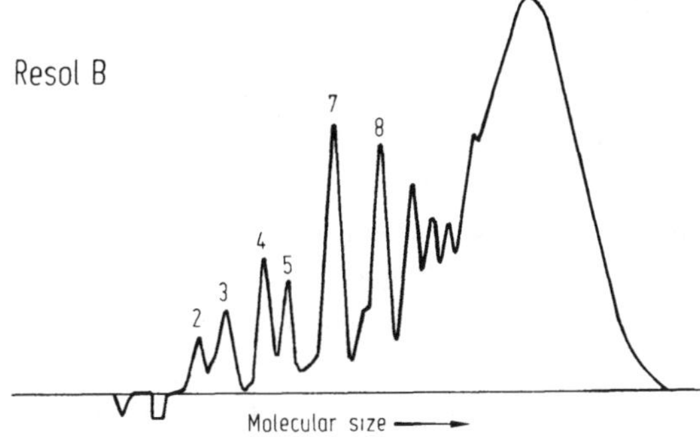

Fig. 4.5. Molecular weight distribution of resols produced under different reaction conditions. *1* = Ph; *2* = *o*-HMP; *3* = *p*-HMP; *4, 5* = 2,4- and 2,6-BHMP; *6* = DPM Der.; *7* = THMP; *8* = DPM Der.

hydroxide, sodium carbonate, alkaline earth oxides and hydroxides, ammonia, HMTA and tertiary amines are the most important catalysts. The catalyst performance, the possibilities of separation, if required, and costs are the decisive criteria for catalyst selection. Sodium hydroxide and carbonate are mostly not separated from the reaction mixture. The oxides and hydroxides of calcium and barium are precipitated as sulfates or carbonates by addition of sulfuric acid or passing of carbon dioxide at the end of the reaction and separated. Tertiary amines, i. e. triethylamine, can be separated by distillation. The catalyst separation is necessary, if dielectric properties, ageing- and humidity resistance are required. The molar catalyst ratio based on phenol as applied in technical resols is very different. It ranges from 1 : 1 to 1 : 0.01.

Exact temperature and time control are required. Adequate vacuum, at least of 50 mbar, and a cooling water temperature as low as possible, in order to maintain the temperature of the resin low (max. 60 °C) and distillation time short, are necessary. The distillation is stopped when the prescribed resin content is obtained. The viscosity of the resin can be regulated, if necessary, by post-condensing at approximately 70 °C. The resin is then quickly cooled down to room temperature. An almost cured resin coat is gradually formed on the surface of the vessel wall, reducing the heat transfer and prolonging the time of distillation remarkably. The reactors and auxiliary equipment must therefore be thoroughly cleaned after a certain number of batches.

The following procedure is recommended for the production of a resol on a laboratory scale[14]: "A 500 ml resin kettle is equipped with a reflux condenser, stirrer, thermometer and siphoning tube leading to a collecting trap for the removal of samples for testing. To the reaction vessel is added 94 g (1 mol) of distilled phenol, 123 g aqueous formaldehyde 37% by weight (1.5 mol) and 4.7 g barium hydroxide octahydrate. The reaction mixture is stirred and heated in an oil bath at 70 °C for 2 hours. Two layers form if stirring is stopped. Sufficient 10% sulfuric acid is added to bring the pH to 6–7. Vacuum is then applied by means of a water aspirator (pressure regulated at about 30–50 mm) and water is removed through the condenser which is now set for distillation. The temperature is not permitted to exceed 70 °C. Samples are withdrawn every 15 min through the vacuum siphon take off and tested for gel time by working with a spatula on a hot plate at 150 °C."

A wide range of resin types is obtained if the type and amount of catalyst and molar ratio as well as the reaction conditions are varied. A low molecular weight liquid phenol resol is used to impregnate paper for laminates in order to reduce their water absorption (Resol A, Fig. 4.5). However, other requirements are established for plywood glues. The resin must wet very well, but must not penetrate too deeply into the porous structure of the wood. This would cause a high glue consumption and render the adhesive process uneconomical. The molecular weight must, therefore, be considerably raised by post reaction (Resol B).

High melting resols of phenol and cresols made with ammonia as catalyst are of special significance, for instance in the manufacturing of laminates, coatings and brake linings. The uniform production of solid resols requires a high measure of experience. After the raw materials are weighed in, the vessel content is carefully heated up to the boiling point and the temperature is maintained for about 50 minutes. Then, the distillation is carried out under vacuum at approximately 50 °C. As soon as water is removed, the temperature of the highly fluid resin is gradually raised under vacuum up to 95 °C, whereby the MP of the resin gradually rises due to the condensation reaction. As soon as the desired MP is reached, which is limited by the reactivity of the resin, the resin is quickly discharged and cooled. Since the MP-determination takes considerable time, the end point is determined by experience. If the exact point of time is missed, the temperature and MP of the resin continues to increase due to the exothermic reaction. It becomes a very tough and hard mass which can only be removed from the vessel by the use of a pneumatic hammer.

References

1. Biesterfeld & Stolting: DE-PS 758 436 (1939)
2. Rütgerswerke AG: DE-PS 965 923 (1954)
3. Societa Italiana Resine S. P. A.: US-PS 3 687 896 (1972)
4. Perstorp AB: US-PS 1 460 029 (1976) Deutsche Texaco AG: DE-OS 25 38 100 (1977)
5. Lemer, F., Greth, A.: Phenolharze. In: Ullmanns Encyclopädie d. techn. Chem., Vol. 13, 3. Ed. München: Urban und Schwarzenberg 1963
6. Manegold, E., Petzold, W.: Kolloid.-Z. *94*, 284 (1941)
7. Jones, T. T.: J. Soc. chem. Ind. (Lond.) *65*, 264 (1946)
8. Vlk, O.: Plaste u. Kautschuk *4*, 127 (1957)
9. Donndorf, R. et al.: Werkstoffeinsatz und Korrosionsschutz in der chemischen Industrie, Leipzig: VEB Deutscher Verlag für Grundstoffindustrie, 1973
10. Dunn, C. L., Liedholm, G. E.: Oil Gas J. *51*, 560 (1952)
11. Sandler, S. R., Caro, W.: Polymer Syntheses, Vol. 2, Chapt. 2. Phenol-Aldehyde Condensations. New York: Academic Press 1977
12. Wegler, R. Herlinger, H.: Polyaddition u. -condensation v. Carbonyl- u. Thiocarbonylverbindungen. In: Houben-Weyl: Methoden d. Org. Chem. Vol. XIV/2 P. 272–291; Chem.. Abs.: Phenol-Condensation Products, 8th Collect. Vol. 1967–1971, 9th Collect. Vol. 1972–1976
13. Euteco S. P. A.: Euteco Continuous Process, Technical Bulletin
14. Sorenson, W. R., Campbell, T. W.: Preparative Methods of Polymer Chemistry, 2nd Ed. New York: Interscience 1968
15. Jones, T. T.: J. appl. Chem., *2*, March 1952, 134
16. Čiernik, J.: Chemický prûmysl, *25/50*, 260 (1975)
17. Čiernik, J. Fiala, Z.: Chemický prumysl, *25/50*, 374 (1975)

5. Physiology and Environmental Protection

In order to judge the risks connected with the handling of phenolic resins, a clear distinction must be made between phenols, phenolic resin prepolymers and cured phenolic resins. Apart from constitutional characteristics, the MW is of great importance regarding the physiological effects. The physiological activity of phenolic prepolymers depends upon the content of free phenol and formaldehyde. Cured phenolic resins are entirely harmless. The FDA permits articles molded from phenolic resins to come into contact with food.

5.1. Toxicology of Phenols

Mononuclear phenols are protein degenerating and highly toxic. The oral LD_{50}-value (rats) is 530 mg/kg[1]. Human skin, which has come in contact with phenol, first becomes white, subsequently red and wrinkled; a strong burning sensation is quickly perceived. Longer contact destroys the skin tissue. Solid and liquid phenol are absorbed by the skin very quickly and cause very severe damage. Contact with large amounts leads to death through paralysis of the central nervous system. Minor intoxications lead mostly to damage of the kidneys, liver and pancreas. If phenol is inhaled or swallowed, local cauterizing occurs and headaches, dizziness, vomiting, irregular breathing, respiratory arrest and heart failure are the results.

The destructive effect of phenol on the human skin is reduced by introduction of lypophilic groups (methyl-, higher alkyl- or chloro groups). The neutral molecules are far more active than the corresponding ions. The biological activity of phenols is the result of their ability to alter biological structures, i.e. cracking of bacterial cell walls. The disruptive effect on cytoplasmic membranes and cell walls develops, it is believed, by the creation of pores large enough to permit cytochromes to diffuse out[2]. Cresols are similar to phenol in their action, but less severe in their effects (Tab. 5.1). Chlorophenols are not used for resin production.

The surface activity of alkyl phenols leads to their concentration on the cell surface, but does not explain the destructive effect to the cells. The bacterial and anthelmintic action of phenols is also influenced by soaps, which are usually used to solubilize phenols in water for use as disinfectants.

5.2. Toxicology of Formaldehyde

Formaldehyde in an aqueous solution is a protoplasm poison with a cauterizing and protein degenerating effect. The use of aqueous formaldehyde to preserve medical or biological preparations is well-known. Formaldehyde is believed to injure bacteria by reacting with the amino groups of proteins, which are thereby changed in nature and action[2]. Formaldehyde in the organism is quickly oxidized to formic acid which is partly separated by the urine.

Formaldehyde in the form of gas or aerosol – the effect of both is comparable – is very irritating to the mucous membranes. The pungent smell is noticeable even at concentrations below 1 ppm. The MAK-value is 1 ppm[1a]. Formaldehyde is a dangerous material to work with and has received the same rating as phenol.

Concentrations of formaldehyde up to 10 ppm cause conjunctivitis within a few minutes, as well as rhenitis with anosmia and pharyngitis. It can be observed that one can get used to formaldehyde to a certain extent. At 10–15 ppm dispnoe, cough and pneumonia develop[1,3].

5.3. Environmental Protection

The environmental policies of progressive industrial nations require not only the return of used media to the environment in a treated condition, but aim to work with these media without causing damages or injuries, i.e. new production processes must be developed which prevent contamination of the environment in the first place. An example is the legislation against environmental contamination in West-Germany. Aims and instruments of environmental policies are set forth in the "Bundesimmissionsschutzgesetz" (Federal Immission Protection Law), 1974[4]. The "Technische Anleitung zur Reinerhaltung der Luft" (Technical Instructions on Clean Air Maintenance), 1974, describes the minimum requirements for plants and their operation and a series of limiting values for emissions and immissions[5]. According to these requirements organic compounds of Class I – phenol and formaldehyde amongst others belong to this class – must not exceed a mass concentration of 20 mg/m^3 at a mass flow of 0.1 kg/h and more. In order to protect the waterways from contamination, the "Wasserhaushaltungsgesetz", 1976, was passed[6]. The law called "Wasserabgabengesetz", 1976[7], includes a scale of fees determined by the quantity of waste water emitted and the amount of injurious substances (according to the COD and BOD) in the water, and deposits as well.

Permissible levels of phenols in waste water have been established in USA by the Environmental Protection Agency (EPA) in the Federal Register. These guidelines generally establish phenol levels of 0.1 mg/l for the Best Practical Control Technology Currently Available (BPCTCA) for 1977, and 0.02 mg/l for the Best Available Control Technology Economically Achievable (BACTEA) for 1983[8].

Even relatively low concentrations of phenols below 10.000 μg/l water are fatal for several kinds of fish after 1–3 days. Lower concentrations are at least injurious and deteriorate the taste of the fish flesh considerably so that it is unfit for human consumption[9]. Phenols in chlorinated water lead to the formation of chlorophenols which will impart objectionable taste and odor to water even in quantities below 0.01 mg/l.

For example, in West-Germany the following requirements must be met when waste water is allowed to flow into the local waters: content of free phenols maximum 0.5 mg/l; temperature maximum 30 °C, pH 6.5–8.0. Furthermore, the total quantity of waste water allowed to flow in within 24 hours is limited to 75 m^3.

Table 5.1 Phenols in Water: Threshold Odour and Taste Concentration and Acute Fish Toxicity; According to F. Dietz, J. Traud[9a)]

	Threshold odour concentration µg/1	Threshold taste concentration µg/1	Acute fish toxicity LC_{50} µg/1
Phenol	300	4000	5000
o-Cresol	3	1400	2000
m-Cresol	2	800	6000
p-Cresol	2	200	4000
2,3-Xylenol	30	500	–
2,4-Xylenol	500	400	28000
2,5-Xylenol	500	400	2000
3,4-Xylenol	50	1200	4000
3,5-Xylenol	–	5000	50000
p-tert.-Butylphenol	30	800	–
Resorcinol	2000	6000	–
o-Chlorophenol	0.1	10	8000
m-Chlorophenol	0.1	50	3000
p-Chlorophenol	0.1	60	3000

Under certain circumstances waste water may be allowed to flow into the municipal biological water treatment plants together with the household sewage[10)]. In any case, the municipal requirements have to be adhered to. In general, the waste water must be as follows: temperature maximum 30–35 °C, pH 6.0–9.0, content of phenols or phenol prepolymers maximum 100 mg/l. In addition, the composition of the waste water must be such that neither the biological processes nor the plant operation are affected.

The disposal of solid phenolic waste is also regulated by the federal legislation. Waste materials containing injurious substances are to be disposed off only at official disposal places[11)]. Flammable waste materials are preferably destroyed within the plant in appropriate incinerators. Used phenolic resin coated sand can be disposed off at official disposal places without problems. Published results of the behaviour of foundry waste sand at disposal places[12)] show that the phenol quantities which may be eluted of cured resin bonded sands are even lower than amounts found for household waste under similar circumstances.

The regulation "Verordnung über gefährliche Arbeitsstoffe"[13)], covering the handling of injurious working materials, 1975, contains requirements for the rating, packing, marking and preparation of dangerous working materials. Dangerous or injurious materials in the sense of this regulation are basic and auxiliary materials and their formulations (blends, mixtures and solutions), if they are explosive, flammable, toxic, detrimental to the health, caustic or irritating. To indicate these properties, official warning symbols are to be put on packaging and containers. The scope of this list is in accordance with the requirements of the European Community of 1967 including their alterations and supplements of May 21, 1973, as well as the EEC rules and regulations for solvents. Phenols and phenolic resins, formaldehyde, the solvents methanol, propanol, toluene and some others which are used to produce resinous solutions, are also governed by these regulations.

All phenolic resins containing more than 5% free phenol must be designated "poisonous" by a skull. Phenolic resins containing 1–5% free phenol are considered detrimental to the health and are to be marked with a "St. Andrew's cross" (Andreaskreuz). Formulations are not considered toxic if the amount of free phenol is below 0.2%. The label on the packaging or containers must also show the name of the producer and the kind of toxic components and must include warnings of the special dangers involved and safety measurements to be taken.

During the handling of phenolic resins, sensitive persons may succumb to dermatological diseases. To prevent such reaction, it is advisable to treat the hands, arms, and other parts of the body which might be exposed to phenolic resins with an appropriate protective cream and to wear rubber or plastic gloves during work. After work, hands and arms are to be washed with a special soap and again treated with protective cream.

Furthermore, special attention should be given to clean working conditions and effective ventilation in the working rooms. The MAK-value (maximum concentration at place of work) is 5 ppm for phenol and 1 ppm for formaldehyde. Additional technical safety data for phenol are given in Chapter 2, Table 2.

5.4. Waste Water and Exhaust Air Treatment Processes

There are no universal solutions for waste problems for plants working with phenol and phenolic resins. The choice of the optimum process requires an individual analysis of the kind and amount of injurious substances as well as the structure of the plant and laboratory performance tests. Occasionally, a combination of different processes may be feasible. Such processes are microbial degradation, thermal combustion, physical and physico-chemical scrubbing, chemical oxidation or resinification reactions and adsorption methods.

5.4.1. Microbial Transformation and Degradation

The breakdown of aromatic compounds is an important step in the natural carbon cycle and many micro-organisms: eubacteriales, pseudomonas, actinomyceatas, endomicetas, higher funghi – are capable of breaking down aromatic substrates[14]. The essential step required for biological degradation is the conversion of the aromatic compound to an *ortho-* or *para*-dihydroxybenzene structure. The enzymes responsible for this hydroxylation have the character of mixed function oxidases or dioxygenases. The first steps of the three possible oxidative cleavage reactions of *o-* and *p*-dihydroxy compounds are shown in the formulae (5. 1/5). In the case of 1,2-

dihydroxybenzene, *ortho*- or *meta*-cleavage (5.1, 5.2) may occur. *Ortho*- and *para*-hydroxybenzoic acids (5.3, 5.5) may be formed as intermediates during the degradation of phenolic resin prepolymers.

ortho cleavage (5.1)

meta cleavage (5.2)

p-dihydroxybenzene cleavage (5.3)

(5.4)

(5.5)

The aliphatic mono- and dicarboxylic acids formed are further converted to 3-oxohexanedioic acid, which is taken up in the Krebs cycle, or to fumarate, pyruvate, acetaldehyde and acetoacetate. After this, the degradation to CO_2 and H_2O follows.

Certain kinds of microorganisms are able to live and cause degradation in water containing up to 1,000 mg/l of phenol[15]. They are most active at temperatures between 25–35 °C. Further essential prerequisites are a sufficient content of nutritive substances (N, P) and oxygen, pH between 7.5–8.5 and the absence of heavy-metal ions (< 5 mg/l). In order to provide the nutritive substances, it is advantageous to treat the waste water together with household sewage. Ammonium phosphate is most frequently used as a nutritive compound. The effectiveness of the biomass increases with time up to a limiting value because biological selection processes take place and the resistance and degradation ability of some kinds of microorganisms increase. The basin must have an effective aerating- and circulating system so that dissolved oxygen is always available in excess. The Unox process[16] uses oxygen instead of air in order to reach a higher oxygen level.

Biochemical degradation is the most used and most effective process for treating waste waters containing phenol. Final effluents in the range of 0.1 mg/l are reported. In order to ensure that feed and environmental conditions for the biomass are constant, properly designed equalization systems are required for optimum efficiency. Particular problems arise if plants are operated discontinuously or 5 days a week[17].

5.4.2. Chemical Oxidation and Resinification Reactions

In chemical oxidation processes phenols are normally destroyed to form intermediate non-toxic compounds (not CO_2 and H_2O), and so only a certain decrease in COD will result. The removal of phenol may reach final levels of less than 1 mg/l or >99% according to the ratio of chemicals used[17].

$$C_6H_5OH + 7\,O_2 \longrightarrow 6\,CO_2 + 3\,H_2O \qquad COD = 2.38\ mg\ O_2/mg\ phenol \qquad (5.6)$$

Hydrogen peroxide[18] in the presence of small amounts of iron-, manganese-, chromium- and copper salts is an effective oxidizer of phenols (and other organics). The temperature has little effect on reaction rate and conversion, a pH in the range of 3–5 is most effective. Hydrogen peroxide may be used to treat concentrated wastes high in phenols or for pretreating of high-phenol waste before biological treatment to obtain uniform phenol levels.

Ozone is a more effective oxidant than hydrogen peroxide. Lower amounts are normally applied as necessary for complete destruction to carbon dioxide and water. The selectivity of ozone is low, operating at pH values of 11.5–11.8 appears to result in preferential oxidation of phenol[19]. Ozone is often used in the final treatment step leading to very low phenol levels (lower than 0.1 ppm).

Sodium hypochlorite or chlorine dioxide, which is the oxidizing agent, will oxidize phenols to benzoquinones (pH 7–8). At pH above 10 further oxidation to maleic acid and oxalic acid will occur; chlorophenols are not formed. Chlorine is not used because of the formation of chlorophenols which are more toxic and have more objectionable taste and odor than the original phenols. Potassium permanganate or potassium dichromate are also effective oxidants; however, the handling of the precipitated sludge can be a serious problem.

Resinification reactions followed by precipitation of the polymeric material can be used for waste waters which contain phenol, phenol prepolymers and formaldehyde by adding sulfuric acid or ammonia and reacting at higher temperature. Ferric chloride or aluminium sulfate are recommended as precipitants. The deposits are burned in most cases. It is customary in the plywood-, particle board- and fiber board industries to acidify the waste waters with aluminium sulfate up to pH 4. By this method, the resinous components precipitate almost completely as a deposit which settles well and is filterable, especially if the precipitation occurs at elevated temperatures. Afterwards, the water must be neutralized with caustic lime (pH 6.5–8.0) and the calcium sulfate which is formed, filtered.

5.4.3. Thermal and Catalytic Incineration

The treatment of exhaust air by oxidation – thermal or catalytic incineration – is taken into consideration if the recovery of the solvents is not feasible or uneconomical. The organic components are oxidized to CO_2, CO and water. The catalytic incineration occurs at temperatures between 350–400 °C. Metal oxides, preferably, however, elements of the platinum group on different supports are used as catalysts. Catalysts are very sensitive. Sulfur-, phosphorus-, halogen-, silicon-, arsenic compounds and many others lead to catalyst poisoning[20, 21].

In principle then, catalytic incineration is only preferred if the exhaust air contains only minor concentrations of organic substances (<3 g C/m^3), if catalyst poisons are absent and if the content of dust in the air is small. The presence of all these prerequisites is very rare in plants working with phenol or phenolic resins, so that catalytic incineration does not have any importance in these industries. Their exhaust air is rich in combustible materials, due to the solvents used (varnish industry, production of laminates) or has a high solids content (production of mineral wool). The phosphoric acid esters which are part of many formulations act as catalyst poisons. They are transformed to phosphorus-V-oxide which may react with the catalyst support (for instance with Al_2O_3 forming $AlPO_4$), thus leading to deactivation[22]. The low molecular phenol alcohols carried along may cause resinification and formation of coke on the surface of the catalyst.

In thermal incinerators, the exhaust gases are burned in metal chambers at 700–800 °C. Gas or oil fired jet or area burners are used. The effectiveness of thermal incinerators, which of course depends on the temperature, is considerably higher (residual content of carbon < 5 mg/m^3 at 750 °C) than that of the catalytic process. This process results in a larger consumption of energy so that the utilization of the exhaust gas heat is the decisive economical factor. The principle of thermal incineration and an incinerator are shown in Figures 5.1 and 5.2[23].

This process is used in almost all plants producing technical laminates as well as in larger varnish drying equipments. In both cases, phosphoric acid esters and a relatively large quantity of solvents are contained in the exhaust air. The combustion heat is utilized to preheat the exhaust air as well as to heat the circulating air for the drying process, occasionally also for the preparation of hot water. The exhaust air, after appropriate adjustment, may be led directly into the dryer.

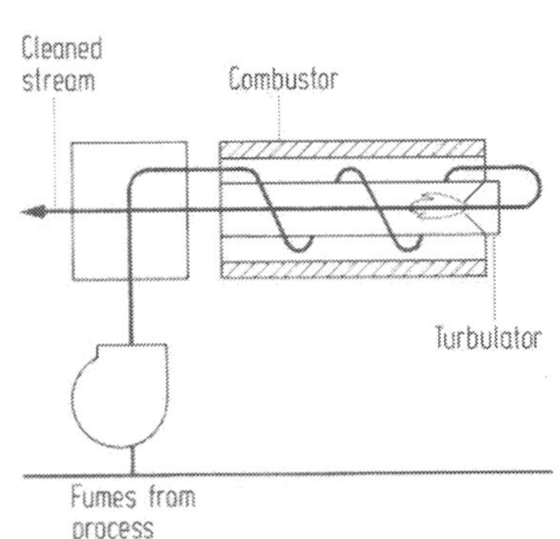

Fig. 5.1. Thermal incinerator with heat exchanger (Drawing: UOP-KAVAG, D-6467 Hasselroth)

Fig. 5.2. Overall view of a thermal incinerator (Photo: UOP-KAVAG, D-6467 Hasselroth)

The combustion is not limited to exhaust air. The combustion of waste water offers the only economical solution in many instances. The liquid waste which must be atomizable at temperatures less than 200 °C is sprayed into the combustion chamber either immediately or, in the case that the liquid is not able to sustain combustion, after the addition of supporting fuel, and burned. Theoretically, a mixture of about 18% phenol and 82% water would be self-sustaining when burned at 760 °C[17]. However, the solubility of phenol in water is limited to about 10% at ambient conditions, and therefore effective mechanical mixing is necessary at such high phenol levels.

5.4.4. Extraction Processes and Recovering

The coking plants in the Ruhr area have been confronted with the problem of the recovery of phenol from waste waters since their beginning (Chapter 2). Over the years, different extraction processes have been developed. Some of them are used to a great extent for the dephenolization in coking plants today[15, 24, 25]. The effectiveness of the benzene/caustic process according to Pott and Hilgenstock, which uses benzene as a solvent and effects the recovery with an aqueous sodium hydroxide solution, can be increased to 98% by the use of multi-stage Podbielniak countercurrent extractors. However, it is not possible to lower the content of phenol below 20 mg/l. Using the Phenosolvan process developed by Lurgi[26], which employes diisopropyl ether as extractive solvent in a 10-stage counter-current extractor, the phenol content can be reduced to values around 10 mg/l. Neither process is effective enough to achieve the limit of 0.5 mg of phenol per liter. Therefore, these extraction processes are only important for the recovery of phenol in coking plants, coal liquefication plants and installations producing phenol and phenolic resins.

5.4.5. Activated Carbon Process

Activated carbons have been developed for the treatment of waste water and exhaust air. They have a high adsorption capacity for phenols, high stability and abrasion resistance[27]. A polymeric adsorbent, also highly effective in removing phenol and similar compounds from aqueous waste yielding an effluent with generally less than 1 ppm of phenol has been recently developed[28]. Other examples of adsorbent media are aluminium oxide, silica gel and zeolites. Generally, these processes consist of three process steps. During the pretreatment, the waste water or exhaust air is cleared of solid substances. In the adsorption step, the organic substances are adsorbed by an activated carbon bed by a purely physical process based on electrostatic interaction. The adsorption process may be accompanied by chemisorption and capillary condensation.

In the regeneration step, the desorption and reactivation of the activated carbon take place. Either diluted caustic, solvents (methanol, acetone) or inert gas mixed with steam are used to desorb phenols. A sulfuric acid treatment is recommended to recover phenol from the sodium salt. If solvents are used, the treatment is done by distillation. Regeneration with inert gas/steam necessitates a post combustion of the exhaust gas.

5.4.6. Gas Scrubbing Processes

Gaseous components are absorbed in a liquid either due to a physical solution process or by chemical reaction. Both mechanisms may be applied to exhaust gases containing phenol. The exhaust gases may also contain solids and droplets[29]. The principle of a continuous absorption plant is that the finely distributed washing agent and the exhaust gas pass each other in the absorber in a countercurrent. The loaded water can either be regenerated or, in some cases, returned to the resin application process in a closed loop thus omitting further addition of water. Otherwise, the exhaust air problem would become a waste water problem.

This process is extremely suitable in cases where relatively small amounts of injurious substances are to be removed from relatively large amounts of air. For example, this is the case in the production of mineral wool products. Here, this process would be the only practicable solution. The essential figures for the absorption are the mass transfer constant, the diameter of the drops, the exchange area and the time of direct contact. Common absorbers are spray tower washers, filter washers and plate columns.

References

1. Sax, N. J.: Dangerous Properties of Industrial Materials van Nostrand Rheinhold Co.: 1975
1a. Deutsche Forschungsgemeinschaft: Maximale Arbeitsplatzkonzentrationen 1978. Boppard: H. Boldt-Verlag 1978
2. Albert, A.: Selective Toxicology. London: Methuen & Co.: 1968
3. Kimmerle, G.: Toxikologie. In: Formaldehyd. Ullmanns Encyklopädie d. Techn. Chem. Vol. 11, 4 Ed. Weinheim: Verlag Chemie 1976
4. Gesetz zum Schutz vor schädlichen Umwelteinwirkungen durch Luftverunreinigung (Bundes Immissionsschutzgesetz-BImSchG) vom 15. März 1974, BG-Bl. I, S. 721 (1974); BG-Bl. I, S. 1253 (1976)
5. Erste Allgemeine Verwaltungsvorschrift zum Bundes-Immissionsschutzgesetz (Technische Anleitung zur Reinhaltung der Luft; TA Luft) vom 28. August 1974, GM-BL. Nr. 24, S. 426 (1974)
6. Gesetz zur Ordnung des Wasserhaushaltes (Wasserhaushaltsgesetz) vom 26. April 1976, BG-Bl. I, Nr. 48, S. 1109 (1976)
7. Gesetz über Abgaben für das Einleiten von Abwasser in Gewässer (AbwaG) vom 13. September 1976, BG-Bl. I. S. 2721 (1976)
8. Development Document for Effluent Limitations Guidelines for Petroleum Refining, U. S. Environmental Protection Agency, EPA 440/1-74-014-a, April 1977.
9. Liebmann, H.: Handbuch der Frischwasser- und Abwasserbiologie, Bd. II. München: Oldenburg 1960
9a. Dietz, F., Traud, J.: gwf-Wasser/Abwasser *119* (6) 318 (1978)
10. Hinweise für das Einleiten von Abwasser aus gewerbl. und industr. Betrieben in eine öffentliche Abwasseranlage, ATV-BDI-KfK 1970, BDI-Drucksache Nr. 90, BDI, 5 Köln 51
11. Gesetz über die Beseitigung von Abfällen (Abfallbeseitigungsgesetz-AbfG) vom 5. Jan. 1977, BG-Bl. I Nr. 2, S. 41 (1977)
12. Bradke, H.-J., KLein, T.: Deponierverhalten und Verwertung von Gießereisanden, Teil I und II: Laborauslaugungen von Gießereisanden zur Ermittlung der eluierbaren Stoffe und des voraussichtlichen Deponierverhaltens. IWL, 5 Köln 51, 1975
13. Neufassung der Verordnung über gefährliche Arbeitsstoffe vom 8. September 1975, BG-Bl. I, Nr. 107, S. 2493 1975

14. Kieslich, K.: Microbial Transformations of Non-Stereoid Cyclic Compounds. Stuttgart: Thieme 1976
14a. Rozovskaya, T. I., Lazareva, M. F.: Mikrobiologiya *40,* 370 (1971)
15. Sierp, F.: Die gewerblichen und industriellen Abwässer, Berlin/Heidelberg/New York: Springer 1967
16. Martin, P., Schönfelder, H., Tischer, W.: gwf, Wasser/Abwasser *116,* 272 (1975)
17. Lanuette, K.: Chem. Engng.-Deskbook issue, Oct. *17,* 99 (1977)
18. Kibbel, W. H., Jr., Raleigh, C. W., Shepherd, J. A.: Hydrogen Peroxide for Industrial Pollution Control, 27 th Ann. Purdue Indl. Waste Conf. Part 2, West Lafayette, Ind., 1972.
19. Nebel, D., Gottschling, R. D. Holmes, J. L., Urangst, P. C.: Ozone Oxidation of Phenolic Effluents. Welsbach Ozone Systems Corp. 1976
20. Quillmann, H.: Chem. Ind. *25,* 497 (1973)
21. Rüb, F.: Wasser, Luft u. Betrieb *19,* 702 (1975)
22. Kirchner, K., Angele, B.: Chem.-Ing.-Techn. *49,* 243 (1977)
23. UOP-KAVAG, UOP Thermal Incineration Systems, Techn. Bulletin
24. Wurm, H. J.: Chem.-Ing.-Techn. *48,* 840 (1976)
25. Hahn, E., Herpes, E. Th., Neff, I.: Chem. Ind. *28,* 591 (1976)
26. Lurgi GmbH: Dephenolization of Effluents by the Phenosolvan Process. Techn. Bulletin
27. Jüntgen, H., Klein, J.: Energy Sources *2* (4) 311 (1976)
28. Fox, Ch. R.: Hydrocarbon Proc., July 1976, P. 109
29. Reither, K.: Wasser, Luft u. Betrieb *10,* 3 (1974)

6. Analytical Methods

Phenolic resins are produced in a discontinuous batch process. There is a multitude of analytical methods[1, 2] for controlling the production and final quality of the resin. Quality control is performed by the producer and rechecked to a limited extent by the consumer. The most frequently controlled properties are dry resin content, reactivity, monomer content and viscosity. Control methods are determined by individual requirements in the different fields of application; resin efficiency for instance is important in the production of mineral fiber insulations. Analytical methods differ from country to country due to differences in national standards so that the comparability of the results is limited. The expenditure for quality control of thermosets is very high. Therefore, apart from the basic requirements of analytical procedures like reproducibility, accuracy and sensibility, the economy, meaning the grade of automation of the method, is of increasing importance. Thus, the analytics of phenolic resins is in the process of being revised, because the physical methods of investigation (GC, HPLC) coupled with data processing are applied to an increasing extent. Also customers working with phenolic resins have often to decide which analytical method will be the best for batch to batch uniformity control with regards to information quality, quantity and economy. In most cases, chemical methods do not come into question because the analytical laboratory equipment as well as skilled personnel are not at one's disposal. Furthermore, their information quantity is limited, they allow only single point determinations.

In spite of the relatively high cost of automatized equipment, the physical methods – e. g. gel permeability chromatography (GPC) and high performance liquid chromatography (HPLC) – may be the best ones because a multitude of characteristic properties and values like free phenol content, MWD and content of additives are displayed visually in one recorder trace.

6.1. Monomers

6.1.1. Phenols

Phenols in general undergo a number of reactions resulting in colored products which have been used in the past for the qualitative identification of phenolic resins. The ferric chloride test is probably the most widely known identification reaction[3, 4]. Qualitative color tests are no longer used.

Koppeschaar Method

The determination of free phenol is mostly performed according to the Koppeschaar method[5] as described in ASTM D 1312. Phenol is separated by steam distillation and then converted to tribromophenol by use of a 0.1 N potassium bromide-bromate solution as the source of bromine. Excess of bromine is determined by the addition of potassium iodide, followed by titration of the liberated iodine with sodium thiosulfate solution. The selectivity of the method is limited; other bromine reactive compounds e. g. furfural or furfuryl alcohol, cannot be distinguished.

Colorimetric Determination

At low phenol levels the quantitative determination is performed colorimetrically. The phenol obtained by steam distillation of an aqueous solution is converted into an azo dye by the addition of *p*-nitroaniline and a sodium nitrite solution, which is then determined colorimetrically either directly or after extraction with butanol.

Also the 4-aminoantipyrine (1-phenyl-2,3-dimethyl-4-aminopyrazolone-5) method (ASTM D 1783-70) is widely used, e. g. for the determination of phenols in industrial waste waters, despite some shortcomings of the method[2, 6]. 4-Aminoantpyrine reacts with phenols in the presence of ferricyanide ions in alkaline solution resulting in quinoide-type compounds (6.1). *para*-Alkylphenols give negative or weak results.

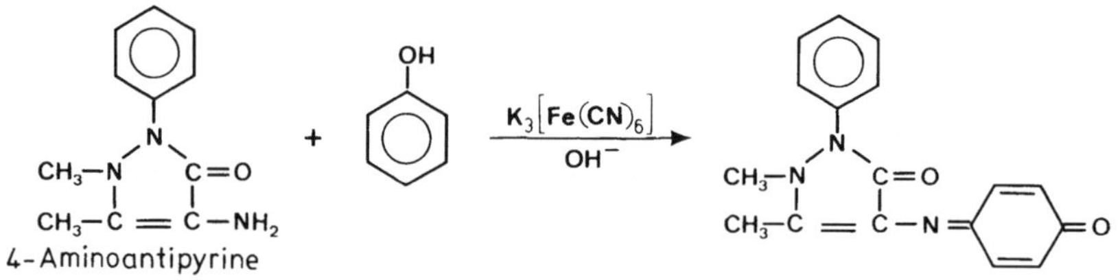

4-Aminoantipyrine + phenol $\xrightarrow[OH^-]{K_3[Fe(CN)_6]}$ quinoide-type compound (6.1)

MBTH · HCl (6.2)

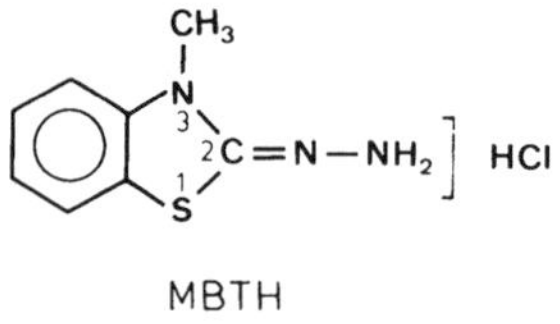

Some disadvantages of 4-aminoantipyrine seem to be avoided by use of MBTH (2-(2-hydrazono-2,3-dihydro-3-methylbenzthiazol hydrochloride).

Gas-chromatographic Determination

Gas-chromatographic determination is being used to an increasing extent for the determination of free phenols[7] in phenolic resins. This method is very sensitive and has the advantages that the distribution of cresol and xylenol isomers can be determined quantitatively and that it is not disturbed by other bromine-reacting compounds. An internal standard compound is added for the quantitative determination.

The resolution and information obtained by this method can be increased considerably by conversion of phenols and methylolphenols to their trimethylsilyl-derivatives. The intramolecular interaction is reduced and the volatility increased considerably by etherification of the hydroxyl group, e.g. with N,O-bis-(trimethylsilyl) trifluoracetamide.

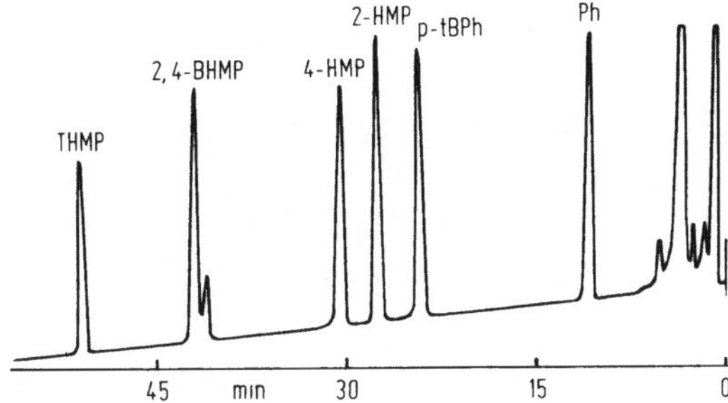

Fig. 6.1. Gas chromatographic determination of phenol alcohols analyzed in form of their trimethylsilyl ether derivatives.
Ph = phenol; *p*-tBPh = *p*-tert. butylphenol; HMP = hydroxymethylphenol; BHMP = bis(hydroxymethyl)phenol; THMP = tris(hydroxymethyl)phenol

The separation of *m*- and *p*-cresol, 2,4- and 2,5-xylenol and *m*-ethylphenol, *p*-ethylphenol and 3,5 xylenol and of methylolphenols as well is possible by this technique.

6.1.2. Formaldehyde

The (free) formaldehyde[8] is converted quantitatively into the corresponding oxime by the addition of hydroxylamine hydrochloride. In the reaction an equivalent amount of HCl is set free, which is determined alkalimetrically.

$$CH_2{=}O + [NH_2{-}OH]\,HCl \rightarrow H_2C{=}N{-}OH + H_2O + HCl \qquad (6.3)$$

A simple and highly sensitive colorimetric identification test is based on the formation of a specific blue-violet color when formaldehyde reacts, in the presence of strong acids, with Schiff's fuchsine-bisulfite reagent.

6.2. Nitrogen and Water

Total Nitrogen

According to Kjeldahl (1883) organic substances containing nitrogen are destroyed by heating with concentrated sulfuric acid. The carbon is oxidized to carbon dioxide while organically bound nitrogen is converted quantitatively into ammonium sulfate[9]. The decomposition is facilitated by the addition of dehydrating substances like potassium sulfate or catalysts like copper(II)-sulfate.

The weak base ammonium hydroxide is then displaced from the ammonium salt solution by adding sodium hydroxide, that means it is removed as gaseous ammonia. The amount of ammonia is determined quantitatively by absorption in a measured amount of acid of known content.

HMTA in Powdered Resins

Two methods are used. In the first method, the powder resin is dissolved in a butanol/ethylglycol mixture and titrated either potentiometrically or by addition of thymol blue as indicator with 0.1 N hydrochloric acid. According to the second method, the determination is performed in acetone solution with 0.1 N perchloric acid, also in acetone, potentiometrically.

Water

The water content of liquid resins is determined by the Karl-Fischer-titration method (1935) according to DIN 51 777. This method is based on the oxidation of iodine. The determining solution – a solution of 1 mol J_2, 3 mol SO_2, 10 mol pyridine (Py) and 50 mol water-free methanol – contains the oxidizing agent J_2 as well as the reducing agent. The redox reaction, however, is only released by the missing component of the oxidizing system, the water contained in the test solution:

$$3\,Py + SO_2 + J_2 + H_2O + CH_3OH \rightleftharpoons 2\,[PyH^+ + J^-] + PyH^+ + CH_3{-}O{-}SO_3^- \quad (6.4)$$

The equation shows that 1 mol of iodine is equivalent to 1 mol of H_2O. The titration is performed according to the Dead-Stop-technique.

According to the carbide method, calcium carbide is reacted with water yielding acetylene.

$$CaC_2 + 2H_2O \rightarrow CH{\equiv}CH + Ca(OH)_2 \quad (6.5)$$

The reaction is performed in a closed vessel connected to a precision manometer. The increase of pressure is strictly proportional to the water content.

6.3. Physical Properties

Content of Resin

2–3 g of resin in a porcelain cup or sheet metal lid are heated to 150 °C (30 min) in a dryer to constant weight. The cup is cooled in a desiccator and weighed.

Viscosity

The most important processes are:

The determination with the forced ball viscosimeter (DIN 53015). The principle of measuring is the rolling and sliding movement of a ball in a sloping, cylindrical tube which is filled with the liquid to be tested.

The use of a cone-plate-rotational viscosimeter (DIN 53229, Paragraph 8.1). The measuring device consists of a plane plate on which a cone is rotating. The substance fills the space between the plate and cone. The time necessary for one measurement is 1–2 minutes.

The viscosity cup (Ford cup, DIN 59211), the Ubbelohde capillary viscosimeter (DIN 53177) and the rotational viscosimeter (DIN 53229, Paragraph 8.2) are also used.

Water Dilutability

Water dilutability plays a role in the processing of many phenolic resins. When diluting with water, two points can be stated, the first turbidity (opalescence) and the heavy turbidity (milky or precipitation). It is the heavy turbidity, above all, which is important in dilutability determination.

A measured quantity of liquid resin is diluted with water drop by drop and shaken until a permanent turbidity is achieved. The water dilutability, which is influenced by the content of alkali and solvents, mainly depends upon the grade of condensation of the resin and represents a good criterion for the molecular weight.

Specific Gravity

The SG is determined with the pycnometer at 20 °C according to DIN 53217.

Refractive Index

The resin content of a liquid resin can be determined very easily and quickly by the determination of the refractive index according to DIN 53491 with an Abbe-refractometer. The conversion of phenol with formaldehyde at resin production is followed by the determination of the refractive index.

Melting Point

The MP is determined by the capillary method according to DIN 53181. Since phenolic resins are not uniform compounds with a well-defined MP, it is more reasonable to speak of a melting range which generally is within 3 °C.

Bulk Weight and Stamp Volume of Powder Resins

The bulk weight is determined according to DIN 53468. A 100 ml measuring cylinder, which is cut off at the 100 ml mark is used. The powder is poured in from a height of 2 cm above the rim of the cylinder. The liter weight of the powder is an approximate criterion of the fineness of grain and should fluctuate only within very narrow limits.

At the stamp volume (DIN 53194), the reproducibility is somewhat better. Special stamp volumeters with a design fixed by standards serve for the determination.

Sieve Analysis

A complete sieve analysis to determine the grain size distribution, which in accordance with DIN 53734 is to be performed with an air blast sieving equipment, is too expensive for quality control. Therefore, in most cases only the residue on the 0.06 mm sieve (10,000 mesh per cm^2) is determined according to DIN 1171. For example, for powdered resins to be used for the manufacturing of grinding wheels, this residue should not exceed 1–2%.

6.4. Reactivity

Determination of B-Stage Time

B-stage time means the time which the resin needs to change from the liquid (A-stage) to the rubber-like B-stage at a distinct temperature. The determination is performed on an iron plate with a diameter of 200 mm and a height of 20 mm. Round cavities with a diameter of 20 mm and 5 mm deep are arranged on a concentric circle with a diameter of 130 mm. A hole for the thermometer is drilled in the center of the plate. The iron plate is put on an electrically heated plate with exact temperature control. 0.5 g of liquid phenol resin or 0.15 g powdered resin are filled into the cavity where the temperature is held at 130 or 150 °C and slightly stirred with a thin glass rod. As soon as a certain degree of three-dimensional cross-linking is reached, the resin shows a rubber-like condition and is unable to flow. The time in seconds until this stage is reached (B-stage) is the measurement for the curing rate.

Gel Time

A weighed amount of resin is filled into a test tube and dipped into an oil bath at 130 °C. The increase in viscosity is identified with a glass rod formed like a piston. As soon as the gel point of the resin is reached, the glass rod is more or less stuck. This moment in time can be automatically recorded (DIN 16945).

Flow Distance

Information about flow and reactivity of solid resin-HMTA mixtures is obtained by determining the flow distance in an oven (according to Norton) at 125 °C. 0.5 g of resin are pressed to a pellet with a diameter of 12.5 mm and 4.8–5.2 mm height on a pellet press. The flow distance is determined on a hot, inclined glass plate. A tilting device along with the glass plate is preheated to the required temperature in a drying cabinet which is set at a temperature of 125 °C. The pellet is placed on the horizontal glass plate which is slanted after three minutes to form an angle of 60 ° with the base plate of the device. The whole device is left in the drying cabinet in this position for 30 minutes. The length of the flow in mm is the measurement for the melt viscosity and curing rate. Resins of high molecular weight and high reactivity do not flow and only a blowing up of the pellet is observed. In this case, the diameter of the pellet is expressed as flow distance.

Differential Scanning Calorimetry

Differential scanning calorimetry (DSC) gives a measure of the difference in the rates of heat absorption by a sample with respect to an inert reference (maintained at the same temperature) as the temperature is raised at a constant rate. This contrasts with conventional Differential Thermal Analysis (DTA) in which the differential temperature caused by heat changes in the sample is monitored[10]. A recorder prints out the data as a trace of dH/dt versus temperature. The integral of this function is the reaction enthalpy. The difference between the heat development ΔH of an uncured sample and a partly cured sample is the measurement of the degree of cure[11]. In the absence of a chemical reaction, second order transitions are indicated as a discontinuity in the thermogram caused by change in the specific heat.

By assumption of a constant specific heat, further kinetic data, for instance cure rate and activation energy, can be approximately determined[12–14].

At phenolic resins cure, endothermal processes (evaporation of water, formaldehyde and ammonia) superimpose exothermal chemical reactions, thus complicating the evaluation considerably. However, techniques have been developed which enable a heating up under pressure.

DTA and TG analysis have been used by Orrel and Burns[15] to investigate the phenol-HMTA and novolak-HMTA reactions. TG diagrams of different phenolic resins are shown in Chapter 7, Figure 7.1.

6.5. Chromatographic Methods

Liquid Chromatography

The technique of high performance liquid chromatography HPLC[16] has been used relatively seldom for analytical characterization of phenolic prepolymers up to date[17, 18] although it offers certain unique advantages for the molecular weight range up to approximately 1,000. It is applicable to thermally unstable materials and allows the separation of compounds, the physical behaviour of which is determined by intermolecular interactions. The problem of appropriate separating solvent-systems for phenolic resins has not yet been satisfactorily solved.

The reversed-phase HPLC was used to separate phenols, alkylphenols and dihydroxybenzenes by Engelhardt et al.[19, 20].

The order of elution in RP-techniques is opposite to that in chromatography with polar stationary phases – the more polar the compound, the earlier it is eluted. Dichlorodibutylsilane on silica (5–10 μm) was used as stationary phase. Methanol

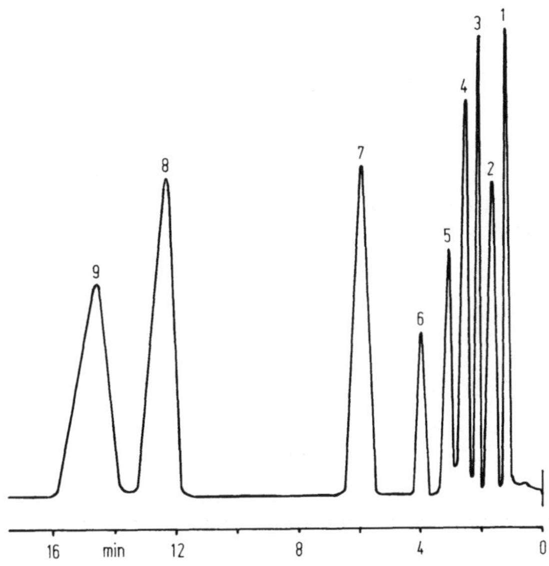

Fig. 6.2. Reversed-phase HPLC separation of phenols[20]. Stationary phase: Si-100-C_4. Eluent: water/methanol. Samples: *1* = methanol; *2* = hydroquinone; *3* = resorcinol; *4* = catechol; *5* = orcinol; *6* = phenol; *7* = *p*-cresol; *8* = *m*-xylenol; *9* = *o*-xylenol

or acetonitrile or mixtures of these solvents with water qualified as eluents. This process is supposed to be the best and fastest method of determining phenols and phenol alcohols.

Gel Permeation Chromatography (GPC)

Gel permeation chromatography is a special type of liquid-solid chromatography that separates molecules according to molecular size. The filling in the column is a material with pores and channels of approximately the size of the molecules to be separated. Smaller molecules can more easily penetrate into the pores and remain in the stationary phase longer. The amount of equipment is comparable to that of HPLC. Gels are not used anymore today because of their low pressure stability[16]. Polystyrene, PVAc or silica gels of very narrow sieve fractions around 5 μm are used as they are obtained, for instance, by sedimentation in liquids.

GPC-chromatograms have been used several times in this book to explain the MWD[21]. A home-made apparatus was used including a differential refractometer. PVAc was applied as stationary phase, DMF was used as solvent.

Paper and Thin Layer Chromatography

The two-dimensional paper and thin layer chromatography[7, 22] are appropriate for the qualitative identification of phenols, phenol alcohols and di-nuclear prepolymers as well as for the approximate quantitative determination. The statistical deviations are within the range of 5–15%. Formely, these methods were frequently used to determine kinetic data (see Chapter 3). A methanol/chloroform mixture, for instance, is used as eluent in the first direction and benzene/MEK/diethylamine in the second. The visualization is performed by spraying with diazotised *p*-nitroaniline. Today, the faster and highly automated high-pressure chromatographic methods will be used to an ever increasing extent.

Pyrolysis/Gas Chromatography

Information on structure of phenolic resins is obtained by pyrolysis and gas chromatographic determination of the fission products. This method is chosen for cured phenolics which cannot be studied by other methods due to their insolubility and infusibility. Past experience has shown that the best results are obtained by stepwise pyrolysis at 300 and 800 °C.

6.6. Spectroscopy

Nuclear Magnetic Resonance Spectroscopy NMR

The structure of transient molecules involved in the resin formation[23] and structure of phenolic prepolymers[24, 25] can be quite completely analyzed by NMR[26]. It was shown that quantitative estimation could be performed for the *ortho/para* hydroxymethyl ratio and the ratio of *p-p, o-p* and *o-o*-methylene linkages in bisphenols. There is a clear difference between NMR spectra of resols and novolaks; the presence of alkyl substituents is also clearly indicated.

Infrared Spectroscopy

The infrared spectra of phenol-formaldehyde resins were studied by many authors. Hummel[27, 28] published a multitude of typical infrared spectra. *Ortho-ortho* linked novolaks show a strong band at 13.3 μm, *ortho-para* novolaks show bands of almost equal intensity at 13.3 and 12.2 μm[29]. HMTA is evident from a sharp doublet band at 9.9–10 μm. In the vicinity, however, the typical hydroxymethyl absorption appears at 9.9 μm. The resol structure is further indicated by strong absorption at 3 μm (hydroxyl group) and 6.9 μm[1]. The dibenzyl ether band appears at 9.5 μm. Etherified resols show an absorption at 9.2 μm. IR-spectra are also very qualified for the identification of additives; the identification, however, requires highly skilled personnel.

References

1. Haslem, J., Willis, H. A., Squirell, D. C. M.: Identification and Analysis of Plastics, 2. ed. London: Iliffe Books 1972
2. Ettre, L. S., Obermüller, E.: Phenols. In: F. D. Snell, L. S. Etre (ed.): Encyclopedia of Industrial Chemical Analysis, Vol. 17, New York: Interscience
3. Wesp, E. F., Brode, W. R.: J. Amer. Chem. Soc. *56*, 1037 (1934)
4. Soloway, S., Wilen, S. H.: Analyt. Chem. *24*, 970 (1952)
5. Koppenschaar, W. F.: Z. analyt. Chem. *15*, 233 (1876)
6. Feigl, F., Anger, V.: Analyt. Chem. *33*, 89 (1961)
7. Harborne, J. B.: Chromatography of Phenolic Compounds. In: E. Heftmann (ed.): Chromatography, 3. ed. New York: Van Nostrand Reinhold 1975
8. Bleidt, R. A.: Formaldehyde. In: F. D. Snell, L. S. Etre (ed.): Encyclopedia of Industrial Chemical Analysis, Vo. 13. New York: Interscience
9. Pohloudek-Fabini, R. Beyrich, Th.: Organische Analyse. Leipzig: Akademische Verlagsges. 1975
10. Wendlandt, W., (ed): Thermal Methods of Analysis, 2. ed. New York: John Wiley 1974
11. Nachtrab, G.: Kunststoffe *60*, 261 (1970)
12. Kay, R., Westwood, A. R.: Europ. Polymer J. *11*, 25 (1975)
13. Kamal, M. R., Sourour, S.: Polymer Engng. and Sci. *13*, 59 (1973)
14. Taylor, L. J., Watson, S. W.: Analyt. Chem. *42*, 297 (1970)
15. Orrell, E. W., Burns, R.: Plastics & Polymers *36*, 469 (1968)
16. Engelhardt, H.: Hochdruck-Flüssigkeits-Chromatographie. Berlin, Heidelberg, New York: Springer 1975
17. Sebenik, A., Lapanje, S.: J. Chromatography *106*, 454 (1975)
18. Karch, K.: Ph. D. Thesis, University of Saarbrücken, Saarbrücken 1974
19. Engelhardt, H., Weigand, N.: Analyt. Chem. *45*, 1149 (1973)
20. Karch, K., Sebestian, I., Halasz, I., Engelhardt, H.: J. Chromatography *122*, 171 (1976)
21. Knop, A., Trapper, W.: Unpublished.
22. Pastuska, G., Petrowitz, H. J.: Chemiker-Ztg. *86*, 311 (1962)
23. Kopf, P. W., Wagner, E. R.: J. Polymer Sci.-Polymer Chem. Ed. *11*, 939 (1973)
24. Woodbrey, J. C., Higginbottom, H. P., Culbertson, H. M.: J. Polymer Sci. Part. *A 3*, 1079 (1965)
25. Hirst, R. C., Grant, D. M., Hoff, R. E., Burke, W. J.: J. Polymer Sci. Part. *A 3*, 2091 (1965)
26. Bovey, F. A.: High Resolution NMR of Macromolecules. New York: Academic Press 1972
27. Hummel, D. O.: Polymer Spectroscopy. Weinheim: Verlag Chemie 1976
28. Hummel, D. O., Scholl, K.: Infrared Analysis of Polymers, Resins and Additives. Carl Hanser/Verlag Chemie 1973
29. Bender, H. L.: Mod. Plastics *30*, 136 (1953)

7. Degradation of Phenolic Resins by Heat, Oxygen and High Energy Radiation

7.1. Thermal Degradation

For particular applications, the serviceability of the polymer, depending on temperature, is the focus of interest. The general term polymer thermal stability is not always exact. It is mostly used with reference to the alteration of one or more chosen physical properties dependent upon temperature and time. Another characterization of the stability is derived from the ability of the polymer to degrade. TGA appears to have only a restricted value, however, in relating relative performance in the actual environment, which is far different from laboratory conditions. Some phenolic resin types with relatively better TGA, e. g. PF-resins containing naphthol – or *p*-phenylphenol units, gave inferior ablative performance, indicating that the mechanical strength of the material also plays an important part[1].

In general, all thermochemical and physical impacts must be taken into consideration. PF-resins are known to be very temperature resistant plastics and to give high yield of strong char at pyrolysis. They are, therefore, used for ablative materials, friction materials, grinding wheels, and HT-resistant moldings. The thermal degradation of PF-resins[1, 2] can be clearly divided into three stages, indicated by weight loss and volume change[3].

In the first stage, up to 300 °C, the polymer remains almost intact. The quantity of gaseous components released in this stage is relatively small (1–2%). Mainly water and the remaining monomers, phenol and formaldehyde, which had been enclosed during cure, diffuse out.

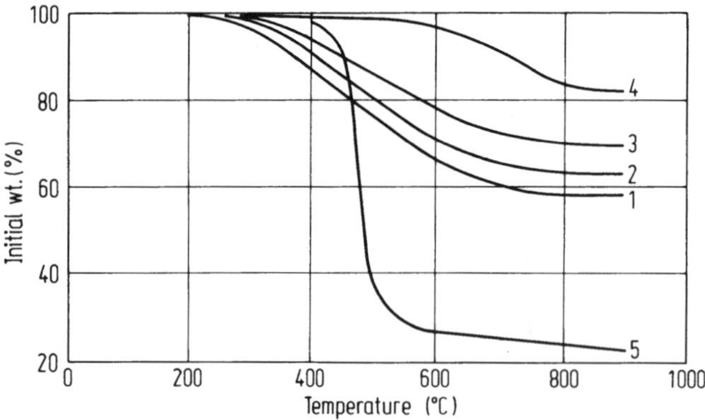

Fig. 7.1. Thermogravimetric analysis (N_2, 15 °C/min) of different phenolic resins *1–3*, Poly-*p*-phenylene *4* and Polycarbonate *5*.
1 Phenol-novolak resin, 10% HMTA; *2* Phenol novolak/resol resin 60:40, 6% HMTA; *3* Boron-modified phenol resin (18% B); *4* Poly-*p*-phenylene; *5* Polycarbonate

The decomposition starts at approximately 300 °C. Up to 600 °C the main quantity of gaseous components is split off. The reaction rate reaches the maximum within this range. In this second stage, water, carbon monoxide, carbon dioxide, methane, phenol, cresols and xylenols are split off. In this degradation stage random chain scission occurs; no depolymerization takes place. At the same time, a strong increase of ketone- and carboxyl groups can be observed by IR. The shrinkage, however, is relatively low. The internal porosity increases clearly and so the density decreases. It is assumed that a coalescence of chains to form ribbons occurs in this stage.

In the third stage above 600 °C, CO_2, CH_4, H_2O, benzene, toluene, phenol, cresols and xylenols are split off. The high shrinkage is characteristic; the density increases and the permeability decreases considerably. The electrical conductivity increases.

The mechanism of the thermo-chemical degradation was thoroughly investigated by Conley[4)] and Gautherot[5)]. It was stated that there is always a thermo-oxidative process taking place regardless of whether the pyrolysis reaction occurs in oxidative or inert atmosphere. The high oxygen content of phenolic resins is the reason for this process. The reaction steps described below have been proposed by Conley[4)].

The methylene bridge is the thermodynamically most stable cross-link in PF-polymers and therefore the prevailing one in cured resins. The degradation process is therefore dependent upon the stability and concentration of the dihydroxyphenylmethane units so that different resins show an almost similar behaviour if the curing process is complete.

The first step of the thermo-oxidative degradation is the formation of hydroperoxide followed by decomposition to dihydroxybenzophenone and benzhydrole structures.

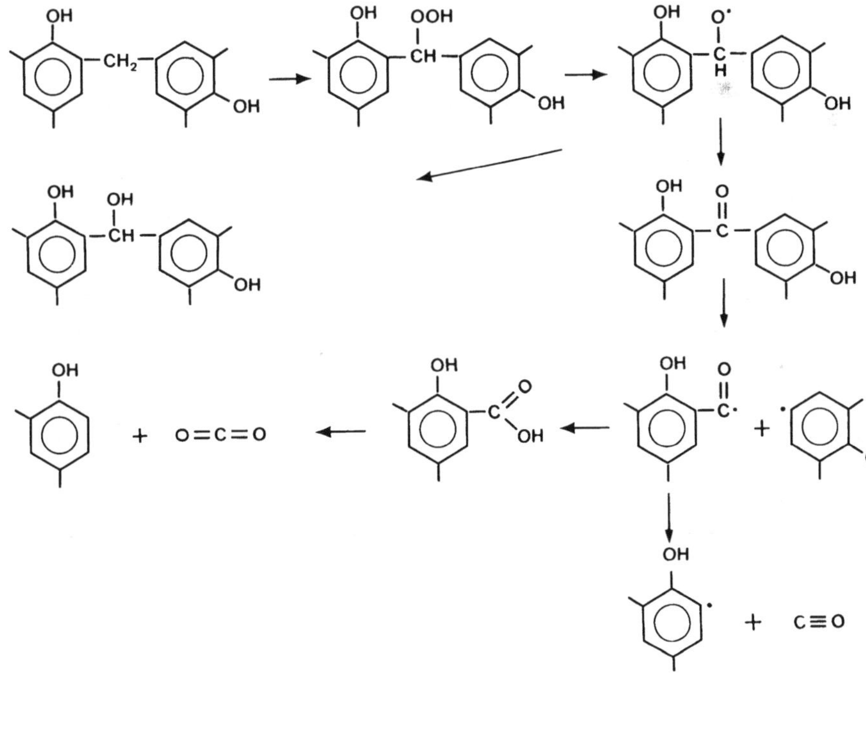

(7.1)

(7.2)

(7.3)

(7.4)

The sequence of this reaction is in accordance with the observed increase in IR-absorption at 3.0 μm (hydroxyl group) and 6.05 μm (carboxyl group) and the identification of volatile products. The decarboxylation and decarbonylation reactions start at approximately 300 °C and 500 °C respectively.

A second fragmentation reaction and hydrogen abstraction leads to cresol and methane formation. The formation of methane occurs in increasing amounts above 400 °C.

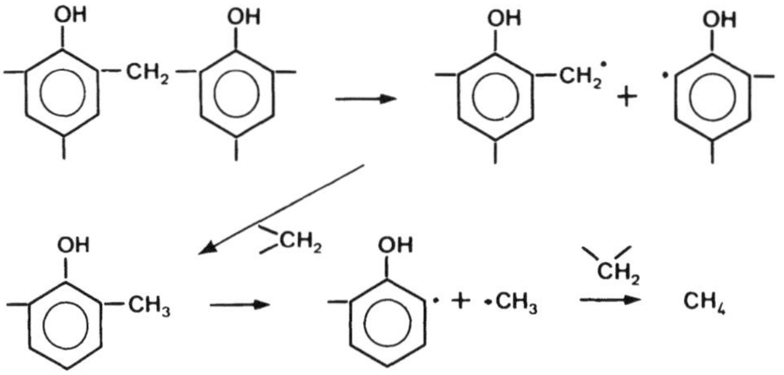

(7.5)

(7.6)

For the formation of benzene, toluene and benzaldehyde the splitting-off of the hydroxyl group is a necessary requirement.

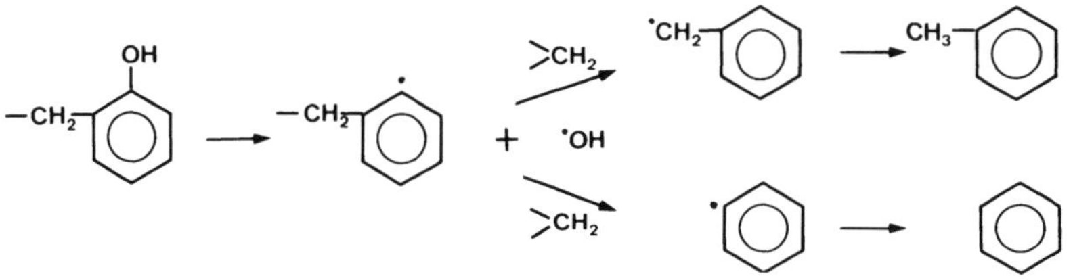

(7.7)

(7.8)

The splitting of phenols to aromatic hydrocarbons is a rather difficult operation and requires special reaction conditions. The formation of aromatic hydrocarbons is observed in metal casting with phenolic resin bonded sand molds.

7.2. Oxidation Reactions

Phenolic compounds can easily be oxidized to higher MW materials containing phenolic and quinoide structural elements. The first step in this one electron abstraction reaction is the formation of relatively stable phenoxy radicals[6)]. Their lifetime can vary from 10^{-3} seconds, in the case of the non-substituted aryloxy radical, up to hours or days by suitable substitution in 2, 4, 6 position due to increased resonance stabilization and steric hindrance. Some of these aryloxy radicals are highly colored (and paramagnetic) as monomers.

Phenoxy radicals, in general, may undergo a variety of reactions depending on substitution and on the presence of other compounds. Reacting with other radicals C-C or C-O coupling may occur, this can be simply illustrated with *p*-cresol as an example (7.9, 7.10).

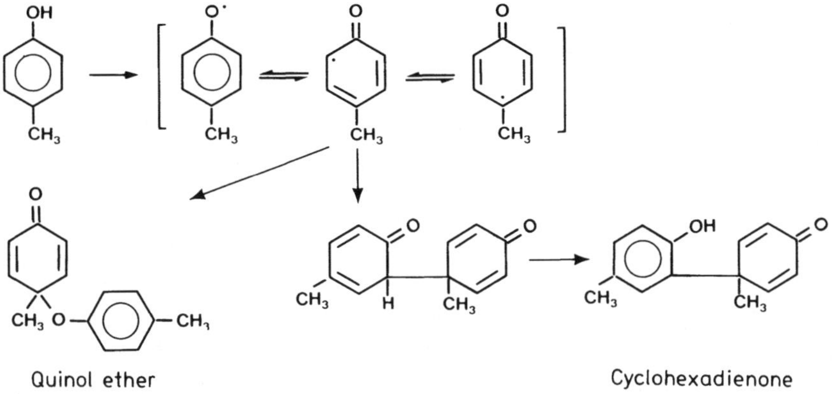

(7.9)

(7.10)

Quinol ether

Cyclohexadienone

C–C coupled dimers can be further oxidized to the corresponding diphenoquinones (7.12). Thus, also additional cross-links are formed with thermal aging. Often, difficult to identify polymeric materials are formed as well.

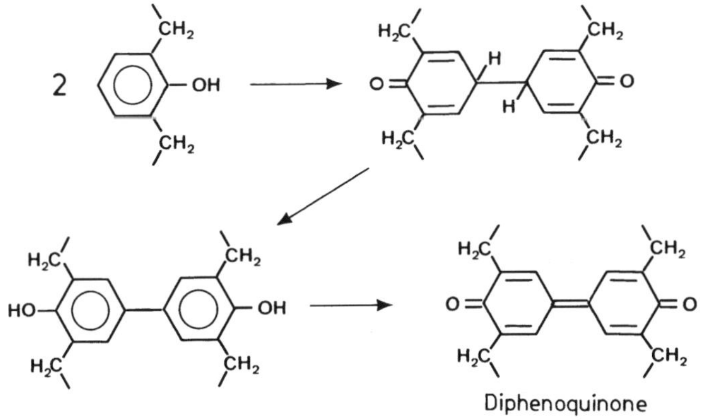

(7.11)

(7.12)

Diphenoquinone

Therefore, there are two weak points in cured phenolic resins when exposed to thermo-oxidative stress: they are the methylene bridge and the phenolic hydroxyl group. The thermal resistance can be improved by cross-linking phenols with heteroatoms or transformation of the phenolic group (see Chapter 8).

7.3. Degradation by High Energy Radiation

Phenolic resins and phenolic resin molding compounds may be of considerable interest for use in electrical components in nuclear power equipment and high voltage accelerators, components of equipment for handling radioactive materials, electrical and structural elements of space vehicles and as protective coatings in nuclear power plants. In these applications high thermal resistance is required apart from resistance to high energy radiation.

When high energy radiation, including γ- and x-rays, neutrons, electrons, protons and deuterons, passes through matter, a strong interaction either with the nucleus or with the orbital electrons occurs, leading to dissipation of a large fraction of the incidental energy[7, 8]. The final result of such interaction in polymeric materials is the formation of ions and radicals followed by rupture of the chemical bonds. At the same time, new bonds are formed, followed by cross-linking and degradation at different rates. The difference in the corresponding rate constants determines the resistance to high energy radiation. The rates of degradation are much lower for polymers containing aromatic rings because of resonance stabilization of the transient species. In general, rigid molecular structures, i. e. thermosettings, are more resistant than flexible thermoplastic and elastomeric structures. TGA gives quite a good, but only qualitative indication of the radiation resistance of the material concerned. Steric effects also have considerable influence on the probability of ion- or radical recombinations. Comparable to the effect of antioxidants, some substances added in minor quantities have a well defined stabilizing effect, but the mechanism of this stabilization is not properly understoood. Radiation resistance is usually improved by the addition of mineral fillers[12]. However, there are known additives which speed up the deterioration. They are called radio-sensitisers. An important example of a material inducing this effect in phenolic resins is cellulose.

An overall picture of the radiation resistance of different thermosetting materials is shown in Figure 7.2.

According to this figure, phenolic resins reinforced by glass- or asbestos fibers can be regarded as very radiation resistant synthetic materials. Non-filled phenolic resins have a relatively low radiation resistance. The oxygen content of phenolic resins has a considerably negative effect on the radiation resistance.

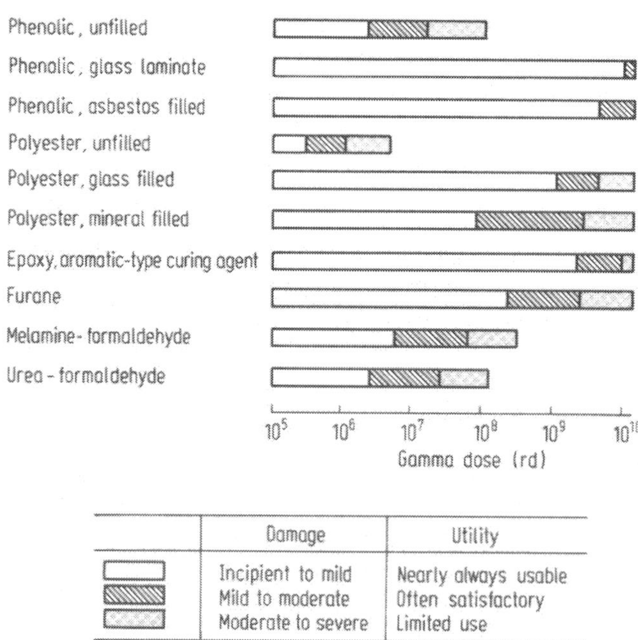

Fig. 7.2. Radiation resistance of thermosetting resins[9]

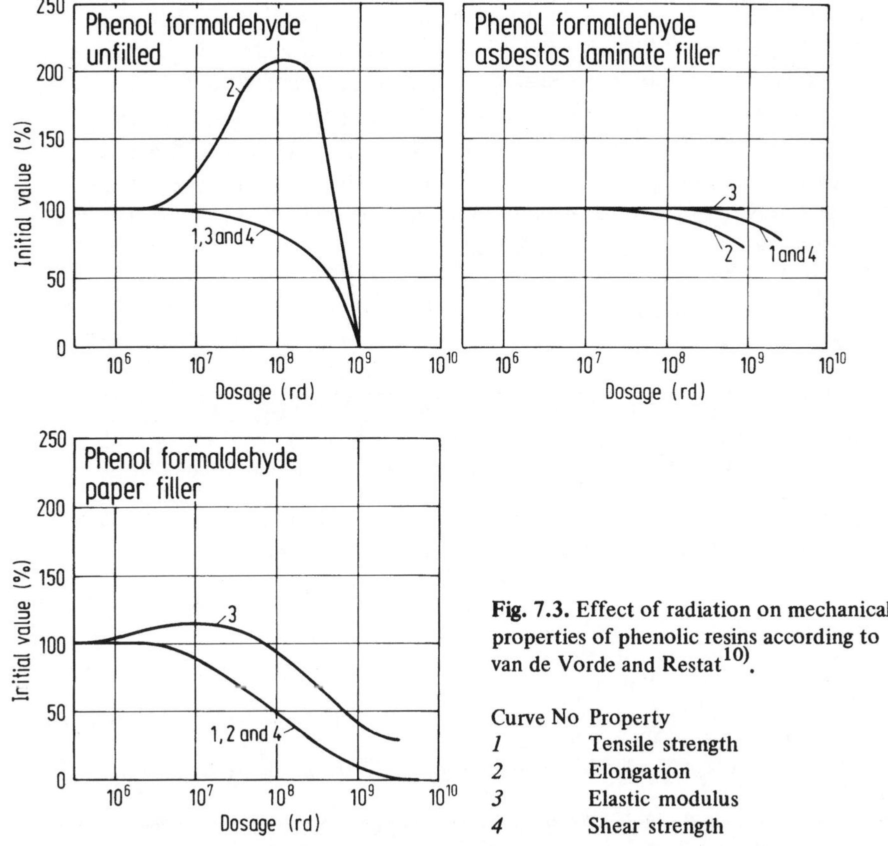

Fig. 7.3. Effect of radiation on mechanical properties of phenolic resins according to van de Vorde and Restat[10].

Curve No	Property
1	Tensile strength
2	Elongation
3	Elastic modulus
4	Shear strength

The main causes of failure of electrical insulations in nuclear radiation environments are primarily loss of mechanical strength or the evolution of gases, rather than changes in electrical properties[7]. Van de Voorde and Restat[10] have reported about the influence of high energy radiation on the mechanical strength of phenolic resins dependent on the reinforcement as shown in Figure 7.3.

Table 7.1. Amount of gaseous degradation products released from phenolic molding powders at high energy radiation impact compared with other plastic materials[11]

Polymer	Filler	Gas evolved cm^3/g Radiation dose Mrd				
		50	100	200	500	1000
Phenolic resin	Wood flour	0.58	1.07	2.09	4.91	8.85
Phenolic resin	Asbestos	0.11	0.18	0.39	1.07	1.97
Polyester	Mineral flour/ cellulose	1.19	2.22	4.00	10.52	19.40
Polyamide 6	–	1.32	2.39	4.68	10.75	–
Polycarbonate	–	1.12	2.00	3.60	8.09	13.69
Polyethylene terephthalate	–	0.22	0.40	0.72	1.58	–
Polystyrene	–	0.07	0.11	0.17	0.30	0.42

Reference literature also includes the development of gases by synthetic materials under the influence of high energy radiation[11]. The behaviour of phenolic resins compared to some competitive plastic materials is described in Table 7.1.

Principally, gas evolution in dependence on the type of filler, proceeds parallel to the loss of mechanical strength, as shown in Figure 7.2. A chemical identification of the gaseous degradation products is not included in this study. However, they should consist mostly of hydrogen and minor quantities of methane and higher hydrocarbons, carbon monoxide and carbon dioxide.

References

1. Marks, B. S., Rubin, L.: ACS-Organic Coatings and Plastics Chemistry, 28/1, 94 (1968)
2. Goldstein, H. E.: ACS-Organic Coatings and Plastics Chemistry, 28/1, 131 (1968)
3. Jones, R. A., Jenkins, G. M.: Volume Change in Phenolic Resin During Carbonization, Carbon 76, Baden-Baden 27.6.–2.7.1976
4. Conley, R. T.: Thermal Stability of Polymers, Chap. 11, New York: Marcel Dekker Inc. 1970
5. Gautherot, G.: Contribution a l'étude de la dégradation des résines phénoliques. Office national d'études et de recherches aérospatiales 1969
6. Mihajlovič, M. L., Cekovič, Z.: Oxidation and Reduction of Phenols. In: S. Patai (ed.): The Chemistry of the Hydroxyl group, New York: Interscience 1971
7. Van de Voorde, M. H.: Effects of Radiation on Materials and Components: CERN 70–5, European Organization for Nuclear Research, Geneva 1970
8. Parkinson, W. W.: Radiation Resistant Polymers. In: Encyclopedia of Polymer Science and Technology, Vol. 11, New York: Wiley 1969
9. Batelle Memorial Institute, Columbus: Radiation-effects, State of the Art, 1965–1966, REIC Report 42 (1966)
10. Van de Voorde, M. H., Restat, C.: Selection Guide to Organic Materials for Nuclear Engng. CERN 72–7, European Organization for Nuclear Research, Geneva 1972
11. Morgan, J. T., Stapelton, G. B.: Gas Evolution from Plastic Materials by High Energy Radiation, Sc. Res. Council RL-74-021, Rutherford Laboratory, 1974
12. Gilfrich, H. P., Wilski, H.: Chemie-Ingenieur Technik *42,* 19 (1970)

8. Modified and Thermal-Resistant Resins

Polymers are considered to be thermally stable, if they resist the influence of temperature in an inert atmosphere without changing their properties significantly as indicated in Table 8.1.

Table 8.1. Guidelines for the rating of plastics as thermal resistant[1]

°C	Hours
175	30,000
250	1,000
500	1
700	0.1

Table 8.2. Thermal resistance of phenol resols in dependence on temperature and time[2]

°C	Time
1,000–1,500	Seconds
500–1,000	Minutes
250– 500	Hours
$<$ 200	Years

Plastics which meet these requirements have the following structural characteristics:
- high proportion of heterocyclic or aromatic rings
- high bond energies between the atoms
- high cohesive strength between the polymer chains
- oxidation resistant bonds.

Phenolic resins possess these prerequisites to a considerable extent. Although non-modified, inorganically filled phenolics are already considered temperature resistant, their thermo-oxidative resistance can be further improved by chemical modification[3]. The weak point of phenolic resins is the oxidative susceptibility of the phenolic nucleus and the methylene linking group. Poly-*p*-xylylene is known for its excellent high temperature stability.

$$-CH_2\left[-C_6H_4-CH_2-\right]_n \qquad (8.1)$$

The following methods are suitable to improve the thermo-oxidative resistance of phenolic resins:
1. Etherification or esterification of the phenolic hydroxyl group,
2. Complex formation with polyvalent elements (Ca, Mg, Zn, Cd, ...)
3. Replacement of the methylene linking group by heteroatoms (O, S, N, Si, ...)

The most important general modification reactions in phenolic polymer chemistry are the etherification and C-alkylation (Friedel-Crafts) reactions[4]. Both reactions are commonly used to enhance flexibility and compatibility with polymers and solvents and to adjust reactivity and performance. Because of the strong nucleophilic activity of the phenol, mild catalysts and operating conditions are usually employed for alkylating them with olefins. However, the ready formation of ethers, especially at mild conditions, and the tendency of the hydroxyl group to complex the catalyst must be considered. Diisobutylene, terpenes and tung oil are the olefins most frequently used (see 13.3.3, 17.1).

8.1. Etherification Reactions

The hydroxymethyl group in phenols and phenol prepolymers can easily be etherified (8.2) with alcohols because of their tendency to form hydroxybenzylcarbonium ions. High hydroxymethylated phenols and an excess of alcohol are used to avoid the self-condensation reaction.

In general the reaction is performed at pH 5–7 and at temperatures between 100–120 °C with monoalcohols like methanol, butanol and isobutanol. Butanol is most frequently used; the water formed is separated at azeotropic conditions at an excess of butanol[5]. Such etherified resols show a higher solubility in aromatic solvents and improved flexibility. They are used mainly in coatings, impregnating resins for electrical laminates and adhesives. Their reactivity, however, is reduced. Also polyhydroxy compounds are recommended for flexibilization, e.g. glycol, glycerine, polypropylene glycols, hydroxypolyesters and polyvinylacetals.

On the other hand, the etherification of the phenolic hydroxyl group (O-alkylation) (8.3) leads to improved alkali resistance. Better flexibility, light fastness and with allyl compounds, enhanced air drying properties are also obtained. The reactivity of phenol ethers towards formaldehyde is strongly reduced in comparison to phenols. Therefore, the resol is first prepared and the phenolic hydroxyl group etherified (alkylated) with stronger electrophiles, e.g. allyl chloride, alkyl bromides, alkyl sulfates, epichlorohydrine and epoxide compounds in the presence of sodium hydroxide. In general, mild conditions are necessary to avoid polymer formation and C-alkylation.

Novolak resins can also be O-alkylated for higher performance. Resins with improved light fastness, which are soluble in alkalies, are obtained by O-alkylation with monochloroacetic acid.

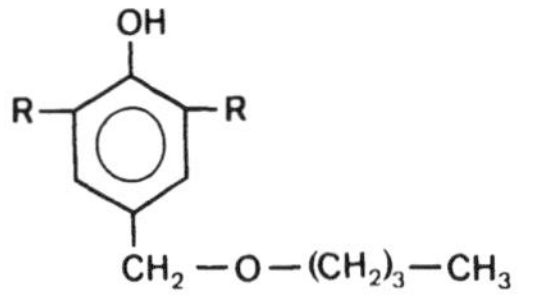

OH; R, R; $CH_2-O-(CH_2)_3-CH_3$ (8.2)

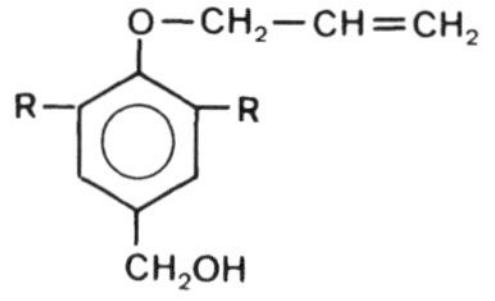

$O-CH_2-CH=CH_2$; R, R; CH_2OH (8.3)

R = H, CH_2OH, CH_2X

Allyl prepolymers, produced by O-alkylation with allyl chloride (General Electric, Methylon® resins), are used as additives for can and drum coatings, and for electrodepositive paints based on epoxy-, polyvinylacetal- or polybutadiene resins because of their increased thermooxidative stability, excellent resistance to chemicals and relatively good flexibility[6].

8.2. Esterification Reaction

The esterification of phenol novolak resins with inorganic polybasic acids, such as phosphoric and boric acid, or the reaction with phosphorus oxyhalides are of particular importance in increasing the heat and flame resistance of phenolic resins. Because of the high OH-functionality of novolaks, the reaction with polyfunctional compounds often leads to gelation. Linear polymers have been obtained using bisphenols and bi-functional derivatives of phosphoric acid or phosgene[7]. Dannels and Shepard have shown[8] that "high ortho"-novolaks can be quite extensively esterified with bi-functional compounds, while novolaks with random isomer distribution gel at a low level of esterification. Similar behaviour is oberved at the reaction with boric acid or diphenyl silyl dichloride. The intermolecular (cross-linking) reaction 8.4 takes place predominately with random novolaks[9], whereby with ortho novolaks the intramolecular etherification seems to be the preferred reaction yielding 8-membered rings (8.5).

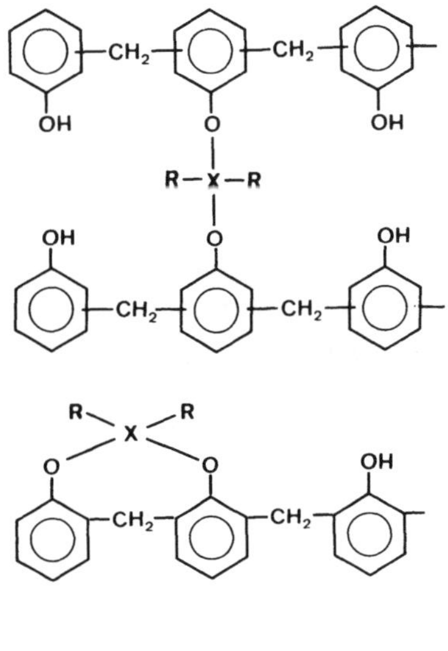

(8.4)

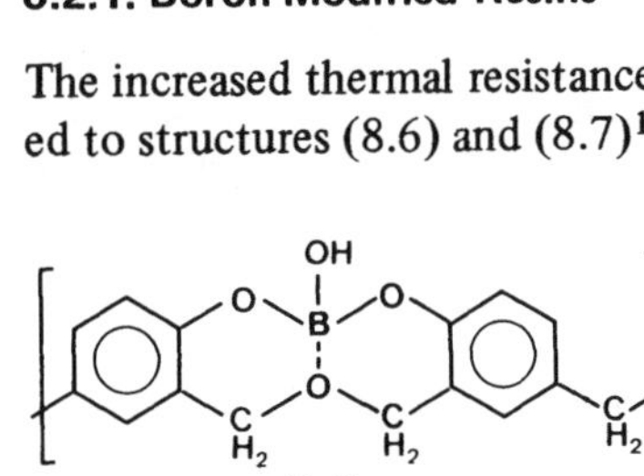

(8.5)

8.2.1. Boron-Modified Resins

The increased thermal resistance of phenolic resins modified with boron is attributed to structures (8.6) and (8.7)[10].

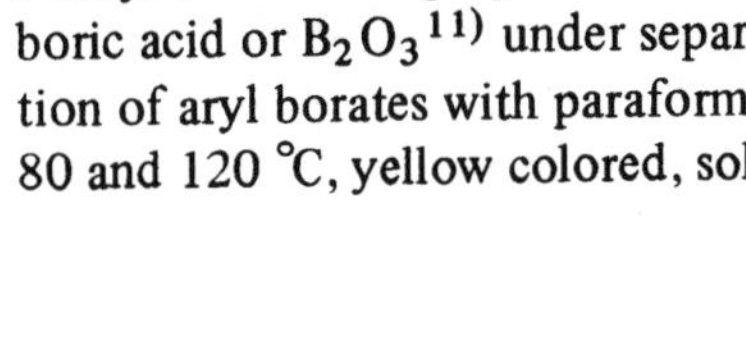

(8.6) 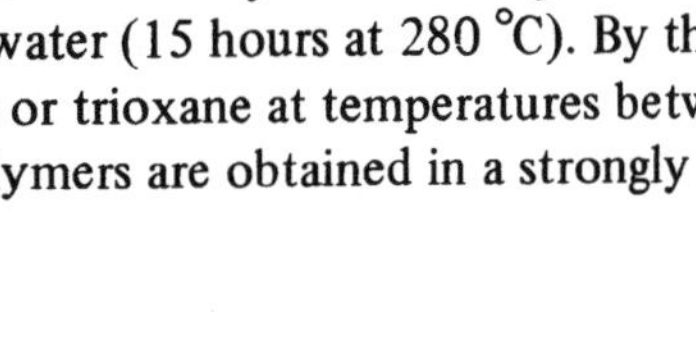 (8.7.)

Phenyl borates are prepared first according to Huster by reaction of phenol and boric acid or B_2O_3[11] under separation of water (15 hours at 280 °C). By the reaction of aryl borates with paraformaldehyde or trioxane at temperatures between 80 and 120 °C, yellow colored, solid prepolymers are obtained in a strongly exothermal reaction, which are formulated in the usual manner to molding compounds with HMTA, fillers, etc. Relatively high temperatures of 200 °C are required for curing in order to obtain the desired high temperature resistance. The curing can be performed at lower temperatures, between 100–120 °C if epoxy compounds are used instead of HMTA[12].

Ablative materials with outstanding temperature resistance are obtained from boric acid modified *p*-aminophenol/formaldehyde resins[13]. These resins are prepared by heating 3 mol *p*-aminophenol with 1 mol boric acid in boiling xylene. Water is distilled off as an azeotropic mixture. The formed tris-p-aminophenyl borate is soluble in water and colored blue. After further reaction with trioxymethylene or formaldehyde at 70 °C by addition of an acidic catalyst (3 h), a red solid resin is obtained. The curing occurs with HMTA. The weight loss of such resins is very small at extreme temperature stress. Above 2,500 °C boron nitride-like structures are obtained.

Although boron-modified phenolic resins are recommended for many different applications, they are at present only used for the manufacturing of brake linings and ablative materials.

8.2.2. Silicon-Modified Resins

The addition of silicon compounds for the improvement of the thermal resistance of phenol-formaldehyde resins was recommended as early as 1941 by E. G. Rochow[14]. Since then, a series of patents have been registered[15–18]. Although the thermal resistance is improved by this modification, silicon modified phenolic resins have as yet been very little used. The reason for this might be the high price of suitable organic silicon compounds. A phenolic resin becomes twice to three-times as expensive, if only 10% of a silicon compound is added. The addition of smaller quantities does not bring significant improvement in thermal resistance. The modification can be performed by chemical reaction of silicones or siloxanes containing reactive groups with phenolic compounds or by a mixture of the components.

$$\left.\begin{matrix} -\overset{|}{\underset{|}{Si}}-X \\ -\overset{|}{\underset{|}{Si}}-OR \\ -\overset{|}{\underset{|}{Si}}-OH \end{matrix}\right. + HO-C_6H_4-R \longrightarrow -\overset{|}{\underset{|}{Si}}-O-C_6H_4-R + \left.\begin{matrix} HX \\ ROH \\ H_2O \end{matrix}\right. \qquad (8.8)$$

During the reaction of siloxanes (X = halogen, hydrogen) with phenolic compounds as O-H acidic components, a Si-O-bond is the result[19]. The reaction is not limited to monomeric compounds; prepolymers also open up a lot of variation possibilities. The reaction conditions are chosen in such a way that the co-condensation of both systems is preferred instead of the self-condensation reaction. The formation of block polymers cannot be completely avoided. The reaction between phenol or phenol novolaks and methoxyphenylpolysiloxanes (8.9) according to the

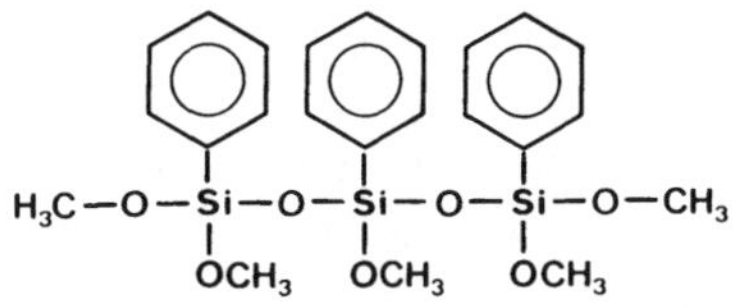

$$H_3C-O-\underset{OCH_3}{\overset{C_6H_5}{Si}}-O-\underset{OCH_3}{\overset{C_6H_5}{Si}}-O-\underset{OCH_3}{\overset{C_6H_5}{Si}}-O-CH_3 \qquad (8.9)$$

general formula (8.8) under splitting-off of methanol is the simplest to be performed. The incorporation of silicon atoms in the polymer chain instead of the methylene linkage is obtained by reaction of *p*-silylphenols, which are available by hydrolysis of phenoxysilanes, with formaldehyde[15].

The addition of silanes as adhesion promoters is widespread, for instance in the manufacturing of mineral wool mats or foundry sands[20]. The added quantities are, however, very small, mostly far below 1%, so that such resins cannot be designated as silicon-modified resins. Amino-functional silanes are used in preference, for instance γ-aminopropyltrimethoxysilane, less frequently epoxy functional ones like glycidoxipropyltriethoxysilane.

8.2.3. Phosphorus-Modified Resins

Phosphorus-modified novolaks are obtained by esterification of novolaks with phosphoric acids or by reaction with phosphorus oxychlorides. The reaction with bifunctional phosphorus oxychlorides (8.10) is performed at 20–60 °C in dioxane[13, 21, 22].

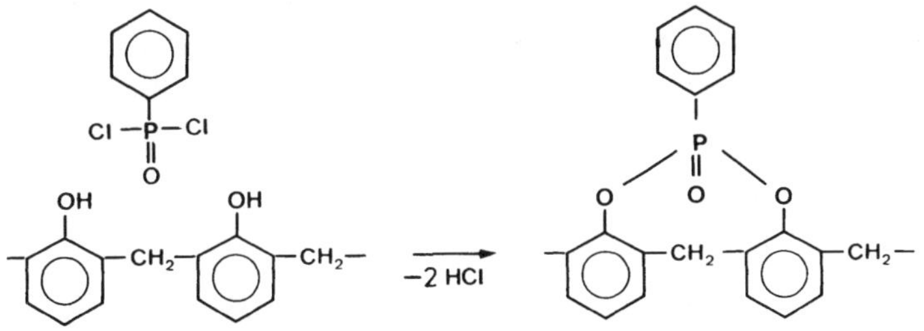

$$C_6H_5-P(=O)Cl_2 + \text{—}C_6H_3(OH)\text{—}CH_2\text{—}C_6H_3(OH)\text{—}CH_2\text{—} \xrightarrow{-2\,HCl} \text{—}C_6H_3(O\text{—})\text{—}CH_2\text{—}C_6H_3(O\text{—})\text{—}CH_2\text{—} \;(\text{O—}P(=O)(C_6H_5)\text{—O bridge}) \qquad (8.10)$$

With *ortho*-linked novolaks, cyclic structures prevail (8-membered ring); at normal isomer distribution, intermolecular esterification is preferred[8]. In this case, early gelation may occur. Although phosphorus-modified resins exhibit excellent heat resistance in oxidizing media and outstanding flame resistance, they are relatively rare on the market. At thermal stress in reducing medium phosphine is formed[13].

8.3. Heavy Metal-Modified Resins

Heat- and flame resistant resins are obtained by the reaction of phenols or phenolic resins with metal halides (molybdenum trichloride, titanium tetrachloride, zirconium oxychloride, tungsten hexachloride), metal alcoholates (aluminium trimethoxide, titanium tetramethoxide) or metal- organic compounds (acetylacetonates). If such metal containing phenolic resins are subjected to high temperatures, they decompose considerably slower than conventional resins. Besides, it is assumed that metal carbides are formed with the carbon of the resins. The resins are deeply colored and may contain up to 20% of ionic bound metal[13]. Red-colored titanium-modified resins for instance are obtained by heating phenol and titanium tetramethoxide under separation of methanol at 80 °C for 1 hour and following prepolymer formation by reaction with paraformaldehyde.

Otherwise[23], titanium tetrachloride is dropped into a phenol melt, and the hydrogen chloride which is formed is removed by passing nitrogen through and raising the temperature up to the boiling point of phenol. The intensely red tetraphenyltitanate (MP 130 °C) is converted in the second stage by reaction with trioxymethylene into a solid prepolymer (MW ca. 500; soluble in dimethylformamide), which can be cured with 10% of HMTA. The reaction of phenolic resins with Ca-, Mg-, Zn- and Cd-oxide and their importance for the formulation of adhesives is described in Chapter 17.

8.4. Nitrogen-Modified Resins

Some polycondensation products of formaldehyde with aromatic amines are characterized by great thermal stability. When heated to 330 °C, they loose 15–30% of their weight, but when heated further to 900 °C, they form a nonvolatile residue constituting up to 65% of the weight of the orginal polymer. The most thermostable are polycondensation compounds of formaldehyde with *p*-aminophenol[24, 25].

By reacting aniline with formaldehyde in a strong acidic medium, *p*-aminobenzylalcohol is obtained which under separation of water forms chain-like polymers[26-29].

$$H_2N-C_6H_4-CH_2OH \longrightarrow -NH-C_6H_4-CH_2-NH-C_6H_4-CH_2-NH-C_6H_4- \quad (8.11)$$

The chains are cross linked with methylene groups to some extent. A co-condensation reaction with phenols is easier to perform in general[30, 31]. Methylenediphenyldiimide is formed from aniline and formaldehyde in alkaline solution. At neutral or weak acidic pH resin-like products which contain azomethine groups, result. They are soluble, brittle and non-resistant to, temperature[27].

Aniline-formaldehyde resins or phenol-aniline mixed resins have been used in the past to produce laminates and molding compounds for electrical engineering. The resins possess higher tracking resistance and favorable electrical properties[26, 32]. Aniline resins do not have any technical importance today. The reasons for this are difficulties in processing, low flow and physiological hazards which are connected with the use of aniline. Aniline is extremely easily absorbed by the skin and even in minor concentration acts as a strong blood pigment poison which cancels the ability of haemoglobin to transfer oxygen because, it interacts competitively in the redox process. Formerly, aniline was said to cause cancer. However, this is to be attributed to higher aromatic amines which were contained as contaminants[33].

The addition of urea or UF-resins, which is performed on a large scale to modify foundry resins, effects a reduction of the temperature and humidity resistance. Melamine, dicyandiamide and sulfonamides may also be used to produce nitrogen-modified resins[27].

8.5. Sulfur-Modified Resins

The direct reaction of phenols and sulfur occurs relatively easily in the presence of alkaline catalysts at temperatures between 130 and 230 °C. The results are liquid or solid resins with an unpleasant hydrogen sulfide odour. The softening range depends upon the kind of phenolic component and molar ratio. The simplest compound which results from this reaction is the dihydroxydiphenylpolysulfide, which is either further reacted with formaldehyde or can immediately be crosslinked with resols[27, 34]. Similar structures can be obtained if phenols are first reacted with aldehydes and then with sulfur and alkali hydroxides. Phenolic resins modified with sulfur do not have any technical importance. Worth mentioning are the high plasticity and relatively high solubility in water.

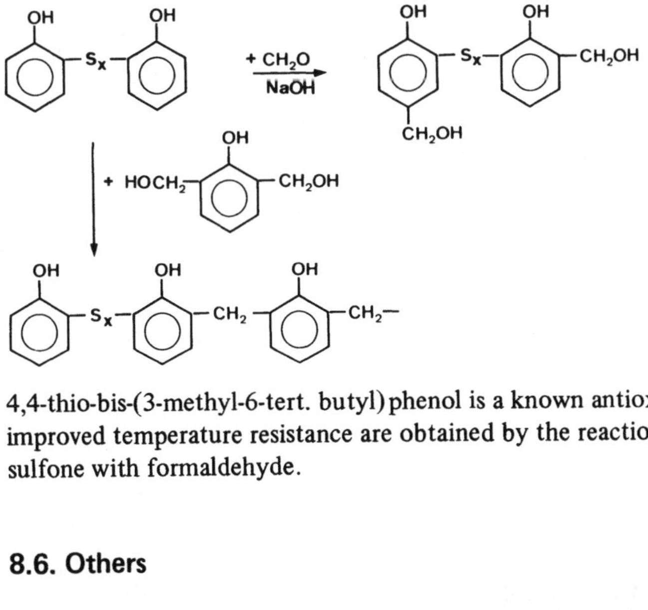

(8.12)

(8.13)

4,4-thio-bis-(3-methyl-6-tert. butyl)phenol is a known antioxidant. Resins with improved temperature resistance are obtained by the reaction of dihydroxydiphenylsulfone with formaldehyde.

8.6. Others

Condensation resins of an aralkylether and phenol (8.14) (Xylok, developed by Midland Silicones Ltd.) take an intermediate position between phenolic- and poly-*p*-xylylene resins. The considerably increased temperature resistance compared to the one of nonmodified phenolic resins[35, 36] is the result of the reduced oxidation susceptibility of the methylene linkage.

Polynuclear aromatic and heterocyclic compounds (naphthalene, carbazol) also react easily with *p*-xylylene-dimethylether. The prepolymers can be cured with HMTA (10–15%) or with (cycloaliphatic) epoxy resins. Xylok resins are recommended for the production of class H glass laminates or in combination with

$$C_6H_5OH + CH_3O{-}CH_2{-}C_6H_4{-}CH_2{-}OCH_3 \xrightarrow[-CH_3OH]{SnCl_4} {-}C_6H_3(OH){-}CH_2{-}C_6H_4{-}CH_2{-}C_6H_3(OH){-} \quad (8.14)$$

p-Xylylene dimethylether Xylok prepolymer

phenolic resins for molding compounds. Due to a cost four to six times higher than that of phenol resins, they have not gained technical importance. Minor quantities are used for the manufacture of friction linings.

Polydehydration of difunctional phenols in the presence of zinc chloride under pressure above 220 °C yields polyhydroxyphenylenes (8.15), benzyne and hydroxybenzyne being intermediates[37].

$$HO{-}C_6H_4{-}OH \longrightarrow \left[C_6H_3(OH) \right]_n \quad (8.15)$$

Linear, thermoplastic phenolic resins (polyhydroxystyrene) are obtained by polymerization of 4-hydroxystyrene (8.16) which can be cross-linked either with HMTA or preferably with epoxy resins[38].

$$CH{=}CH_2{-}C_6H_4{-}OH \longrightarrow {-}CH(C_6H_4OH){-}CH_2{-}CH(C_6H_4OH){-}CH_2{-}CH(C_6H_4OH){-}CH_2{-} \quad (8.16)$$

Polyhydroxystyrene/epoxide combinations are recommended for the production of multilayer circuit boards.

References

1. Levine, H. H.: Ind. Engn. Chem. *54,* 22 (1962)
2. Techel, J.: Plaste u. Kautschuk *10,* 137 (1963)
3. Bachmann, A., Müller, K.: Plaste u. Kautschuk *24,* 158 (1977)
4. Olah, G. A.: Friedel Crafts Chemistry. New York: Wiley 1973
5. Lemmer, F., Greth, A.: Phenolharze. In: Ullmanns Encyclopädie d. techn. Chem. Vol. 13, 3. Ed. München: Urban und Schwarzenberg 1962
6. General Electric Co: Methylon Resins. Technical Bulletin, US-PS 257 330 (1951)
7. Wright, H. R., Zentfmann, H.: Chem. Ind. (London) *244,* 1101 (1952)
8. Dannels, B. F., Shepard, A. F.: Inorganic Esters of Novolaks. Journal of Polymer Sci.: Part. A-l, Vol. 6, 2051 (1968)
9. Cass, U. E.: US-PS 2 616 873
10. Hoechst AG: DE-OS 24 36 358 (1974)
11. Dynamit Nobel AG: DE-PS 1 233 606 (1960)

12. Dynamit Nobel AG: DE-OS 2 214 821 (1972)
13. Nord-Aviation Société Nationale de Constructions Aéronautiques: DE-AS 1 816 241 (1968)
14. General Electric Co.: US-PS 2 258 218 (1941)
15. Dow Corning Corp.: DE-PS 937 555 (1956)
16. Stenbeck, G.: DE-OS 1 694 974
17. Westinghouse Electric Corp.: US-PS 2 836 740 (1958)
18. Dow Corning Corp.: US-PS 2 842 522 (1958)
19. Noll, W.: Chemie und Technologie der Silicone. Verlag Chemie, Weinheim: 1968
20. Union Carbide Corp.: Technical Bulletin
21. Helferich, B., Schmidt, K. G.: Chem. Ber. *92,* 2051 (1959)
22. General Electric Co.: GB-PS 1.031 908 and 1.031 909 (1966)
23. Hitachi Chem. Co. Ltd.: J-PS 7658493 (1974)
24. Sergeev, V. A., Korshak, V. V., Kozlov, L. V.: Plasticheskie Massy *3,* 57 (1966)
25. Korshak, V. V., Vinogradova, S. V.: Russian Chem. Rev. *37,* (11) 885 (1968)
26. Frey, K.: Helv. chim. Acta *18,* 491 (1935)
27. Scheiber, J.: Chemie und Technologie der künstlichen Harze. Stuttgart: Wissenschaftl. Verlagsges. 1943
28. Ellis, K.: The Chemistry of Synthetic Resins. New York: Reinhold 1935
29. Scheuermann, H.: Anilinharze. In: Ullmanns Encyclopädie d. techn. Chem. Vol. 3, 3 Ed. München: Urban und Schwarzenberg
30. Fibre Diamond: FR-PS 644 075
31. E. I. du Pont de Nemours & Co.: US-PS 2 098 869
32. Imhof, A.: Kunststoffe *27,* 89 (1939)
33. Oettel, H.: Anilin, Toxikologie. In: Ullmanns Encyclopädie d. techn. Chem. Vol. 7. 575, 4. Ed. Weinheim: Verlag Chemie 1974
34. Cherubim, M.: Kunststoff Rundschau *13,* 235 (1966)
35. Midland Silicones Ltd.: Xylok 210, Technical Bulletin
36. Harris, G. J.: Werkstoffe u. Korrosion *22,* 227 (1971)
37. Jones, J. I.: J. Macromol. Sci.-Revs. Macromol. Chem. *C 2* (2), 303, (1968)
38. Maruzen Oil, Jap.: Technical Bulletin

9. Composite Wood Materials

Phenol resin bonded wood materials – particle boards (PB), plywood, fiber boards (FB) and glued wood construction elements – are used for outdoor construction and in high humidity areas because of the high water and weathering resistance of the phenolic adhesive bond and the high specific strength.

The basic idea for the use of these materials[1–7] is

- to compensate the anisotropy in strength of natural wood,
- to use wood of minor quality and wood waste of the wood working industry,
- to obtain better and more economical possibilities for the production of shaped parts.

The competitiveness and development of the wood working industry are of utmost importance for the development of thermosetting plastics. This industry is the largest consumer of urea-, melamine- and phenol resins. Approximately 85% of UF- and more than one quarter of PF-resin production is used for wood materials. The building and furniture industry are the main markets for composite wood materials.

Although the high growth rates of PB in Europe up to 1973 – they were influenced by the late start of large scale production – belong to the past, future growth rates are forseen to be at least above average[8, 9].

The plywood consumption in Western Europe is not comparable to the PB-market[10]. From 1960 to 1973 the rate of growth amounted only to 6% or less per annum (FAO). There is some replacement of urea glued plywood by thin particle boards in countries with an advanced particle board industry. Furthermore, plywood production has shifted to a great extent to North America, Africa and East Asia where the production is favored by large wood reserves or economic advantages. The veneer yield of high diameter tropical trees is 50%, far higher than the yield of the local red beech. The effect of lower raw material price is often increased by lower labor costs in these countries[10]. The development of core plywood, on the other hand, shows positive prospects.

A steady or only slow growing market can be assumed for fiber boards, and a clear shifting of the production to East European countries is observed[11]. Growth rates above average are predicted for glued wood structural elements.

The situation in the USA is entirely different from the European one. There, plywood takes first place with a market share of about 58% followed by fiber boards (23%) and particle boards (19%) (Table 9.1). However, due to rising costs and limited supply of peelable wood, also there is a clear trend toward PB's because of the higher utilization of raw wood in comparison to that of plywood, where it is only around 45%. Since also wood of minor quality can be used, the material costs as well as the production costs are lower for particle boards.

Table 9.1. Production of composite wood materials for some countries in 1975.
Source: "1975 Yearbook of forest products" FAO, Rome 1977

Production 1975 1,000 m^3	Particle board	Plywood	Fiber board
West-Germany	5.444	400	319
France	1.964	551	241
Finland	693	415	249
Italy	1.500	400	148
UK	532	25	47
Europe	19.074	3.468	4.244
USA	4.964	15.134	6.085
Canada	566	2.044	763
Japan	645	6.174	597
Korea Rep.	52	1.436	16
USSR	3.085	2.142	1.912
World	30.188	34.473	14.862

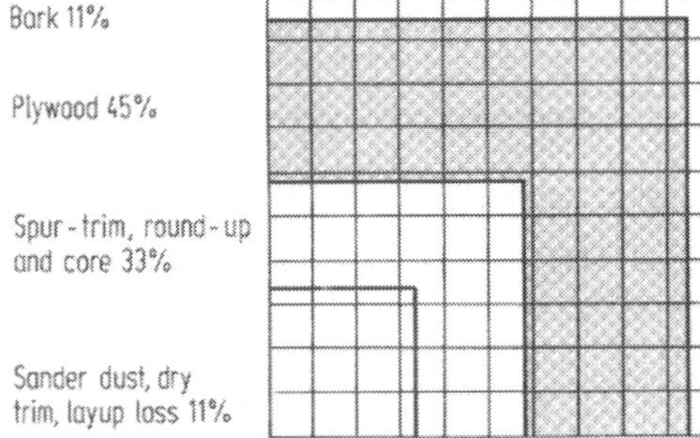

Fig. 9.1. Softwood lumber and residue yields from veneer logs.
(G. A. Koenigshof, 11th Particleboard Symposium, Washington 1977)

9.1. Wood

The main chemical components of wood[2, 4] are cellulose (40–45%), hemicellulose (20–30%) and lignin (20–30%). Further, fluctuating portions of water, different resins, fats (mono-, di- and triglycerides), waxes, tannins and sterols are found. Conifers in particular contain low molecular compounds like mono- and diterpenes which are extractable by organic solvents.

Lignin is a highly polymer substance[12] of phenolic structure. This is the reason for its reactivity to formaldehyde. Lignin is formed from coniferyl alcohol (9.1) and the related sinapin- and cumaryl alcohol by a dehydrogenating, oxidative coupling mechanism (9.2) including free radicals, which does not lead to a uniform constitution.

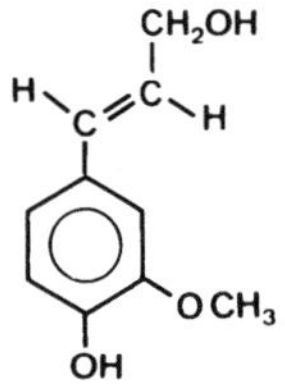

(9.1)

Conyferyl alcohol

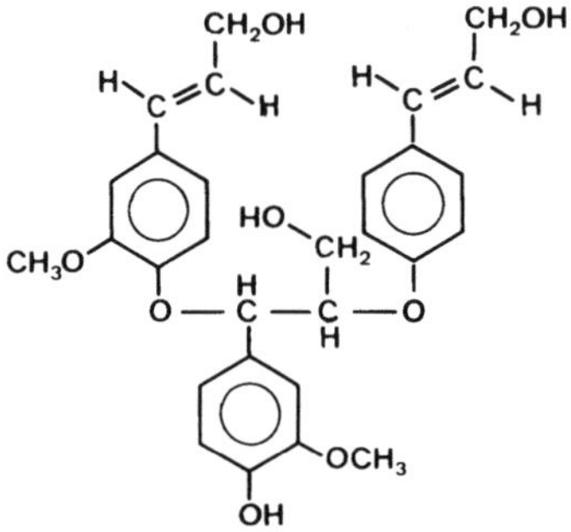

(9.2)

Lignin prepolymer

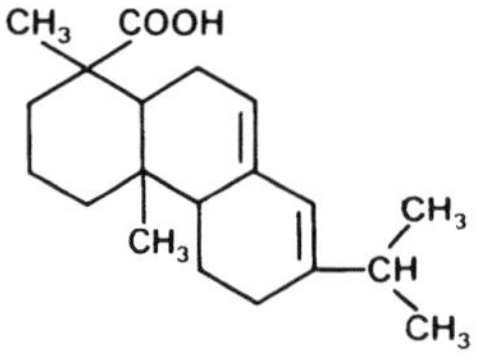

(9.3)

Abietic acid

The resins found in conifers, called rosin, can easily be extracted with organic solvents. Rosin consists of approximately 90% free rosin acids, the remainder being neutral substances like oxidized terpenes and minor quantities of esters and anhydrides. Between the rosin acids there exists a structural equilibrium with respect to the position of the double bonds, leading to laevo-pimaric, neo-abietic, palistric and abietic acid. Abietic acid (9.3) is the main component of different commercially obtainable rosins.

Wood contains a series of reactive compounds and functional groups, so that apart from hydrogen bond formation and acid-base reaction between rosin acids and sodium phenoxide, a number of chemical reactions with PF-condensation products may occur. However, it is not possible to separate the reaction products without destroying them. Therefore, the extent of chemical reaction under conditions existing at PB-pressing and veneer gluing can not be reliably determined.

9.2. Residues of Annual Plants

Some residues of annual plants became an essential lower priced raw material for ligno-cellulosic fibers instead of wood[13]. The most important substitute for wood in particle board and fiber board production in the European countries are flax residues of the production of flax fibers. Bagasse, the residue of sugar cane after the extraction of sugar, is also very important. The constitution of bagasse is similar to that of wood. Further examples are rape-seed and rice straw.

The substitution of wood by annual plants results in reduced tensile and flexural strengths, due to lower cellulose content, higher swelling and water absorption, attributable to the higher hemicellulose content. Lignin is relatively highly hydrophobic and acts against swelling. In 1973, according to the FAO, approximately 4% of the world particle board was made of annual plants, mainly flax shives (81%) and bagasse (13%). These agricultural raw materials have similar importance for the production of fiber boards. The high expenditures for collecting, transport and storage are the main reason for their limited use.

9.3. Adhesives and Wood Gluing

All large scale wood gluing processes such as plywood and particle board manufacturing are dominated by the costs of the adhesives[14]. Higher costs of the latter must be compensated by processing advantages or significant quality improvements. The applied thermosetting adhesives are listed in decreasing order of consumption:
- urea formaldehyde and melamine formaldehyde resins
- phenol resins (including resorcinol resins)
- diisocyanates
- lignin sulfonates

Further, PVA-dispersions, or very seldom, acrylics, rubber, epoxies, animal glue, casein and starch may be used.

An important characterization of composite wood materials follows in accordance with their use in exterior and interior grades. Comparative water resistance of various wood adhesives under well controlled laboratory conditions, reflecting

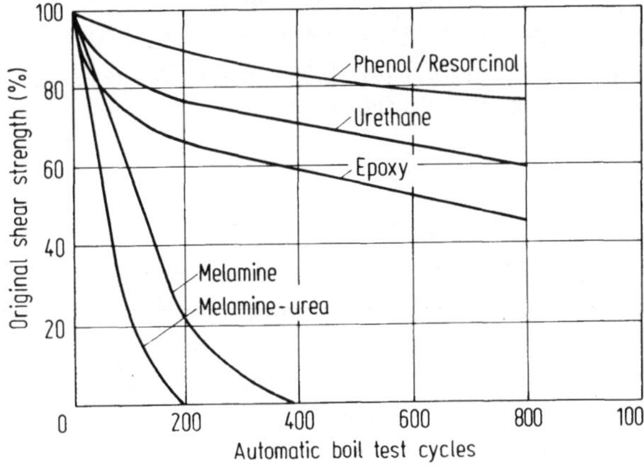

Fig. 9.2. Durability of various adhesive types by boil-dry cycling according to Kreibich and Freeman[15]

outdoor performance, was estimated by Kreibich and Freeman[15]. The results are indicated in Figure 9.2.

The strength of the adhesive bond formed by thermosetting resins can be attributed to several factors[2]. Due to the high polarity and low viscosity the adhesive penetrates deeply into the micro-pores of the wood resulting in a mechanical anchorage initially. Due to their high polarity, thermosetting resins are known to form very strong hydrogen bonds with hydroxyl groups. Strong dipole-dipole interactions and van der Waals forces are developed. Chemical reactions further occur between wood components and PF- and UF-resins or urethanes according to the chemical constitution and temperature. The cohesion force of the cured resin surmounts the tensile strength of the wood. Roughness and cleanness of wood surface have a considerable influence on bond strength.

Phenol Resins

Phenol- and resorcinol formaldehyde resins exhibit strong adhesive bonds with excellent durability under water and humidity exposure[16, 17]. They are therefore used for exterior grade plywood, block board, PB and FB. Formaldehyde emission during production as well as from the finished boards is very low.

Urea and Melamine Resins

Urea-formaldehyde resins are by far the most used wood adhesives. Their main advantages are low costs and high curing rates. Since the urea bond is not resistant to hydrolysis, these resins can only be used for interior application. A further serious disadvantage is the considerable separation of formaldehyde during the production process and in later use. The water resistance can be improved to a limited extent by addition of melamine but not enough to be suited for outdoor application, as the experience with "shipyard" boards has shown[18]. A further improvement has been achieved by additional modification with phenol[19]. Even though such UMP-F boards are used as alternates to PF-boards in Scandinavia and France, and an exceptional approval was achieved for use in West-Germany, the long term exposure shows an unsatisfactory performance, so that their suitability for outdoor application is controversial[17]. An advantage of UMP mixed resins compared to PF-resins are the simpler production switch to UF-boards and the lack of sodium hydroxide and thereby the color fastness of the top layer. Besides, the poorer outdoor performance, the low water dilutability, the necessity of high gluing factors and the narrow limits of the applicable humidity of the chips are disadvantages.

Diisocyanates

Apart from PF-resins only diisocyanates received the V 100 approval[20]. They are remarkable because of the chemical reaction between the isocyanate group and hydroxyl groups of the cellulose. PB bonded with isocyanates have an excellent modulus of elasticity[3]. If diisocyanates are rarely used, it is because of the high costs and the fact that isocyanate bonded chips adhere strongly to pressing plates. They can, therefore, only be used in the middle layer (Novopan AG). Mostly 4,4'-diisocyanatodiphenylmethane is used as the bonding agent.

Lignosulfonates

Lignosulfonates are obtained by the reaction of wood pulp with an aqueous solution of sulfur dioxide and a sulfurous acid salt, usually calcium bisulfite. Several extraction and transformation processes have been developed for the production of highly active forms of lignosulfonic acids. The major quantity of purified lignosulfonates is used in oil well drilling mud formulation. Further uses are the pelletizing of animal feeds, in the mining industry as ore-flotation agents, as dispersants and as Portland cement additives. Especially at times of phenol shortages lignosulfonic acids and ammonium lignosulfonates in combination with phenolic resins are recommended to make low cost binding[21–23] and laminating resins. However, lignosulfonates have not yet been successful as glues for PB's.

Bark Extracts

The use of the phenolic components of bark extracts for the preparation of PB and plywood adhesives in Australia and South Africa has been reported[24]. Bark contains a much higher portion of extractable compounds than wood. The most important are monomeric poly-phenols, tannins and similar phenolic acids. These compounds (9.4, 9.5) are highly reactive to formaldehyde[25]. Solvents for these substances are hot water, aqueous sodium carbonate solution or ethanol.

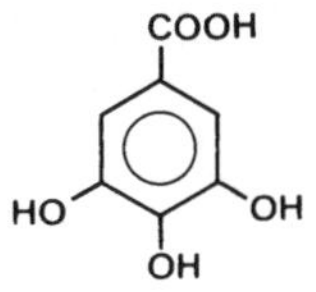

(9.4)

3,4,5-Trihydroxybenzoic acid
(Gallus acid)

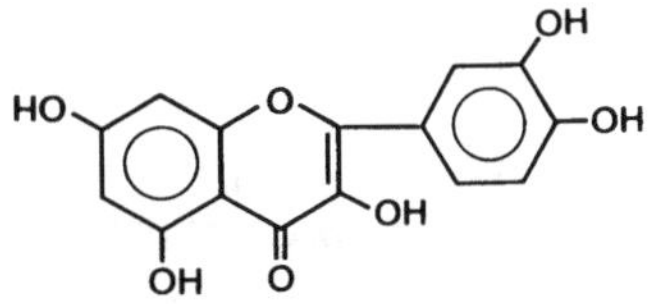

(9.5)

Quercetin

Also finely ground bark itself, applied as adhesive in combination with formaldehyde, leads to acceptable tensile strengths, however, without reaching V-100 values. On the other hand, such acidic phenolic compounds may cause problems at gluing chestnut, oak or eucalyptus wood with PF-resins. Due to their acidic character – the aqueous extract has a pH up to 3.2 – it is assumed that the cause is the buffer action of the salt formed from a strong base and a weak acid. Further addition of sodium hydroxyde to the glue can compensate this effect[26, 27].

9.4. Physical Properties of Composite Wood Materials

A compilation of the physical properties of several composite wood materials is shown in Table 9.2.

Wood as a natural polymeric organic material is an ideal construction material. It has a high modulus of elasticity (MOE) in fiber direction at a low specific gravity (SG). Strength, in addition, unusual for an organic material, is temperature independent over a wide range. In this respect, wood is far superior to synthetic organic polymers. The thermal conductivity is low. Wood is therefore an excellent insulating material. The remarkable anisotropy in strength, high water absorption and swelling are negative effects.

9.5. Particle Boards

Wood waste of the wood working industry induced the idea of PB. The industrial production started relatively late because parallel to a reasonable theoretic and practical conception (Fahrni, Kollmann, Klauditz), highly efficient machines, mechanization and automatization of the production methods had to be developed (Himmelheber-Steiner/Behr/Novopan AG)[1, 3, 6]. The production was started in 1941 by the Torfit-Werke AG, Bremen. Due to experiences gained in the gluing of wood, phenolic resins were used. Soon the cheaper and faster curing urea resins almost entirely displaced the phenolics. Today, 60–70% of the PB's are used for furniture production. After entering in the construction market, the share of PF bound PB's has become considerably higher since 1963. At the moment, it is approximately 10%. PF-boards are also used in the ship building industry and in the construction of freight cars (Fig. 9.3).

In general, the PB-industry tends more and more to produce higher priced semi-finished and finished parts from PB's. They are cut to size, faced with laminates and decorative papers (see Sect. 12.4) and fabricated into furniture as well.

9.5.1. Wood Chips, Resins and Additives

Apart from wood chips and resin, a number of additives like water repellants, extenders, fire retardant agents and fungicides are used to improve special properties.

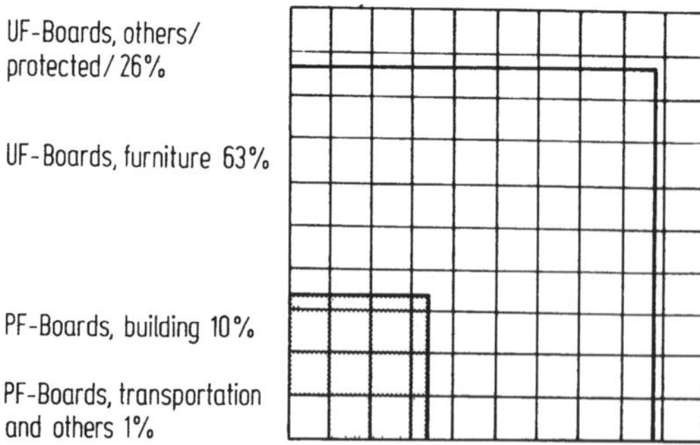

Fig. 9.3. Application of particle boards in West Germany in 1976

Table 9.2. Technical properties of composite wood materials according to Deppe[28)]

Type of board		Plywood		Fiber board		Particle board	
Properties		Veneer plywood	Core plywood	HDF	MDF	Furniture V 20	Construction V 100
Thickness	mm	12–15	16–19	4	6–19	19	19
Specific gravity (SG)	kg/m^3	550–700	450–650	900	680–750	620	700
Flexural strength ∥	N/mm^2	60–100	10–15	40–60	20–40	19	22
Flexural strength ⊥	N/mm^2	20–40	–	43	18–35	18	20
Modulus of elasticity (MOE) ∥	$N/mm^2 \cdot 10^3$	7.0–12.0	1.0–1.5	–	2.0–2.2	3.0	4.5
Modulus of elasticity ⊥	$N/mm^2 \cdot 10^3$	1.5–3.0	–	–	2.0–2.1	2.8	3.3
Internal bond (IB)	N/mm^2	–	–	0.45	0.18–0.70	0.38	0.45
Internal bond, wet	N/mm^2	–	–	0.10	0.02–0.20	–	0.20
Internal bond, dry again	N/mm^2	–	–	0.35	0.15–0.65	–	0.32
Thickness swelling, 2 h	%	–	–	12	3–12	4	8
Thickness swelling, 24 h	%	–	–	18	6–17	11	14
Thickness swelling, 120 h	%	–	–	22	22–80	16	20
Water absorption, 24 h	%	–	–	45	15–35	–	–
Water absorption 240 h	%	–	–	70	–	–	–
Swelling (lengthwise)	%	0.1–0.2	0.2–0.3	0.2–0.4	0.2–0.3	0.2	0.3
Shrinkage	%	0.2	0.2	0.3	0.3	0.2	0.2

Fig. 9.4. Application of PF-resin bonded particle boards in prefabricated houses.
(Photo: Deutsche Novopan GmbH, D-3400 Göttingen)

Wood Chips

A PB consists of up to 90% wood chips or equivalent wood particles. While the use of a certain kind of wood is more or less a matter of economy, the form of the chips and the structure of the board are decisive factors with regard to the quality. Therefore, all kinds of wood which grow in Europe are employed – for instance, fir, Scotch pine, red beech, poplar, birch, etc., either solely or in combination. Larger portions of bark result in considerable strength reduction. Efforts have also been made to use shavings or saw dust. Optimum properties are obtained from chips which have parallel and smooth boundary surfaces leading to high adhesive strength at low binder consumption. This desirable chip form is obtained by cutting machines, on the other hand, chopping and tearing action in mills yields irregular splintered particles.

Resins

According to present knowledge, the use of PF-resins for the production of weather resistant PB is a necessity[29]. This statement is supported by around 35 years of experience in the field of plywood glues. Lately, diisocyanates have been recommended with success for this type of application.

UMP-resins have also been recommended for outdoor application[19] but they have shown some clear deficiencies with regard to longterm weather resistance.

At the end of the fifties mostly phenol-cresol resins or xylenol resins were used. Molding times within the range of 1 min/mm thickness at 155–160 °C pressing temperature were necessary. Therefore, the output of a plant was approximately 50% of the usual production if UF-resins were used. In the sixties, attempts were made to shorten the molding time of PF-boards. They were successful by increasing the quantity of sodium hydroxide and formaldehyde molar ratio. However, the high alkali content could result in high water absorption, thickness swelling, discoloration and staining of light color facings because of the high hygroscopicity of sodium

Table 9.3. Characterization of a phenol resin for the production of particle boards V 100[30].

Characterization	phenol resol, water solution
Dry solids	45 ± 1%
Viscosity, 20 °C	130 ± 20 mPa·s
Specific gravity, 20 °C	1.17 g/cm^3
Alkali content (solution)	5%
Dilutability with water	∞ ratio
Storing ability, 20 °C	3–4 weeks
B-time, 100 °C	20 min.
Content of free formaldehyde	< 0.1%
Content of free phenol	< 0.1%

hydroxide, sodium phenoxide and potassium carbonate. The latter is often added to accelerate curing. The pH of such boards (water extract) is within the range of 9–10. Fast curing phenol resins (Table 9.3) having a lower alkali content, and thus without these disadvantages, are currently available[30]. Cresols and xylenols are no longer used.

The resins are obtained by condensation of phenol and formaldehyde in a molar ratio of 1 : (1.8–3.0) in an aqueous solution with NaOH as catalyst. Sodium hydroxide serves as catalyst for the hydroxymethylation reaction and for the required unlimited water solubility of the resin even at a high MW through formation of sodium phenoxides. At the same time, NaOH accelerates the curing rate of the resin considerably.

The extraordinary low monomer content of such resins is remarkable (Tab. 9.3). Their molecular weight distribution and the low mononuclear phenol alcohols content are shown in Figure 9.5.

A faultless wetting of the chips is guaranteed, but excessive resin penetration which would result in an increased binder consumption is avoided. The pressing time is not sufficient to cure the resin completely. During storage in the hot stack, however, the curing continues and board properties are improved. Often catalysts are added to accelerate the curing rate. Organic or inorganic carbonates like potassium carbonate (FR-PS 15 50 87), formamide (DE-AS 21 10 264), lactones (DE-PS 10 65 605), resorcinol, resorcinol resins or mixtures of these compounds are recommended accelerators.

For 3-layer boards it is reasonable to add the accelerator only to the middle layer, resulting in molding times up to 12 s/mm board thickness at a plate temperature of 220 °C.

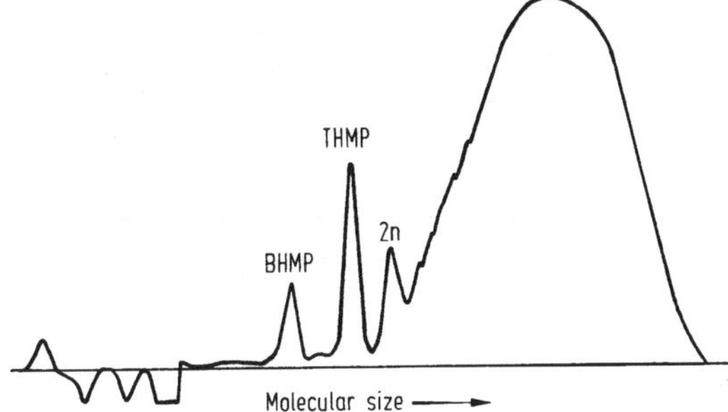

Fig. 9.5. Molecular weight distribution (GPC) of a phenol resin for particle boards
BHMP = bis-(hydroxymethyl)phenol
THMP = tris-(hydroxymethyl)phenol
2 n = dihydroxydiphenylmethane derivative

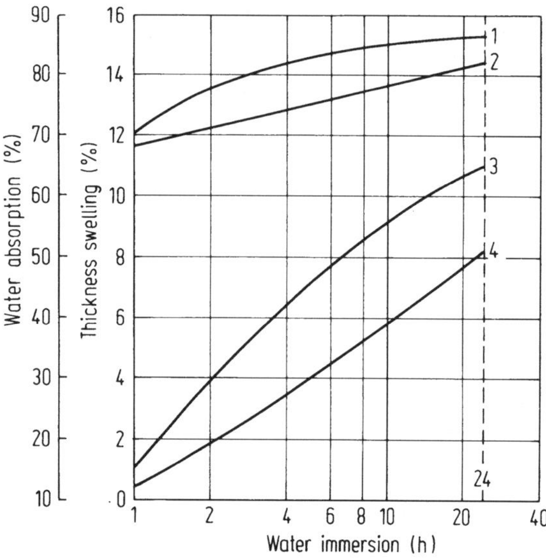

Fig. 9.6. Thickness swelling and water absorption of PF-boards at immersion. 16 mm single layer laboratory PB made of Scotch pine chips, 8% of solid resin referred to dry chip weight. (Drawing: Mobil Oil AG, D-2000 Hamburg). *1* Thickness swelling without additives; *2* Water absorption; *3* Thickness swelling with 1,2% paraffin; *4* Water absorption

Hydrophobic Agents

The swelling of PB's occurs mainly vertically to the board surface due to chip orientation during the molding process. Hydrophobic agents are added to reduce the wettability and water absorption. The amount of bonding agent is, by far, insufficient to coat the chip completely. Paraffins with a MP between 50–60 °C as aqueous dispersions with 30–65% solids content are almost exclusively used as hydrophobic agents[31]. The addition of paraffin is very effective to prevent absorption of liquid water (wetting problem) but is far less effective against damp. The water absorption and deformation processes are therefore only delayed temporarily.

Generally, the paraffin dispersion is mixed homogeneously with the glue. A separate application would be better because of more effective distribution, but is more expensive as far as the equipment needed is concerned. The amount of paraffin used in PF-boards is larger than that in UF-boards and is, in general, approximately 1–1.5% referred to dried wood. A higher percentage of paraffin reduces binder wetting and adhesive strength[31a].

Fungicides and Insecticides

PF-boards offer better resistance to wood destroying insects and fungi than wood. On the other hand, UF-resins further the growth of fungi. Furthermore, the kind of wood, binder content and board density influence the resistance to the attack of fungi. Several cases of damages by fungi in the construction business revealed that additional protective measures are necessary. In West-Germany, PB's resistant to fungi are designated as V 100 G (DIN 68763). Similar boards are made in France according to CTB-G standard. The increased resistance is achieved by the addition of fungicides, preferably tributyl tinoxide or fluorine compounds[32] in quantities up to 1.5%.

Flame Retardants

In West-Germany, standard quality PB's to be used as construction material are classified as B_2 "normal flammability", according to DIN 4102. The requirements of class B_1 can be met by adding flame retardant materials. Ammonium hydrogen phosphate and ammonium polyphosphate are prefered. They may be used in combination with halogen compounds. Boron compounds did not prove to be useful because of lack of compatibility with resols. The material costs are drastically increased by flame retardant additives so that the economy of production is impaired. The addition of inorganic substances like vermiculite or perlite has been recommended. However, board strength is reduced by these materials. Inorganic bonding agents (concrete) and fillers[33)] also increase thermal conductivity so that their use would violate other requirements.

9.5.2. Production of Particle Boards

A description of PB-technology is given in the literature[1, 3, 6, 34–36)]. Only the gluing and pressing processes shall be mentioned briefly as far as they have a decisive influence on the effectiveness of the gluing.

Chip Blending

The chips are dried to a relative humidity RH of 1–5% prior to blending. The chips of the middle layer should have 2% RH. Since the price for bonding agents is a decisive factor in the calculation, a uniform and reproducible glue distribution and a thin as possible spread should be achieved. The resin content in three-layer boards amounts to 10–12% in the surface layers and 7–9% in the middle layer. For single-layer boards the resin content is about 10%, based upon solid resin and dry chips weight. After blending, the chips have a RH of 8–18%. The glue formulation is carried out to a high degree of accuracy by automatically operating equipments[37)]. The homogeneous blend is dosed according to volume and fed continuously into the gluing machine. The formulation and chip size for the middle and surface layers are different, so that separate storage tanks and gluing machines are required for a continuous operation, as shown in Figure 9.7.

In modern high speed chip gluing machines[38)] the glue is uniformly distributed by centrifugal force through a hollow shaft out of several distribution tubes (Figure 9.8). Compressed air is not required to convert the liquid material into very fine drops. The uniform binder spread is achieved by a rolling and wiping effect favored by the high rotation speed. The glue drop is not captured by an individual chip but rolls from chip to chip leaving a thin glue track. The glue is distributed in that way, that only discrete chip areas are covered and glued together.

Gluing with resin powder has been recommended[39, 40)]. Although this method would result in considerable savings of material and energy, the technological problems have not been solved satisfactorily.

Pressing of Particle Boards

The chip mats are predensified in a separate one-daylight press or continuously by rolls at room temperature and at a pressure of 1.0–2.5 N/mm^2. The definite pressing

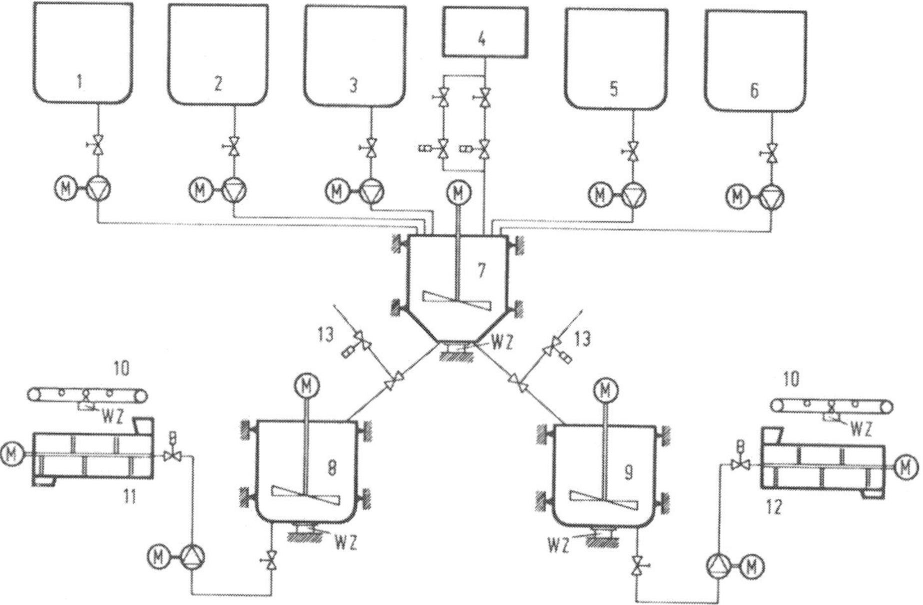

Fig. 9.7. Flow sheet of a PB-glue formulation plant. (Drawing: C. Schenck AG, D-6100 Darmstadt).
1 PF-resin; *2* UF-resin; *3* Paraffin dispersion; *4* Water; *5* Ammonia; *6* Hardener (for UF-boards) or additives; *7* Agitated wessel and scale; *8* Glue formulation, middle layer; *9* Glue formulation, surface layer; *10* Continuous scale; *11* Chip blender, middle layer; *12* Chip blender, surface layer; *13* Compressed air; *M* Motor/Pump

Fig. 9.8. High speed chip blending machine. (Photo: Gebr. Lödige Maschinenbau GmbH, D-4790 Paderborn)

is performed either in an one-daylight press (temperature up to 220 °C) or in a multi-daylight press at 160–190 °C as shown in Figure 9.9.

The pressure applied is, in general, between 1.5–3.5 N/mm^2 according to desired board density. The molding time is between 12–30 sec/mm thickness depending on press plate temperature. The chip mat is a bad heat conductor so that a marked temperature profile is developed in the cross section of the board (Fig. 9.10). The

Fig. 9.9. Hydraulic multi-daylight press for the production of particle boards. (Photo: G. Siempelkamp GmbH & Co, D-4150 Krefeld)

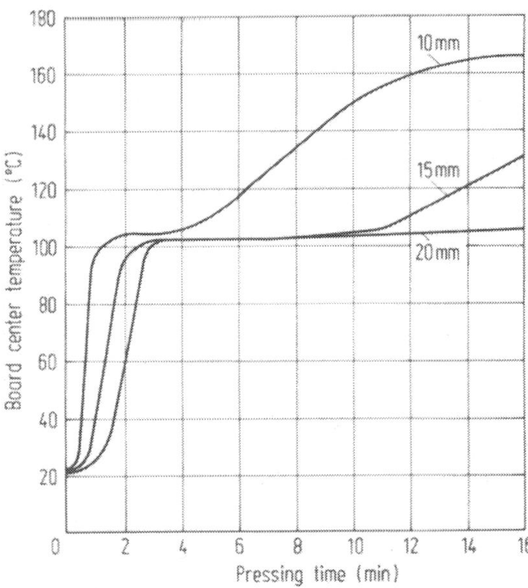

Fig. 9.10. Heating pattern in the center of the particle board dependent upon pressing time and board thickness. Press plate temp. 165 °C, chip humidity 15%, spec. gravity 650 kg/m^3. (Drawing: Bakelite GmbH, D-5860 Iserlohn-Letmathe)

water content of the chips has a decisive influence on the heating rate. Due to the low thermal conductivity of the system, heat is mainly transferred by the steam formed. Sometimes the chip mat is additionally sprayed with water to enhance heat transfer ("steam shock"). Thus, it is understandable that the temperature in the center section hardly exceeds 105 °C. A complete curing is not achieved under these conditions; it is continued during storage in the hot stack. A simple method to de-

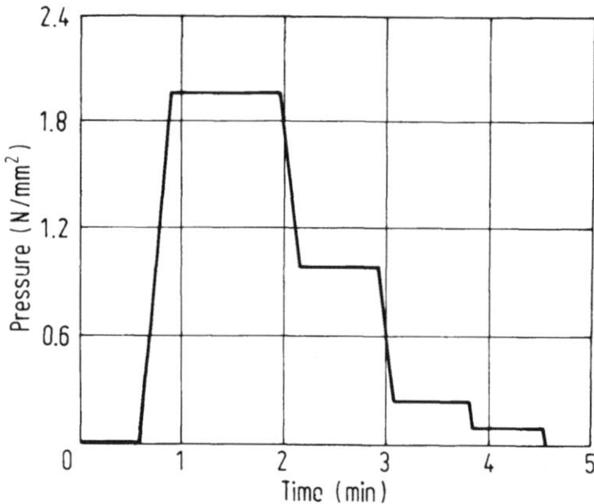

Fig. 9.11. Example of a pressing diagram for the production of particle boards

termine the extent of cure is the evaluation of the internal bond (IB). The IB increases considerably when higher temperature is applied to the board center section. This also applies to the wet IB[41]. About 120 °C in the mid-layer would be enough to cure the resin completely. An even higher temperature has only a small positive effect on board strength. The SG or thickness of the boards are often regulated by stops. Thereby, the effective maximum hydraulic pressure is only effective for a short time. The hydraulic pressure is adjusted to the real course by pressure stages (Fig. 9.11).

When the chip mat is compressed to nominal thickness, the air can escape at the sides. When the temperature in the center layer has risen to 100 °C, water is evaporated vigorously and a considerable gas/vapor pressure is developed. Towards the end of the pressing cycle, the internal pressure drops due to the permeability of the board. Theoretically, the pressing time is completed when the internal steam pressure is lower than the adhesion- and cohesion strength, indicated by the IB[42].

9.5.3. Properties of Particle Boards

The requirements for PB's to be used for furniture are different from those for the construction business[28, 43, 44]. High surface quality is demanded of UF-boards because their bond to various top layers must be flawless and optically perfect. PB's for construction must have high strength (MOE, flexural and tensile strength) and high weather resistance (swelling, IB etc). The SG of a PB (Fig. 9.12) may be affected by production parameters (chip form, RH, press diagram and reactivity of the resin). Even if there are no specifications for PF-boards concerning the roughness of the surface – PF-boards are not covered with decorative facings – a high density of the boundary layer is desirable (sandwich principle, water resistance).

Board strength depends upon chip form, board structure and binder content. Mechanical properties are indicated in Table 9.2. Table 9.3a shows the strength of single-layer boards depending upon resin content. These boards were made without paraffin addition, this explains the relatively high thickness swelling. The influence of paraffin is shown in Figure 9.6.

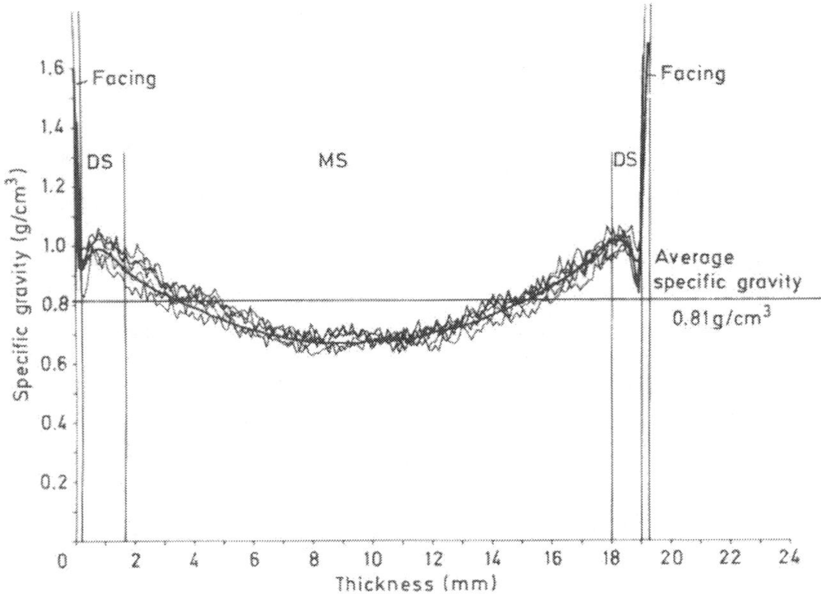

Fig. 9.12. Plotter-drawing of the density profile of a three-layer particle board clad with decorative facings, measured by gamma ray transmission (44). (Kossatz, May, Wilhelm-Klauditz-Institut für Holzforschung, D-3300 Braunschweig)

Table 9.3a. Properties of single-layer particle boards depending on resin content. Chips 80% conifer, d = 20 mm, molding conditions 0.35 min/mm at 160 °C, 2 N/mm^2 pressure

Properties		% Resin content			
		5	7.5	10	12.5
Specific gravity	kg/m^3	650	645	640	660
Internal bond, dry	N/mm^2	0.48	0.57	0.69	0.76
Internal bond, 2^h H_2O/100 °C	N/mm^2	0.04	0.12	0.22	0.27
Thickness swelling	%	28.2	17.7	14.2	11.9

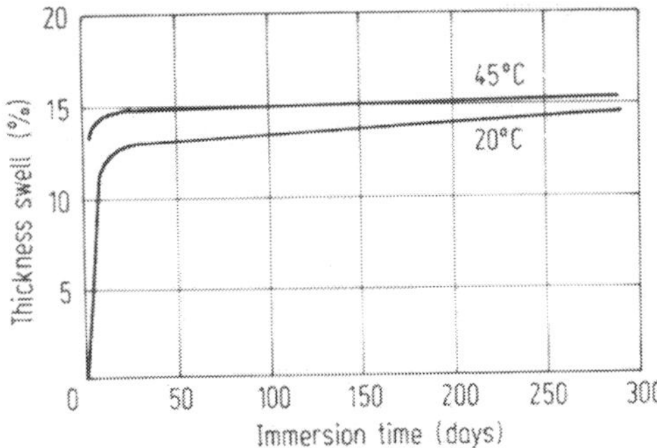

Fig. 9.13. Thickness swelling of a PF-bonded three-layer PB dependent upon time and temperature at water immersion. resin content – top layer 11%; resin content – middle layer 9%; paraffin content 1.2%; pressing time 0.30 min/mm; pressing temperature 170 °C; chips: fir; spec. gravity 665 kg/m^3

The high water resistance of PF-boards[16, 17, 29, 45, 46] is shown in Figure 9.13. Comparable values are found if the boards are boiled in water. After 20 hours at 100 °C the thickness swelling amounts about 15%. Phenolic resin bonded plywood shows a similar behavior.

In the USA and Canada, a special type of board with higher strength (dry) has been developed. These wafer boards are made of large area chips 0.4–0.8 mm thick, 10–15 mm wide and 35–75 mm long. The PF-resin content (5–6%) is considerably lower than in European boards. It is therefore not surprising that after boiling them for only 2 hours the strength is reduced by 50%. The SG of the wafer boards is 650–750 kg/m^3. The flexural strength is supposed to be 22 to 28 N/mm^2, the IB 0.8–1.2 N/mm^2. However, the wafer boards and some special production processes used in Europe are of little importance.

The formaldehyde separation during board production[47] and use[48] will be a very important factor for further market acceptance of different PB-glues. The exceptionally low formaldehyde emission of PF-boards in comparison to UF-bonded boards[49, 50] may favour further market development of PF-boards. Methods for evaluation of the formaldehyde separation of PB's are to be found in the literature[51, 52]. The toxicity of formaldehyde is indicated in Chapter 5.

9.6. Plywood

In general, plywood is a material which is made of at least three layers of wood, mostly veneers, with application of adhesives, pressure and heat. The fiber direction of the individual layers is mostly vertical to each other, but can also be parallel (veneer plywood, wood core plywood, block board, laminar board and com-

Table 9.4. Types of plywood according to DIN 68 705 with respect to their humidity resistance and type of adhesive

Class acc. to DIN 68 705	Type of adhesive	Property	Pre-treatment of samples
IF 20	Urea resins	Resistant to low humidity, usual in closed rooms	24 h in H_2O at 20 ± 2 °C
IW 67	Urea and urea/ melamine resins	Resistant to higher humidity, resistant to water up to 67 °C for a short time	3 h in H_2O at 67 ± 0,5 °C and 2 h in H_2O at 20 ± 3 °C
A 100	Urea/melamine resins melamine resins	Outdoors limited resistance, resistant to cold and hot water	6 h in H_2O at 100 °C and 2 h in H_2O at 20 ± 5 °C
AW 100	Phenol resins Phenol/resorcinol resins Resorcinol resins	Resistant to all climatic influences	4 h in H_2O at 100 °C, 16–20 h in air at 60 ± 2 °C and 4 h in H_2O at 100 °C, 2–3 h in H_2O at 20 ± 5 °C

Fig. 9.14. Phenol resin bonded concrete form plywood. (Photo: Schütte-Lanz, D-6800 Mannheim)

Fig. 9.15. Application of concrete form plywood. (Photo: Schütte-Lanz, D-6800 Mannheim)

posite plywood) or diagonal (star plywood). Plywood is generally classified in interior and exterior grades[1, 5, 10]. Exterior type plywood will retain its glue bond when repeatedly wetted and dryed or otherwise subjected to weather and is, therefore, intended for permanent exposure. The classification in West-Germany according to DIN 68 705 is indicated in Table 9.4. In the USA the interior type plywood is similarly differentiated into three levels according to glue type (interior glue, intermedial glue and exterior glue). Within each exposure capability (exterior and interior) there are a number of panel grades based on the quality of the veneers and the panel construction[53].

Examples for PF-glued plywood application are concrete form plywood, building and construction elements, interior parts in transportation vehicles (trucks, railway cars, passenger cars), boats and ships, airplane construction, containers, foundry patterns, handles for tools and machine parts (made from high-densified plywood).

9.6.1. Resins, Additives and Formulations

Non-modified resols catalyzed by sodium hydroxide are used for weather resistant plywood. The MWD of those resins is shown in Figure 4.5, Chapter 4. The viscosity of the resins varies between 700 to 4,000 mPa·s according to the gluing process – dry and wet. Typical resin properties are compiled in Table 9.5.

Table 9.5. Phenol resol resins for AW 100 plywood manufacture[54]

Resin type		Dry process	Wet process
Dry solids content	%	45 ± 1	44 ± 1
Viscosity at 20 °C	mPa·s	4.000 ± 500	700 ± 200
Alkali hydroxyde content	%	3	7
Gel time at 100 °C	min	20	33
Water solubility		∞	∞
Free phenol content	%	0.1	0.1
Free formaldehyde content	%	0.1	0.1
Storage stability at 20 °C		3 weeks	12 months

As with PB-resins the content of free phenol and free formaldehyde is extremely low and practically non measurable. In order to adjust wetting and avoid excessive penetration and to obtain a uniform joint thickness, diluents and/or fillers are almost always used. While rye or wheat flour is used in combination with UF- and MF-resins, only inert (chalk) or non-swelling fillers like coconut shell flour can be used for exterior grades. Apart from cost reduction, the brittleness of the adhesive joint is reduced by these additives. An excessive quantity of fillers, however, may lead to strength reduction. Examples for the glue formulation are indicated in Table 9.6.

9.6.2. Production of Plywood

Of all the European woods red beech takes the first place in the production of plywood followed by birch, pine and fir; of tropical woods limba and okoume are mainly used. These wood species differ considerably concerning their weather resistance. The RH of the veneers should be between 4 and 8%. This is also valid for the non-coated intermediate plies. Veneers which are too wet or too dry could be the cause of faulty gluing. Too high humidity of veneer favours excessive resin diffusion out of the glue joint or the formation of blisters. Water also reduces the curing rate. Veneers which are too dry draw the water out of the glue and reduce resin flow. The use of hot veneers is not advisable[54–57].

Table 9.6. Glue formulation for the production of weather-resistant plywood[54]

Formulation and processing		Dry process	Wet process		
Phenol resin	pbw	100	100	100	100
Chalk	pbw	–	10	10	10
Coconut flour (300 mesh)	pbw	5–10	8	8	8
Water	pbw	10–20	–	–	–
Hardener (paraformaldehyde)	pbw	–	–	2	2
Accelerator (resorcinol resin)	pbw	–	–	–	6
Glue service life		1 d	1 d	1 d	4 h
Press temperature	°C	130–150	130–140	120–125	100–110

Fig. 9.16. Glue coating machine for veneers. (Photo: R. Bürkle GmbH & Co., D-7290 Freudenstadt)

The glue spread is approximately 160 g/m^2, that means 320 g/m^2 for the usual two-sided coating. A glue coating machine is shown in Figure 9.16. The two-side coated veneer is then covered with two non-coated veneers. Glue spread of 130–140 g/m^2 is often enough for faultless gluing of low porous veneers. A thin glue line reduces costs and may offer technical advantages, for instance minimal joint brittleness and low water penetration into the veneer. However, the risk of gluing defects is increased[56].

After gluing, the veneers are dried in a veneer dryer to a RH between 8–12% (dry process). The drying temperature should not exceed 70–75 °C. At 70 °C the drying lasts approximately 10–15 minutes. The veneers are then cooled by cold air. If resins with a relatively low viscosity are used without the addition of water, the manufacture can be performed without oven drying (wet process). While soft wood veneers can be pressed without a closed or open waiting time for the glued assembled material, waiting times must be observed for at least 5–10 minutes if denser kinds of wood are used. This applies especially to beech.

The closed assembly time can be extended to at least 30 minutes at 20–25 °C; for non-accelerated glues it can frequently be prolonged for several hours. Too strong airing of the adhesive coat results in gluing defects because the resin flow is impaired. The molding temperature is – according to the glue formulation – 100 to 140 °C. The placing in and the closing of the press must be done quickly ($<$ 2 min). Otherwise, the exterior plies start to cure before the maximum molding pressure is reached. The molding pressure is 0.8–1.5 N/mm^2 for softwood (SG up to 0.55 g/cm^3) and 1.5–2.5 N/mm^2 for hardwood (SG above 0.55 g/cm^3). Furthermore, the molding pressure may be modified according to veneer moisture, surface and resin flow.

The actual curing time is approximately 5–6 minutes at 130 °C, and about 3 minutes at 140 °C. The additional heating time depends upon the number and the thickness of the veneers. It is 1 min/mm for boards up to 10 mm thick; for boards over 10 mm of thickness it is 1.5–2 min/mm. Attention must be given to an even warm through especially for multiplex-boards. The curing temperature can be reduced to about 100 °C by the addition of hardeners and accelerators. Gluing with resorcinol resins can be accomplished at room temperature. The glue bond quality (exterior type) is tested by the vacuum-pressure test, boiling test, scarf and finger joint tests and heat durability test[53].

9.6.3. High-Densified Plywood

High-densified plywood is closely related to plywood. It is made by coating and impregnating wood veneers with a higher proportion of resin[58]. It is then pressed under high pressure to laminates with a SG between 1.0–1.4 g/cm^3.

Compressed laminated wood has excellent mechanical strength, water resistance and very good machinability. In the field of mechanical engineering, screws, bolts, drives, cog wheels and parts for weaving machines are made of it. Further fields of application are chair seats, trays, dash boards, knife handles, bearing shells, sliding ledges for conveyer plants etc. Molded parts with excellent electrical properties are made by using phenol resins which are free of inorganic ions. They are used for the construction of transformers and control devices because of their good insulating properties, high specific strength and resistance to transformer oil.

Red beech veneers 0.2 to 1.0 mm thick are mostly used. They are predried to a RH of maximum 8% and the surface is then either coated with the resin or impregnated very homogeneously depending on future use. A resin content up to approximately 20% can be applied by coating. In the dipping (impregnation) process the veneers are flooded with aqueous resin for several hours. A resin take-up of about 30% is obtained. A far better homogeneity can be achieved by the vacuum-pressure impregnation in autoclaves. A maximum resin take-up of 60% can be obtained by this process.

After gluing or impregnating the veneers are dried to 2–10% RH depending on product type in a veneer dryer. A tolerance of ± 0.5% is not to be exceeded. This

Fig. 9.17. Application of high-densified plywood in the seating furniture field. Pagholz-seats made of PF-resin impregnated beech veneers. (Photo: PAG Presswerk AG, D-4300 Essen)

Fig. 9.18. High-densified plywood foundry tools, Delignit. (Photo: Blomberger Holzindustrie, B. Hausmann GmbH & Co. KG, D-4933 Blomberg)

can be achieved by post conditioning. The pressing temperature is generally 140–160 °C. The pressure is guided by the kind of wood, the desired degree of compression and the part shape. In general, it is between 5 and 40 N/mm². For high pressures molds made of high polished chromium-nickel steel are required, while at low pressures molds of aluminium with a modified surface are adequate[59].

9.7. Fiber Boards

As early as 1872 Clay registered a patent which suggested using thick paper of minor quality – "papier mache" – in the construction field and furniture manufacturing. Later, the strength of the paper was improved by additives. This can still be seen in the ISO definition of fiber boards: "Sheet material generally exceeding 1.5 mm in thickness, manufactured from lignocellulosic fibers with the primary bond from the felting of the fibers and their inherent adhesive properties. Bonding materials and/or additives may be added".

A number of classifications of fiber boards can be given according to the appearance, method of production, kind of application and SG[1, 4, 5]. According to SG, five types can be differentiated (Table 9.7).

The market for insulation boards is relatively small and is further losing importance; acoustical boards are more and more replaced by non-flammable, inorganic materials. Medium density fiber boards (MDF), developed in the USA and Scandinavia especially for use in the construction business, may also be used for the manufacture of furniture because of their remarkable homogeneity and smooth surface.

Table 9.7. Classification of fiber boards according to specific gravity[1, 28]

Fiber board		Specific gravity g/cm³
Non-compressed:	Semi-rigid insulation board	0.02–0.15
	Rigid insulation board	0.15–0.40
Compressed:	Medium density fiber board, MDF	0.60–0.80
	Hardboard (high density FB), HDF	0.90–1.20
	Special densified hardboard	1.20–1.45

MDF's of comparable thickness attain the same strength as particle boards. However, their production is more expensive and causes environmental problems. Therefore, they have not yet had success in the West-German construction board market.

In West-Germany, the most important fields of application for hardboards are the furniture industry (about 40%) followed by the manufacturing of doors (about 25%), and the automotive and house trailer industry (about 20%). In the furniture industry, they are mainly used for rear walls and drawer bottoms. For this purpose, the boards are coated; lamination with decorative papers is of minor importance.

Weyerhaeuser Co, USA, developed a special form of a fiber mat which is used for the production of shaped parts. These mats for the "pres-tock" technique contain a higher amount of non-cured PF-resin in combination with colophonium as bonding agent. After steam conditioning the mats are compression molded. In the automotive industry, for instance, these mats are used for switch boards, ceilings or coverings – parts which are mostly clad with plastic foils.

9.7.1. Wood Fibers, Resins and Additives

Only wood of inferior quality and industrial wood is used for the production of fibers. Conifer wood is more suitable for the wet fiber process, because it is easier to dehydrate than wood from deciduous trees. The latter is, however, better suited for the dry process and is preferred in Europe because of its availability. More and more hard- and softwood mixtures are used for both processes due to economic reasons. The bark is not completely removed because presence up to 15% does not affect the properties of the fiber board considerably. Sawdust can also be added up to 30%. In addition non-wood fibrous materials such as residues from annual plants can be used. The fiber efficiency of this material is rather low and the overall properties are inferior to those of wood fibers.

In general, fiber boards can be produced without bonding agents, using only the bonding ability of the fiber. In order to improve the mechanical properties and to reduce the water absorption and swelling, phenol resols, sometimes combined with natural resins like colophonium, are added as bonding agents. Urea resins are used in the dry process for boards which are not subjected to high humidity, for instance for boards to be used in the furniture industry[60]. The portion of synthetic resins in fiber boards – their specification is described in Table 9.8 – is very small. In general, the dry resin content is between 1 and 3% calculated on dry fiber weight. This low resin coat can only be achieved by a special process. The PF-resin is precipitated out of a diluted aqueous solution by acidic chemicals – diluted sulfuric acid or aluminum sulfate – at pH 4 and so fixed to the fiber. MDF's as produced in the USA for building purposes contain a considerably higher PF-resin amount of between 8–10%. The main problem at MDF production is the uniform gluing of the voluminous, non-free-flowing fiber mass[61]. The process is performed with centrifugal gluing machines similar to those shown in Figure 9.8.

Hydrophobic agents like wax and paraffin reduce water absorption and swelling. The amount of wax is normally within the range of 1% based on the weight of the dry fiber.

Table 9.8. Characteristic properties of a phenol resin for fiber boards[62]

Solid resin content	%	41 ± 1
Content of precipitating resin	% min.	35
Alkali content	% max.	6
Viscosity at 20 °C	mPa · s	500 ± 100
B-time in precipitated condition (pH 4) at 100 °C	min.	1.5 – 3
Dilutability with water	–	∞
Storage life at 20 °C	weeks	4

Furthermore, flame retardants, fungicides and insecticides, release agents and other improving materials like drying oils can be added.

9.7.2. Production of Fiber Boards

The technology of fiber board production is described in detail in the literature[1, 4, 5]. The preparation of the fiber material can be performed thermo-mechanically or chemo-mechanically (Table 9.9).

It is obvious that the preservation of the fiber structure and the relation of length to cross section are the factors decisive for the quality and strength of the fiber board. Principally, the wet and dry process are differentiated by the medium for the fiber sheet formation. In the wet process the sheet is made with Fourdrinier

Table 9.9. Processes for the defibration of cellulosic materials[5]

Process	Defibration unit	Defibration conditions	Pulping
Masonite	Mason gun	1. 30–40 s steam, 210 °C and 2.0–2.4 N/mm² pressure 2. 5 s steam, 285 °C (7 N/mm²) 3. explosion-like steam pressure reduction	Thermo-mechanical
Defibrator (Asplund)	Defibrator	2–5 min steam (in the pre-heater), 160–185 °C, 0.6–1.2 N/mm² pressure	Thermo-mechanical
Bauer	Bauer mill with counter running grinding discs	3....10 min steam in the pipe vessel, 0.3–0.6 N/mm² pressure	Thermo-mechanical
Boja-Jung (Biffar)	Biffar mill	Pre-boiling in a globular boiler with diluted sodium carbonate solution 6....8 h at 0.6 N/mm² pressure	Chemo-mechanical

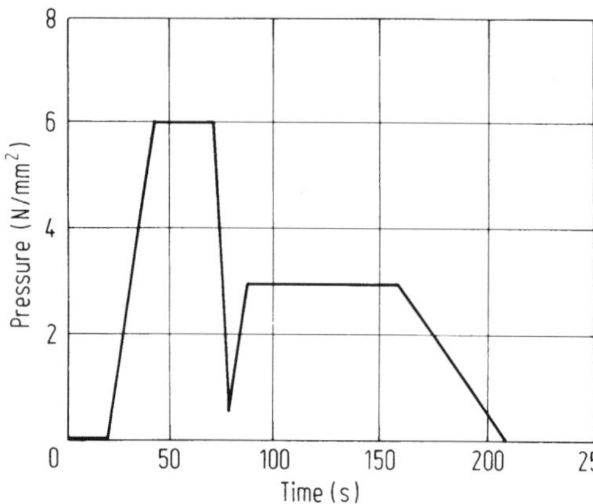

Fig. 9.19. Example of a pressing diagram for production of fiber boards

or cylinder machines from an aqueous fiber suspension similar to the paper manufacturing process.

A dewatering screen is placed under the sheet at molding in a multi-daylight press. Because of this, screen marks develop on one side of the fiber board (1–S–1 = smooth on one side). The pressing time is approximately 2.0–3.5 min/mm of board thickness at temperatures of about 180–200 °C. The pressing diagram includes several pressure stages. In the dry process an air suspension of fibers is used for the sheet formation. The products are still inferior in quality to those obtained by wet felting, but the dry process offers great advantages with regards to environmental problems and lower water consumption. The bonding agents and additives can be added at different process steps, for instance prior to or during the defibration, but also afterwards in special gluing machines, for example, type Drais or Lödige. The pressures and temperatures for semi-dry and dry processes are higher than those for the wet process. They are 7 N/mm^2 and 200–250 °C, respectively.

The physical properties of fiber boards are listed in Table 9.2.

9.8. Structural Wood Gluing

The gluing of wooden construction elements[63)] with cold setting adhesives is gaining more and more importance compared to the conventional techniques using nails and screws. Wood faults can be eliminated by lamination of thin layers; the swelling and shrinkage of the wood can be reduced; and a wide range of design variation is obtainable. The behavior of wooden structures in fire is better than generally assumed. Wood is a very bad thermal conductor (Table 9.2) and, the increase of temperature in the interior of a structural beam is considerably delayed by the formation of a charcoal layer. The type of glue usually used does not have any influence on the flammability.

In Europe, fir wood is preferred for structural elements. The quality of the wood must be thoroughly rechecked. Prior to its use, it is dried at 100 °C and airconditioned. According to exposure capability, cold setting glues based on urea, resorcinol or phenol/resorcinol combinations are used. For interior parts UF-resins are adequate.

Fig. 9.20. Resorcinol resin glued bridge construction. (Photo: Hüttemann Holz GmbH & Co. KG., D-5787 Olsberg)

Fig. 9.21. Resorcinol resin glued structural wood elements of a salt warehouse. (Photo: Hüttemann Holz GmbH & Co. KG., D-5787 Olsberg)

Structural elements subjected to weather or other climatic stresses must be glued with resorcinol/phenol- or resorcinol resins. This is also true for boat construction. In the marine field, the glued parts are also subjected to strong dynamic stresses; the frequent change from wet to dry as well as the attack of salt water have to be taken into consideration[64].

9.8.1. Resorcinol Adhesives

The reactivity with aldehydes is greatly increased by the introduction of a second hydroxyl group into the phenol nucleus[65–67]. Resorcinol reacts with formaldehyde at room temperature even without the addition of catalysts. The rate of reaction is at a minimum at pH 3.5 (phenol 4–4.5) as shown in Figure 9.22.

However, alkaline catalysts are often used to produce resorcinol-formaldehyde resins. Prepolymers are first made at a low formaldehyde molar ratio (< 1). Those

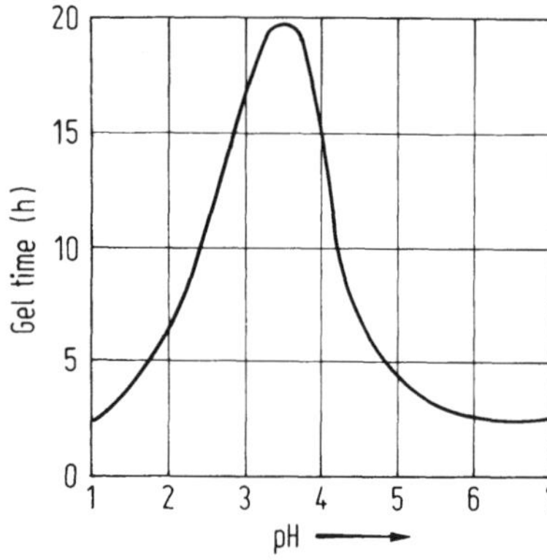

Fig. 9.22. Reactivity of a resorcinol-formaldehyde prepolymer with formaldehyde. Relation between time of gelation and pH[66]

prepolymers, which are completely stable, are cured by the addition of paraformaldehyde or formaldehyde solutions (now called hardeners). A total amount of 1.1 mol of formaldehyde is sufficient for curing, which is performed at ambient temperature and neutral conditions. The application of acids is not required so that the wood is not damaged. The bonds are gap filling and resistant to boiling water (Fig. 9.1), acids, mild alkalis and common solvents. The temperature resistance of resorcinol adhesives is up to approximately 200 °C. The adhesion to various materials is very good so that they can also be used to bond leather, rubber, plastics, ceramics, etc. For the glue preparation, a prepolymer solution (55–60% solids) is mixed in appropriate mixers with the hardener (paraformaldehyde) shortly before application. If chalk is used as filler, it is recommended that the paraformaldehyde is first mixed with the chalk. The service life of the mixtures is approximately 1.5 hours at 20 °C; if chalk is added, up to 2.5 hours, according to the quantity of the hardener. The plies to be glued must have a moisture content between 6 and 15%. The glue is applied on both sides; the coating amounts to 200–400 g/m^2 for soft wood, and 150–300 g/m^2 for hard wood. The glue spread applied with modern glue pouring machines on one side only is approximately 400 g/m^2. The open assembly time is 30 to 45 minutes maximum; the closed assembly time can be extended up to 1.5 hours. The pressure can be applied by clamps or a press. Depending upon the construction and the application, approximately 0.6 to 1.5 N/mm^2 are necessary. The pressing time is about 8–14 hours at 20 °C; the curing, however, is completed after 2 to 3 days. The application of higher temperatures reduces the curing time considerably[68].

References

1. Kollmann, F. P., Kuenzi, E. W., Stamm, A. J.: Principles of Wood Science and Technology, Vol. II, Berlin, Heidelberg, New York: Springer 1975
2. Bosshard, H. H.: Holzkunde, Vol. 3, Basel: Birkhäuser 1975
3. Deppe, H. J., Ernst, K.: Fortschr. d. Spanplattentechnik, Stuttgart: DRW Verlag 1973
4. Lampert, H.: Faserplatten, Leipzig: VEB Fachbuchverlag 1967
5. Kehr, E. et al.: Werkstoffe aus Holz, Leipzig: VEB Fachbuchverlag 1975
6. Deppe, H. J., Ernst K.: Technologie d. Spanplatten, Stuttgart: Holz-Zentralblatt Verlag 1964

7. Froede, O., Witt, H.: Holzwerkstoffe. In: Ullmanns Enzyklopädie d. techn. Chem. Vol. 12, 4. Ed. Weinheim: Verlag Chemie 1976
8. N. N., Holz-Zentralblatt 102, No. 135 (1976)
9. N. N., Holz-Zentralblatt 103, No. 8 (1977)
10. Deppe, H. H.: Holz-Zentralblatt 102, No. 57 (1976)
11. N. N., Holz-Zentralblatt 103, No. 9 (1977)
12. Sarkanen, K. V., Ludwig, C. W.: Lignins. New York: Wiley-Interscience 1971
13. Hesch, R.: Holz als Roh- und Werkstoff *26*, 129 (1968)
14. Bird, F., van den Straten, E.: J. Inst. Wood Sci. (London) *29*, 41 (1971)
15. Kreibich, R. E., Freeman, H. G.: Forest Products J. No. *12*, 24 (1968)
16. May, H. A.: Holz-Zentralblatt *102*, No. 96 (1976)
17. Jellinek, K., Müller, R.: Holz-Zentralblatt *102*, No. 113 (1976); Jellinek, K.: A Comparison Between Particleboards Bonded by Phenolic Resins and Phenol-Modified Melamine Resins. 11th Washington State University Symposium on Particleboard, Pullman, March (1977)
18. Ernst, K.: Holz-Zentralblatt *102*, No. 20 (1976)
19. Clad, W., Schmidt-Hellerau, Ch.: Holz-Zentralblatt *102*, No. 24 (1976)
20. Sachs, H. J.: Holz-Zentralblatt *102*, No. 96 (1976). – Deppe, H. J.: Technical Progress in Using Isocyanate as an Adhesive in Particleboard Manufacture, 11th Washington State University Symposium on Particleboard, Pullman, March (1977)
21. Roffael, E., Rauch, W.: Holz-Zentralblatt *100*, No. 96 (1974)
22. Shen, K. C.: Adhesive Age, Febr. 1976, P. 33
23. Roffael, E., Rauch, W., Bayer, S.: Holz als Roh- und Werkstoff *32*, 225 (1974)
24. Anderson, A. B., Wu, K. T., Wong, A.: Forest Products J. *24/7*, 40 (1976) and *24/8*, 48 (1974)
25. Roffael, E.: Adhesion *11*, 306 (1975)
26. Roffael, E., Rauch, W.: Holz als Roh- u. Werkstoff *32*, 182 (1974)
27. Schmidt-Hellerau, Ch.: Holz-Zentralblatt *94*, 327 (1968)
28. Deppe, H. J.: Möglichkeiten und Grenzen der Weiterentwicklung von Holzwerkstoffen. In: Verbund von Holzwerkstoff und Kunststoff in der Möbelindustrie. Düsseldorf: VDI-Verlag 1977
29. Riehl, G., Rajkovič, E., Kubin, J., Kehr, E.: Holztechnologie *13*, 14 (1972) and *13*, 75 (1972)
30. Bakelite GmbH, HW–2504, Technical Bulletin
31. Amthor, J.: Holz als Roh- und Werkstoff *30*, 422 (1972)
31a. Ranta, L.: Holz als Roh- und Werkstoff *36*, 37 (1978)
32. Metzner, W., Bollmann, H.: Holzschutz. In: Ullmanns Enzyklopädie d. techn. Chem. Vol. 12, 4. Ed. Weinheim: Verlag Chemie 1976
33. Deppe, H. J.: Holz-Zentralblatt No. 49/50, 737 (1973)
34. Hutschnecker, K.: Holz als Roh- u. Werkstoff *33*, 357 (1975)
35. Himmelheber, M., Kull, W.: Holz als Roh- u. Werkstoff *22*, 28 (1969)
36. Hickler, H. H.: Holztechnologie *18*, 84 (1977)
37. Kull, W.: Holz als Roh- u. Werkstoff *34*, 361 (1976)
38. Gebr. Lödige Maschinenbau GmbH, Technical Bulletin; Engels, K.: Holz als Roh- u. Werkstoff *36*, 21 (1978)
39. Luthardt, H.: Holztechnologie *13*, 135 (1972)
40. Deutsche Texaco AG, DE–AS 23 64 251
41. Roffael, E., Rauch, W.: Holz-Zentralblatt *140/141*, 2233 (1973)
42. Denisov, O. B.: Holztechnologie *14*, 43 (1973)
43. Neußer, H.: Holztechnologie *18*, 88 (1977)
44. Kossatz, G.: Holzwerkstoff und Kunststoff – ihre wirtschaftliche Bedeutung für den Möbelbau. In: Verbund von Holzwerkstoff und Kunststoff in der Möbelindustrie. Düsseldorf: VDI-Verlag 1977
45. Gressel, P.: Holz als Roh- u. Werkstoff *26*, 140 (1968) and *30*, 347 (1972)
46. Meierhofer, U. A., Sell, J.: Holz als Roh- u. Werkstoff *33*, 443 (1975)
47. Bernett, J.: Wasser, Luft u. Betrieb *20*, 75 (1976)
48. Frank, M., Thiemann, A.: Kunststoffe im Bau *11*, 36 (1976)

49. Peterson, H., Reuther, W., Eisele, W., Wittmann, O.: Holz als Roh- u. Werkstoff *32,* 402 (1974)
50. Ginzel, W.: Holz als Roh- u. Werkstoff *31,* 18 (1973)
51. Roffael, E.: Holz-Zentralblatt *111,* 1403 (1975)
52. Roffael, E., Melhorn, L.: Holz-Zentralblatt *154,* 2202 (1976); L. Melhorn, E. Roffael, H. Miertzsch: Holz-Zentralblatt *20,* 345 (1978)
53. American Plywood Association, US Product Standard, PS 1–74 (1974)
54. Bakelite GmbH: Bakelite-Harze für wetterfestes Sperrholz AW 100, Technical Bulletin
55. Kreibich, R. E.: Adhesives Age, Jan. 1974, P. 26
56. Reiter, L. L.: Holz- u. Holzverarbeitung *6,* 440 (1975)
57. Neusser, H., Schall, W.: Holzforschung u. Holzverarbeitung *24,* 108 (1972)
58. Wichers, H.: Holzschichtstoffe. In: R. Vieweg, E. Becker (ed.): Kunststoff-Handbuch, Vol. 10 – Duroplaste. München: Carl Hanser 1968
59. Bakelite GmbH: Bakelite-Harze für Preßlagenholz, Technical Bulletin
60. Deppe, H-J.: Holz-Zentralblatt *104,* No. 19 (1978); Wehle, H. D.: Holztechnologie *6,* 37 (1965)
61. Kehr, E.: Holztechnologie *18,* 67 (1977)
62. Bakelite GmbH: Phenolharze für Hartfaserplatten, Technical Bulletin
63. Kollmann, F.: Adhäsion *5,* 134 (1974)
64. Noack, D., Frühwald, A.: Holz als Roh- und Werkstoff *34,* 83 (1976)
65. Rhodes, P. H.: Modern Plastics, Aug. 1947, P. 145
66. Glauert, R. A.: British Plastics, Aug. 1947, P. 233
67. Koppers Co. Inc.: Resorcinol, Technical Bulletin
68. Bakelite GmbH: Resorcinharzleime für den Holzleimbau, Technical Bulletin

10. Molding Compounds

Phenol molding powders, being the first true engineering plastics[1, 2], offer the following key properties:

- high-temperature resistance
- modulus rentention throughout a wide temperature range
- flame and arc resistance
- resistance to chemicals and detergents
- high surface hardness
- good electrical properties
- low costs

Based on these benefits, they are ideal for use within a wide range of applications in household and other appliances, electrical engineering and the automotive industry[3]. Typical examples for appliances are dishwashers, air conditioners, coffee machines, toasters, refrigerators and flat- or steam iron handles. Light sockets, switch and transformer components, blower wheels, relays, connectors, coil forms and wiring devices represent examples in the field of electrical engineering. In the automotive industry, phenolics are mainly used for under-the-hood components such

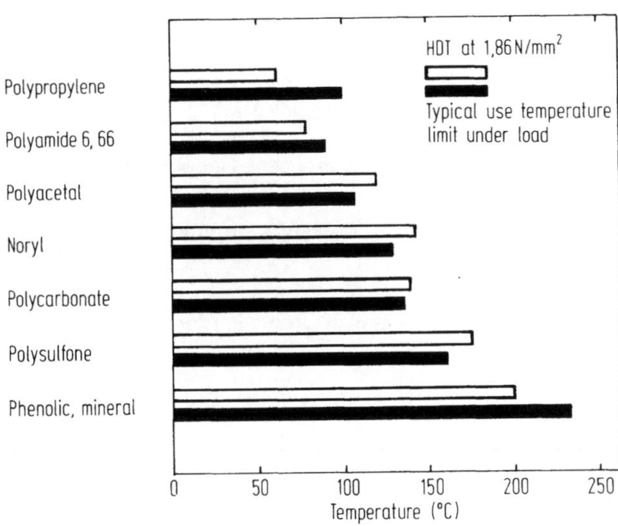

Fig. 10.1. Heat deflection and use temperature of phenolics and other engineering plastics and polypropylene

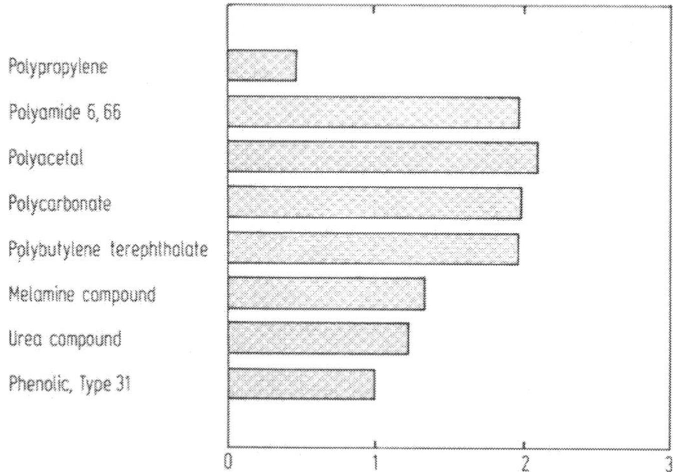

Fig. 10.2. Relative volume price of different engineering plastics and polypropylene, phenol molding powder set equal to one. Situation Jan. 1978, USA

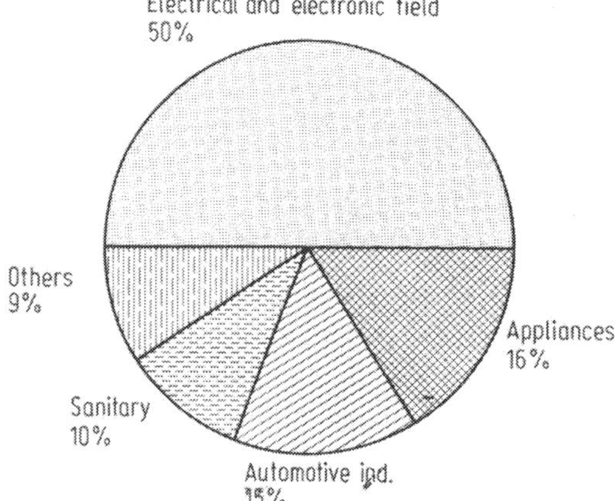

Fig. 10.3. Breakdown of end uses of phenol molding powders[5, 6)]

as distributor caps, coil towers, commutators, fuse blocks, bulkheads, connectors, and brake components.

The molding powder production of some leading countries is summarized in Table 10.1:

Table 10.1. Production of phenol molding powders. x = not available at present time

Production 1,000 to	1975	1976	1977	Ref.
USA	102	155	158	6
West-Germany	37	46	41	7
Great Britain	22.5	23	24	8, 9
France	25	25	26	10
Japan	45	63	x	11

Fig. 10.4. Application of phenol molding powders. (Photo: Bakelite GmbH, D-5860 Iserlohn-Letmathe)

Fig. 10.5. Pot handles made of high-temperature resistant, carbon-filled phenol molding powders. (Photo: Bakelite GmbH, D-5860 Iserlohn-Letmathe)

Tailoring of the material for the injection process, better feeding properties, automation of the injection equipment and warm runner tools will open even more fields of application[12]. New modified phenolics with considerably improved temperature resistance serving at temperatures of approximately 280 °C have been introduced[13] for technical moldings as well as for the manufacturing of pot handles (Fig. 10.5). To avoid hazards at handling asbestos-filled compounds, asbestos-free compounds have been successfully developed. Glass fibers are used mostly in these formulations; however, they lead to increased wear. Combinations of inorganic fillers with low quantities of organic fibers represent a valuable compromise. Dust-free, pelletized general-purpose phenol compounds with improved feeding properties and cleanliness have been offered recently; however, they have as yet met with limited acceptance.

Using carbon and graphite as fillers, compounds with exceptionally low thermal expansion have been developed, which correspond in expansitivity to steel and other metals (Table 10.12). The successful use of phenolics usually derives from a combination of cost savings and improvements in performance or reduced weight. Sheet metal construction often involves an assembly of several parts, spot welding or mechanical fastening together, followed by painting. Injection molding, on the other hand, enables a much more economic "one shot" production process. Weight savings and amount of work required for finishing are the main criteria, which must be considered when comparing molded parts with die casting. Therefore, an increased use of phenol molding compounds is expected in the automotive industry to replace a series of metal parts. Steel disk brake pistons were replaced by phenolic ones in the USA (Chrysler). The pistons, compression molded of glass filled phenolics, offer better performance, half the weight and can be manufactured more economically. In the USA, approximately 2.5 kg of phenolics were used in 1975 models per average car. It is expected that this amount will be doubled by 1980[14].

The high necessity of security and reliability of operation which determine in ever increasing measures the material selection and construction of moldings, appliances and machines will contribute to further market development of thermosetting molding compounds in the electric and appliance industry, because of their high-temperature resistance, flame retardance and infusibility. Since injection molding has broken through as the most economic processing method, the lower impact strength, shortcomings in colorability and the lack of reprocessability of the waste material (sprue, runner and scrap) of phenolics should be mentioned as remaining inhibiting factors against further market acceptance. Two solutions for the waste problem were found recently: warm runner tools and reuse of the pulverized scrap material which is blended with virgin material.

To overcome the shortcomings in colorability, melamine-phenol compounds (MP) were developed. Using light colored fillers like cellulose fibers, relatively bright colored molded parts can be obtained. A further benefit is the improved tracking resistance rendering them especially qualified for the application in the electrical industry and for the manufacturing of household appliances. However, the demand for MP-compounds did not meet the forecasts. Substantial quantities were replaced by polycarbonate and thermosetting polyester compounds. No further growth is to be expected at the present time. The phenolic resin content of these MP-molding powders is rather low, in general, about 10%.

The main competitors of phenolics among plastic materials are polypropylene, polycarbonate, polybutylene terephthalate and thermosetting polyester molding compounds.

10.1. Standardization and Minimum Properties

Due to the multitude of common fillers and reinforcements, standards have been issued to facilitate the comparison of properties and selection of materials.

In general, the following differentiation is common: general purpose, improved impact, improved electrical and heat resistant grades.

In West-Germany, the standardization according to DIN 7708 is far more specific. Further differentiations have been made according to the type of filler, quantity of resin and colour. Minimum requirements for physical properties according to DIN 7708 are compiled in Table 10.2.

Table 10.2. Standardization of phenol molding compounds according to DIN 7708

Type DIN 7708	Fillers/Reinforcing fibers	Flexural strength	Impact strength	Impact strength, notched	Heat resistance (Martens)	Water absorption
		N/mm^2	$N\ mm/mm^2$	$N\ mm/mm^2$	°C	mg
11	Mineral fluor	50	3.5	1.3	150	45
12	Asbestos	50	3.5	2.0	150	60
13	Mica	50	3.0	2.0	150	20
15	Asbestos fibers, long	50	5.0	5.0	150	130
16	Asbestos cord	70	15.0	15.0	150	90
31	Wood flour	70	6.0	1.5	125	150
51	Cellulose fibers	60	5.0	3.5	125	300
71	Textile fibers	60	6.0	6.0	125	250
74	Textile chips	60	12.0	12.0	125	300
83	Textile fibers and wood flour	60	5.0	3.5	125	180
84	Textile chips and cellulose fibers	60	6.0	6.0	125	150
85	Cellulose fibers and wood flour	70	5.0	2.5	125	200

A further differentiation is obtained by "point" types, for example Type 31.5. The point designation is as follows:
0.5 improved electrical resistance
0.8 low content of organic acid ($< 0.18\%$)
0.9 free of ammonia compounds

Table 10.3. Standardization of melamine-phenol molding compounds according to DIN 7708

Type DIN 7708	Fillers	Flexural strength	Impact strength	Impact strength notched	Heat resistance (Martens)	Water absorption	Tracking resistance
		N/mm^2	$N\ mm/mm^2$	$N\ mm/mm^2$	°C	mg	Grade
180	Wood flour	80	6	1.5	120	180	KC 175
181	Cellulose fibers	80	7	1.5	120	150	KC 250
181.5	Cellulose fibers	80	7	1.5	120	150	KC 600
182	Wood- and mineral flour	70	4	1.2	120	120	KC 600
183	Cellulose fibers and mineral flour	70	5	1.5	120	120	KC 600

Four more figures are added to this designation in order to indicate the resin content and coloration, for example: Type 31–1449 acc. to DIN 7708. The requirements for MP-molding compounds according to DIN 7708 are listed in Table 10.3.

10.2. Composition of Molding Powders

Phenol molding powders are heterogeneous compounds. They consist of phenol resin, HMTA and occasionally catalysts, fillers and reinforcing fibers, colorants and pigments as well as various lubricants.

A general purpose, wood flour filled molding powder of Type 31 is composed of the following:

Table 10.4. Typical composition of a general purpose molding compound (Type 31), compression grade

40%	Phenol novolak resin
6%	Hexamethylenetetramine
1%	Magnesium oxide
50%	Wood flour
1%	Lubricants and separating agents
2%	Colorants and pigments

Molding compounds for compression molding have a lower resin content in comparison to injection compounds with a resin level up to 50%. In addition, it is customary that each material is available in at least three qualities regarding the flow: low, medium and high flow.

10.2.1. Resins

Phenol is the dominating resin raw material; technical cresol grades are seldom used. They delay the curing rate more or less according to their content of *o*- and *p*-cresol. The flexibility of the molded parts can be improved by addition of cresols to a certain extent. Novolaks for molding compounds (Table 10.5) are produced mainly by use of oxalic acid as catalyst; hydrochloric acid and phosphoric acid are seldom used.

The P/F molar ratio is within the range of 1 : (0.75–0.85). The MWD of an appropriate resin is shown in Figure 3.5., Chapter 3. Novolak resins with a low free phenol content ($\leqslant 2\%$) and relatively high flow are used in newer formulations.

Resols are only used for special applications in electrical fields where high hydrolytic resistance is required. Ammonia can be split-off from HMTA-cured novolaks under influence of moisture and temperature resulting in corrosion of the current conducting copper and brass parts. The thermal shock resistance of resol based moldings is also better because less gas is released at curing. Resol/novolak mixtures, containing smaller quantities of HMTA, take a middle position.

Table 10.5. Specification of a novolak resin for the production of molding compounds

Melting range	°C	82–95
Viscosity, 50% sol. in acetone	s (Ford 4 mm)	30–45
B-time with HMTA at 130 °C	min	12–16
Free phenol	%	⩽2

Epoxide resins are used to cross-link novolaks instead of HMTA if the release of gaseous compounds on curing must be avoided and strong adhesion to copper is desired. Such compounds may be advantageously used for collectors and similar electrical applications (see Section 3.4.5.). Also melamine-formaldehyde resins are used to cross-link phenolic resins for special applications.

Epoxide-phenol low-pressure molding compounds with selected and silanized inorganic fillers have been developed for the microelectronic field. They are superior to pure epoxies (cured with diaminodiphenylmethane or HET-acid anhydride) regarding dimensional stability, humidity resistance, storage stability and physiological behavior.

10.2.2. Fillers, Reinforcements and Additives

Wood Flour and Cellulose Fibers

Cellulose fillers, wood flour, nutshell flour or cellulose fibers are used in moldings to reduce the shrinkage during cure, to improve the impact strength and as flow control. Wood flour is by far the most used general purpose filler giving reasonably good all-round performance at relatively low costs. Soft wood species like pine, spruce and fir are preferred; although hardwood flour may be used either solely or in combination. A somewhat lower water absorption is obtained with hardwood flour.

The properties of wood flour, prepared by wet grinding in stone mills or hammer mills, are indicated in Table 10.6.

The retention of the wood fiber structure during the grinding operation is decisive for the mechanical properties, especially the impact strength. Smaller granulation yields lower strength; the impact strength is reduced significantly. The reduction in stiffness, however, is not significant.

Table 10.6. Properties of wood flour for molding compounds

Fiber length	20–140 μm
Length/diameter ratio	~2.5 : 1
Apparent density	0.20–0.40 g/cm³
Ash	< 0.4%
pH	4.5 ± 0.5

Table 10.7. Properties of cellulose fibers for molding compounds

α-Cellulose	90% min.
Lignin	0.1% max.
Calcium	0.03% max.
Ash	0.15% max.
pH	5.5

The chemical constitution of wood is briefly described in Chapter 9. The relatively low water absorption of molded parts made of wood flour phenolics indicates that chemical reactions as well as coating and impregnation may occur between the phenol resin, lignin and cellulose.

The substitution of wood flour by nutshell flour results in distinct improvements, mainly flow is enhanced. Flour made of walnut or coconut shells, of stones of apricots or olives, contains a considerable amount of lignin, resins, oils and waxes as well as cellulose (~60%) and pentosans (~8%). These oils and waxes act like internal and external lubricants and therefore improve fluidity and surface quality. At the same time, water absorption is reduced, i.e. the compounds behave as if they had a higher resin content. However, these fillers are not of fibrous structure and so strength is reduced to some extent. Normally, they are used in quantities up to approximately 10%.

Cellulose fibers (Table 10.7) manufactured from carefully debarked softwood logs in a manner used in the paper making industry, are used mainly for light colored melamine-phenol compounds or for general purpose compounds to improve impact strength.

Asbestos

Asbestos is one of the most important fillers for phenolic resins and is used for several applications, e. g. molding compounds, acid and base resistant materials, brake linings and ablative materials, and should be therefore treated in greater detail. Asbestos is a generic term for fibrous silicates mainly occurring in Canada, Rhodesia, China, USSR and Italy. Asbestos fibers possess a high tensile strength with acceptable flexibility and offer exceptional temperature and chemical resistance due to their mineral structure[15, 16]. For their use in molding compounds, the length of the fibers is of the utmost importance. In the most accepted Canadian classification according to the Quebec standard grading test, seven groups, 1–7, with subgradings D, F, K, M, R, T, Z are differentiated. No. 1 fibers are the

Table 10.8. Physical properties of some asbestos fibers in comparison to E-glass fibers[15, 18]

	Chrysotile	Crocidolite	Amosite	Anthophyllite	E-Glass
Chemical composition	$Mg_3[Si_2O_5](OH)_4$	$Na_2Fe_5[Si_8O_{22}](OH)_2$	$(Mg, Fe)_7[Si_8O_{22}](OH)_2$	$(Mg, Fe)[Si_8O_{22}](OH)_2$	
Mohs hardness	2.5–4	4	5.5–6.0	5.5–6.0	6–7
Tensile strength N/mm^2	500–1,000	700–1,000	700–1,000	700–1,000	2,200
Modulus of elasticity N/mm^2	$16 \cdot 10^4$	$19 \cdot 10^4$	$16–19 \cdot 10^4$	$16–19 \cdot 10^4$	$7.3 \cdot 10^4$
Melting point °C	1,520	1,190	1,400	1,470	1,320
Density g/cm^3	2.4–2.6	3.2–3.3	3.1–3.2	2.8–3.1	2.54
pH (in CO_2 free water)	10.3	9.1	9.1	9.4	
Fiber diameter μm	>0.02	~0.5	~0.5	~0.5	5–15
Surface area m^2/g	55	15	10	6	–

longest (crude hand-picked), 4–7 are shorter milled fibers, whereby 7 designates shorts and floats (fine powder). The physical properties of the different types of asbestos are indicated in Table 10.8.

The stiffness of asbestos fibers is superior to that of E-glass fibers, so the Young's modulus of asbestos reinforced plastics can be higher than that of the corresponding material reinforced by glass fibers. Chrysotile, superior to all other types for industrial purposes, accounts for 95% of the world consumption.

The key property for the overall performance is the morphology of the single fibril of chrysotile. This, in turn, depends upon the chemical structure. Chemically chrysotile, sometimes called white asbestos, represents the family of the sheet forming silicates of the serpentine group. Because of the steric hindrance caused by the space-consuming oxygen atoms and hydroxyl groups, these sheets are curved. Therefore, the layers will warp into complete cylinders, and relatively thickwalled tubes with hollow cores are formed.

The overall loss of water by chrysotile is about 14%. The DTA curves show a small endothermic peak at approximately 400 °C. The main endotherm peak appears at 650 °C. The surface of the chrysotile fibriles consists mainly of magnesium hydroxide (brucite layer) and therefore, it is not very resistant to strong acids. Generally, the following decreasing acid resistance is found: anthophyllite > crocidolite > amosite > chrysotile. Aqueous alkalies have little effect on all types of asbestos up to 100 °C.

The hollow fiber structure which offers many benefits regarding mechanical strength, is, on the other hand, a health risk. The risk from fibrous materials is limited to inhalation of the fibers. Recent research on asbestosis has suggested that the fibrosis is caused principally by fibers between about 5 μm and 100 μm long. Asbestos fibers with exceptionally small diameters cause lung cancer (< 5 μm) and mesothelioma (< 2 μm), a relatively rare form of cancer of the outer surface of the lung or intestinal lining, which is fatal[19]. The most convincing evidence of the structural relationship to carcinogeneity is that fine, long fibers of durable materials completely unrelated to asbestos in chemical structure, i. e. glass or aluminum oxide, are similarly carcinogenic in the pleura of the rat. Current medical evidence suggests that the risk is greatest with crocidolite, and apparently less with chrysotile. Legal requirements are imposed upon industries involved in the processing and manufacture of asbestos products. The present acceptable level of air borne asbestos fibers is set in the USA and U. K. for chrysotile at 2 fibers/ml, although 0.5 fibers/ml has been suggested as a new standard (OSHA).

As indicated, the risk is limited to the inhalation of single fibers. Considerably different criteria exist for handling resin-coated and bonded asbestos compounds. Granted a good exhaust system and careful handling of the compounds, their use by no means unfavorably affects the personnel. Asbestos-free compounds having properties very similar to those containing asbestos have been developed. However, they cannot be applied in fields where extreme temperature resistance is required.

Mineral Flour

Generally, mineral flour fillers in thermosetting plastics are used to effect the following improvements:

- reduced shrinkage on cure and reduced exothermic heat during cure
- higher compression strength and stiffness
- higher thermal resistance
- improved flame retardancy
- improved electrical properties
- adjusted flow
- machinability and surface quality
- lower costs

The physical properties of some important mineral fillers are shown in Table 10.9.

Table 10.9. Properties of mineral fillers (* = in CO_2-free water)

	Chemical composition	Density g/cm^3	Mohs-hardness	pH*
Calcium carbonate	$CaCO_3$	2.71	3.0	9–10
China clay	$Al_4(OH)_8[Si_4O_8]$	2.69	2.0–2.5	5–6
Mica (Muscovite)	$KAl_2(OH, F)_2$ $[Al\,Si_3O_{10}]$	2.7–2.9	2–3	7–8
Silica flour	SiO_2	2.65–2.7	7.0	7–8
Talcum	$Mg_3(OH)_2$ $[Si_4O_{10}]$	2.65–2.8	1–2	9–10
Wollastonite	$CaSiO_3$	2.85	5.0–5.5	9–10

The abrasive action of inorganic fillers may cause serious problems. Not only the hardness of the mineral has to be carefully observed, but also the accompanying impurities. The addition of only 1% of an abrasive filler, e.g. quartz, can increase the tool wear tenfold. Furthermore, the different basicity of the fillers may influence the curing rate. Magnesium oxide for instance, is frequently added to accelerate the curing reaction.

Mica, a sheet forming mineral, should be especially mentioned. Muscovite, a calcium-aluminum silicate, is the almost exclusively used mineral species. The mica sheets are flexible and possess outstanding dielectric properties as well as high thermal resistance. Mica filled compounds are used in the electrical engineering field for collectors etc. Besides dielectric strength and thermal resistance, these compounds have low thermal conductivity, low water absorption and very good resistance to chemicals since diffusion processes are delayed considerably due to the sheet form of the filler.

Other Fillers and Fibers

Cotton fibers or chopped cloth are used as reinforcements for materials for the production of large, flat parts with high impact and tensile strength. It is difficult to impregnate these materials uniformly with dry resin and therefore such molding materials are produced by the wet impregnation method using alcoholic novolak solutions or aqueous resols.

In addition, glass fibers which are used to a great extent, a series of organic fibers is used to increase the impact strength. Organic fibers reduce the thermal resistance considerably; therefore, they are mostly added in minor quantities. Polyester-, polyamide- or polyvinyl alcohol fibers have been used with success. Carbon fibers and aromatic polyamide fibers (Kevlar, Arenka) have also been recommended as reinforcing fibers. Some other organic fibers are destroyed or evenly dissolved by phenol at higher temperatures.

The sliding properties of moldings are considerably improved by the addition of graphite or molybdenum sulfide. Such compounds are used to manufacture sliding rings, slide bearings and gaskets. Formerly, graphite was used as an additive to compounds for metal plated parts. Today, however, better plating methods and compounds, free of conductive fillers, are available.

Colorants

The limited colorability of phenolic moldings is caused by the yellow coloration of the cured resin.

Aside from the most important criteria like heat resistance, light fastness, weather resistance, migration resistance, physiological harmlessness, and dispersiveness, which generally must be observed in the choice of the colorant for plastics, the costs of colorants are of the utmost importance. The coloring of the compound is performed in the first phase of the production. Later coloration of molding compounds with a weigh feeder in the injection molding unit have not yet succeeded. Also coating of naturally colored parts like flat-or steam iron handles, side parts of toasters, etc., has not proved practicable since the surface layer will not be scratch resistant and the adhesion not sufficient. Also the electrostatic powder coating is not adequate.

Lubricants and Release Agents

In most cases, a combination of several lubricants is necessary to obtain optimal moldings. Formulations contain up to 1% of lubricants. The friction process is influenced by external and internal lubricants. External lubricants are used to reduce the adhesion to metals. They improve the feeding property of the plasticized material (lower frictional heat development) and act as mold release agents. Internal lubricants affect the melt flow (lower viscosity and injection pressure) and improve the homogeneity of the melt. It is plausible that internal lubricants must have a more polar chemical structure and must be easily soluble in the phenolic resin melt. Examples are fatty alcohols, fatty acid esters or fatty acid amides. Fatty acid salts like calcium or magnesium stearate take a middle position. External lubricants should be non-polar in nature and therefore practically insoluble in phenol resins. This group includes paraffins and waxes.

10.3. Production of Molding Powders

Molding powders are produced by a combination of discontinuous and continuous process steps. A phenol molding compound, for instance Type 31 or Type 12, consists of 6–10 components (Table 10.4). The mixtures are produced discontinuously in appropriate mixing machines and stored in an intermediate bunker. Plasticizing and homogenizing of the mixture, as well as the adjustment of the appropriate condensation grade, can be performed on a two-roll mill or in an extruder. The rolling process can be performed discontinuously or continuously. The rolls are operated at different speeds to effect friction and at different temperatures, approximately 90 and 130 °C. A rough sheet develops on the colder front roll which is homogenized by removing and turning several times, adjusted, and removed as a whole. Otherwise, the mixture is continuously fed to the center of the rolls and displaced to both sides where narrow tapes are cut off by stationary blades. After cooling, the compound is fed to a series of crushers, grinders and sieves where it is reduced and screened to the desired fineness and flowability. The rolling mill process has certain disadvantages. These are high dust load, higher labour costs and higher energy consumption. Inspite of these disadvantages, this process is successful for formulations containing abrasive fillers or changing or small production lots. The maximum throughput for continuous, high speed roll mills is approximately 400 kg/h. The production of molding compounds in extruders, on the other hand, offers economical, dust- and odor-free processing. Efficient homogenizing, which is necessary, can only be accomplished in special extruders, for instance single screw extruders model "Buss-kneader" or in special twin screw extruders with appropriate mixing sections. The throughput of these extruders can be as much as 2,000 kg/h. The use of extruders is limited, for example for compounds containing abrasive fillers. The following finishing and treating of the compound is the same as for the mixing roll process. The production of highly reinforced (high impact) materials is not satisfactory on differential rolls because of the grinding action of the rolls and breakdown of the fibrous reinforcing material (glass fibers, cord or chopped cotton fabrics). In this case, application of sigma blade mixers and impregnation with phenol resin solution and subsequent drying gives good results. With screw type extruders compounds of medium strength can be made. By use of suitable attachments, the molding compound can be delivered in a pelletized, free flowing form.

10.4. Thermoset Flow

When heat and pressure are applied to a thermosetting molding material three processes occur: melting, flowing and gelling. The viscosity change during these phases is a complicated function which depends upon the following variables:

$$\eta = f(\vartheta, t, \tau)$$

the temperature ϑ, the time or curing rate t and shear rate τ. The theoretical treatment of the rheology of thermosets[20, 21] is considerably more difficult than that of thermoplast flow because the time is an additional and essential variable due to the curing reaction which occurs simultaneously. At first, the resin softens under the influence of temperature and the viscosity is reduced. The cross-linking reaction

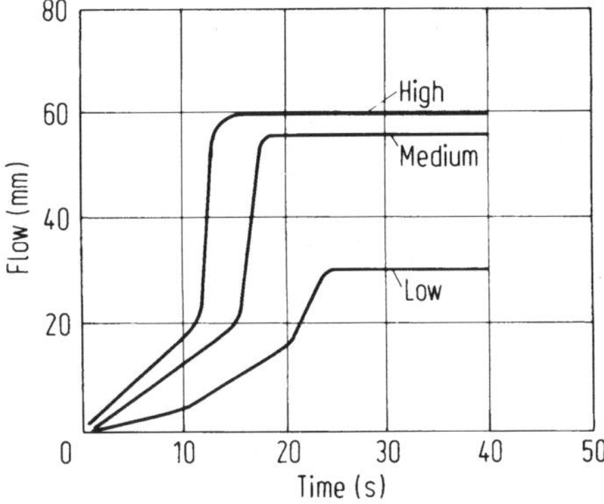

Fig. 10.6. Flow in dependence on time of phenol molding compounds according to DIN 53478[1]

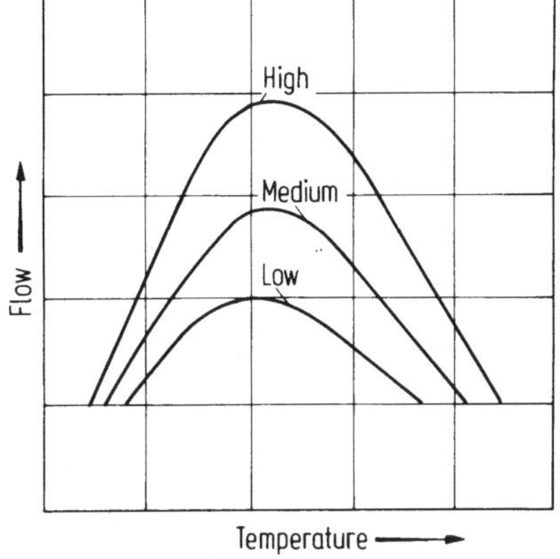

Fig. 10.7. Flow in dependence on temperature of phenol molding compounds according to DIN 53478[22]

causes a viscosity increase at the same time. The superposition of these two exponential functions yields an unsymmetrical parabolic curve. The viscosity also depends upon the shear rate, i. e. thermosetting molding compounds show a non-Newtonian flow. Constant flow parameters are an essential criterion for a trouble-free and economical large scale production. The flow depends upon the resin content and reactivity, the kind and amount of fillers, the lubricants, the free phenol content and especially on the content of humidity, which means also upon the storage conditions and conditioning of the compound. Certain fluctuations from batch to batch are therefore possible. A multitude of processes have been introduced in order to predict the processability of a molding compound[1]. A number of methods describe the processing behavior by means of a molded part, e. g. a plate, cup, stick or spiral. A weighed quantity is molded to a plate in a press and the diameter measured (plate method) or a standard cup is molded and the closing time measured (DIN 53 465). According to ASTM-D-731, the minimum pressure at which the cup mold is just filled is determined. Therefore, several moldings are necessary. In the bar flow test a pellet is pressed into a flow channel from a heated chamber at a definite pressure. The length of the formed stick is the measure for the flow duration. Also here, there are two methods – according to Rossi-Peakes (ASTM-D-569) and DIN 53 478 – which differ from each other by the dimensions of the flow mold.

The EMMI-spiral flow test developed by the Epoxy Molding Materials Institute is similar in principle, but is only suitable for low-pressure molding compounds[21, 22], for instance epoxy-phenolic molding compounds. The flow distance is here enlarged to 262 cm by an Archimedes winding.

All processes described have two disadvantages. The information concerning the total molding process is limited. Furthermore, the results are very dependent upon the testing devices, for instance the roughness of the mold. The results of several testing institutions are only comparable to a certain degree. However, the information is satisfactory as long as the materials are processed by compression molding. For

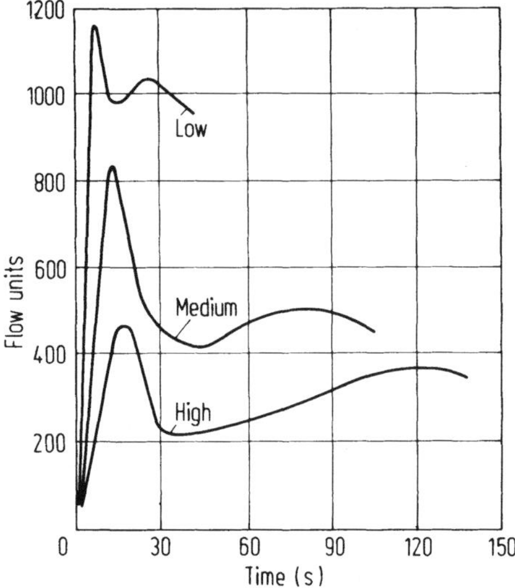

Fig. 10.8. Moment of torsion/time function of several phenol molding compounds determined with the Brabender plastograph[1]

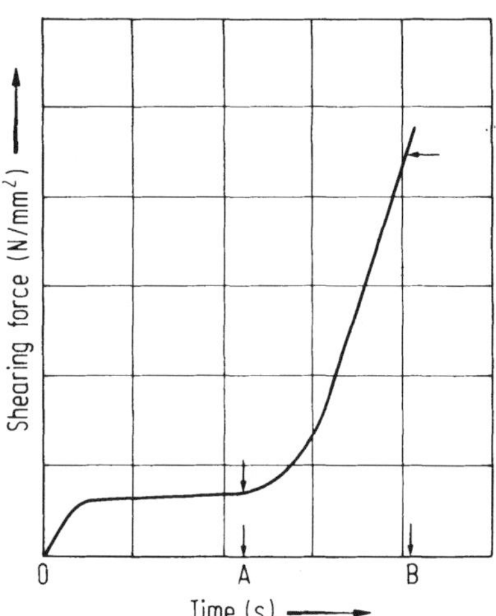

Fig. 10.9. Shearing force/time function of a phenol molding compound determined with a rotational viscosimeter according to Kanavec[25]

injection molding, it is necessary to know the melt viscosity and duration of the melt phase. They can be evaluated in empirical terms with a torque rheometer[24] using sigma-type measuring heads (Brabender plastograph). Torsional moment (viscosity)/time recorder traces of different flow adjustments (soft, medium, hard) are shown in Figure 10.8.

The final status is not determined. The material becomes crumbly very early during the curing process. A further rotation viscosimetric method has been especially successful in East Europe (Kanavec plastograph). A typical shearing force/time function is shown in Figure 10.9. An empirical value B of the shearing force (viscosity) corresponds to the minimum curing time required[25]. The melt viscosity and the duration of the plasticated phase are very accurately determined by both methods.

10.5. Manufacturing of Molded Parts

The most important components of the processing economy are the degree of automation and the cycle time[26]. The molding time is composed of the heating time to mold temperature and the time required for the chemical reaction. The latter is at 160 °C within 5 to 10 seconds. Since the thermal conductivity of molding compounds is relatively low, the heating time is the dominating part. It is therefore reasonable to reduce the molding cycle by preheating of the compound outside of the mold up to a temperature a little below that of the mold. Experience has shown that the economy of a processing technique influences the market development of plastic materials decisively[27–30].

10.5.1. Compression Molding

The preweighed molding compound is fed into the open heated mold, and by closing the mold and simultaneous application of heat and pressure, it is molded and cured. Before, the pelletized compound is preheated by high-frequency radiation to a maximum of 110–120 °C within 60 seconds. The uniform content of humidity of the molding compound is important, since it influences the pelletizing ability and the dielectric loss factor. A further advantage of preheating is the low molding pressure which results in less wear of the mold.

The compression molding process yields stronger moldings than the transfer or injection molding process. Since the required flow of the material is relatively low, there is no fiber orientation or fiber damage. The main disadvantage is the low automation of the conventional compression molding process. The material is difficult to feed, must be premeasured or preweighed and placed manually into the mold. The dust contamination at the place of work is high. The economy of the process can be improved significantly by use of a screw preplasticizing unit[28]. Pelletizing, handling and preheating of the pellets, and manual feeding of the press would be omitted, thereby minimizing dust contamination. A somewhat higher preheating temperature is possible with screw plastification in comparison to HF-preheating.

Newly developed control systems promise to advance compression molding technology[31].

Thin-walled large parts, for example boxes or toilet seats, are manufactured more economically in multicavity molds by compression molding. In addition, the fiber orientation and fiber damage which occur during the injection molding process may be unacceptable. Therefore, the compression molding process will not be completely superseded by injection molding in the future.

10.5.2. Transfer Molding

The portioned, preheated molding compound is pressed (~100 N/mm^2 pressure) from a heated transfer chamber by a plunger through a gate into the closed mold. The compound is preheated in the transfer chamber and, in addition, frictional heat is transferred during the flow through the narrow gate. Metal inserts or pins are not so easily pushed out or damaged due to the low melt viscosity. This process is especially effective for the manufacture of relatively thick-walled parts with different wall thickness. Since the material is subjected to a high pressure drop at the jet, an effective gas release occurs so that less shrinkage will be found in general. Since the material is fed into the closed mold, the metering accuracy is higher, flash is considerably reduced. Some fiber orientation may occur, however, not as much as with injection molding[1, 32].

10.5.3. Injection Molding

The first thermoset injection molding machine was presented in Japan in 1955 at the Second Tokyo International Trade Fair [33]. The real processing by injection molding of thermosets started in the mid-sixties when appropriate molding com-

Fig. 10.10. Injection molding machine for thermosets with process control equipment. (Photo: Bucher und Guyer, Niederwenningen, CH)

pounds were developed[27]. Basic research work on injection compounds had been performed by Bakelite GmbH, West Germany.

The operating sequence of a thermoset injection machine is similar to thermoplast injection molding. Loose material is fed from a hopper, heated and plasticized by the shearing action of the screw. Heat is mainly generated by friction and the temperature is adjusted by hot water. As the preheated material builds up in the nozzle section, the screw is forced back into the barrel. Plastificated and homogenized material is then forced under high pressure into the hot mold by the forward movement of the screw. Further heat is generated by its flow through the narrow nozzle at high speed.

The maximum shot capacity is presently in the range of 2 kg. Another molding technique currently gaining in favour is injection/compression molding. A full shot of material is injected into the mold when the mold is not completely closed. The mold is then closed at full clamping pressure forcing the material to fill the mold cavity. This technique combines the advantages of the compression molding and

Table 10.10. Typical operating conditions for injection molding of phenol molding powders (Type 31)

Barrel temperature	
Middle	65–85 °C
Front	85–110 °C
Nozzle	110–130 °C
Mold temperature	165–200 °C
Screw speed	35–100 rev/min.
Injection pressure	1,200–2,500 N/mm^2
Molding time	15–60 s

the injection molding processes. The orientation of fillers and the strength anisotropy are reduced, but flash is increased. The open mold permits effective escape of gases.

A typical injection compound differs from one for compression molding by a higher resin content (up to 50%) and higher internal lubricant content and therefore by a longer flow. Ideal curing behavior would include a very low reactivity at temperatures up to about 110 °C, i. e. long "barrel" life, and a fast curing rate at temperatures above 140 °C. The use of latent catalysts is mentioned in the patent literature. The molding time may further be reduced by use of higher advanced that means "harder" compounds, which are also predried to a higher extent. The heat input required for the injection process is performed through barrel heating, by transformation of frictional energy into heat, by adiabatic compression and partly by the reaction enthalpy of the curing reaction. Doubtless, the largest part is attributed to temperature increase due to friction. The amount of friction can be increased by higher melt viscosity, by increasing the screw speed (rev/min) and back pressure and by increasing material flow speed, e. g. by injection pressure and/or narrowing the nozzle cross-section. The great advantage of transformation of mechanical into thermal energy is the extraordinary fast and uniform temperature rise. Most effective is the increase of the flow speeed by a narrow nozzle supported by high injection pressure. By reducing the nozzle cross section from 28 mm^2 to 3 mm^2, the molding time of a Type 31 compound is reduced by 45%[35]. Furthermore, the surface quality is considerably improved at the same time.

The material loss with conventional injection molding amounts to 15–20% on an average[36], due to sprues and runners which are not reused. When multicavity tools are used for small moldings, for example in the electrical engineering and electronics fields, the loss may exceed 50%. The material costs represent the major part of the production costs of molded parts. The amount of scrap is reduced considerably by application of warm-runner molding which was developed[37] analogically to hot channel tools for thermoplastics. Sprue and runner scrap is reduced or eliminated in multicavity molding by assuring that the material in the sprue and runner does not cure and can be injected into the mold cavities on the next shot. A third, heat insulated partition is inserted in the tool where the temperature is maintained between 80° and 110 °C by liquid tempering. In this section the material is fed and distributed. Constructional alterations which are not so expensive are the use of a long nozzle with temperature control or of a hot sprue[36]. A closed loop recovery system for reuse of thermoset scrap has been developed in Japan by Meiki & Co, Nagoya. Flash, runners and other discards are pulverized in special mills to 50–100 μm grain size, mixed uniformly with the virgin raw material in a predetermined ratio and then conveyed to the hopper of an injection molding machine[38]. It was shown, that up to 15% powdered phenolic filler can be incorporated without deterioration of part quality.

Further improvements in automation, output, energy saving and part quality can be achieved by process monitoring or process control in closed loops[39]. There is no quantifiable property for the description of part quality. However, it has been observed that more than 90% of all defects in injection molding can be traced to improper mold filling. Overfilling results either in excess flashing or in warping, underfilling yields poor surface quality. It is therefore obvious to use the part weight as criterion of quality. The variations in weight are the result of changes in melt viscosity and material flow back in the barrel. Up to now it has not been possible to develop a back-flow barrier which works faultlessly. The back pressure can also influence the temperature and homogeneity of the melt and the extent of back-flow.

A new development in process control[35] is based on cavity pressure and screw speed control. The control system consists of a quartz crystal load cell[40] in the cavity, i. e. behind one of the ejector pins. Deviations of the desired value are compensated for by adjusting the length of the stroke or the hydraulic pressure.

A further approach uses the small opening of the tool during the injection procedure[41]. The opening distance depending on time (and shot size) can be determined very exactly, and the plasticating volume and charge stroke may be adjusted accordingly on the next shot (System "Volumatic", Bucher und Guyer). The hydraulic system is not affected.

Alternatively, key parameter variations can also be overcome by simpler instrumentation of process monitoring in open loops and signalled change. The examination and the adjustments must then be made by the operator. Monitors now available cover single and multiple parameter measurement in conjunction with visual display or recording systems. Single point monitors observe cycle time or injection rate, or in case of multiparameter systems, injection rate, injection pressure and holding pressure[42].

10.6. Selected Properties

Phenolic moldings offer remarkable advantages compared to other engineering plastics in fields where higher temperature stress under load is to be expected. Data on the resistance to high energy radiation are mentioned in Section 7.3. Figures for properties and strength under physical and environmental stresses are available in the literature[43–47]. Minimum requirements according to DIN 7708, measured on standard specimens, are included in Table 10.2; the effective attainable properties exceed these limits more or less. The overall chemical resistance is considerably higher in comparison to thermoplastic materials. Indications are given in Chapter 19.2.

10.6.1. Thermal Resistance

The following Figures 10.11–13 demonstrate the influence of the filler and material pretreatment on strength and modulus at elevated temperatures for the high temperature resistant PF-compounds which have been developed recently.

10.6.2. Shrinkage and Post-Mold Shrinkage

For producing parts precisely to size, a number of considerations concerning elements governing thermosetting part shrinkage must be made before the designation of tool dimensions. Otherwise, a considerable amount of reworking or trial and error mold designing will be necessary. The shrinkage of a molded part is divided into

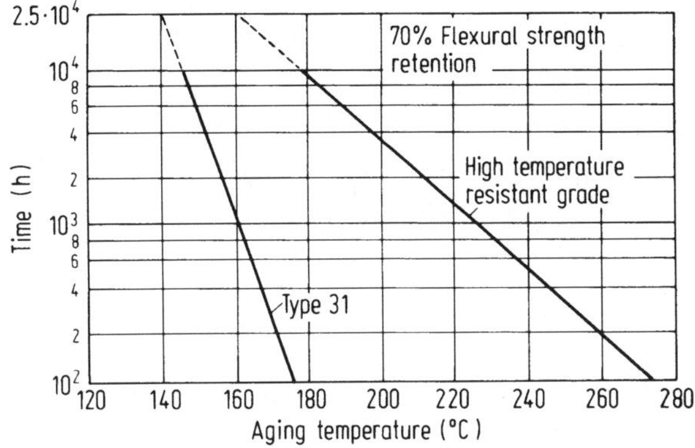

Fig. 10.11. Temperature/time limit functions according to DIN 53446, VDE 0304 for general purpose and high-temperature resistant grade phenolics; 70% flexural strength retention

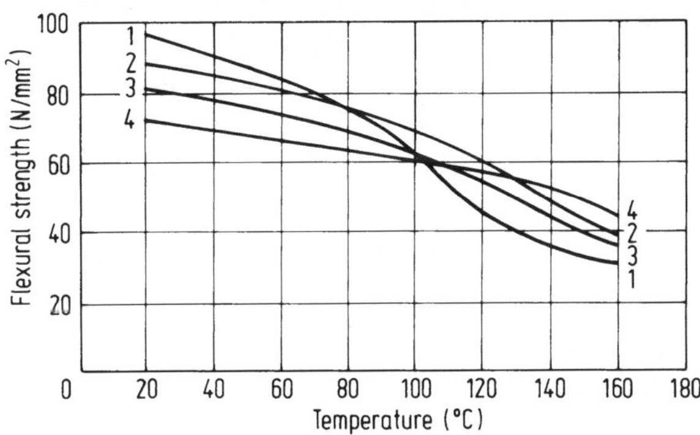

Fig. 10.12. Flexural strength in dependence upon the temperature, type of filler and pretreatment of the molding material[48].
1 Type 31, wood flour; *2* as *1*, preheated for 10 min. at 110 °C; *3* Type 12, asbestos; *4* HT-resistant type, carbon filled

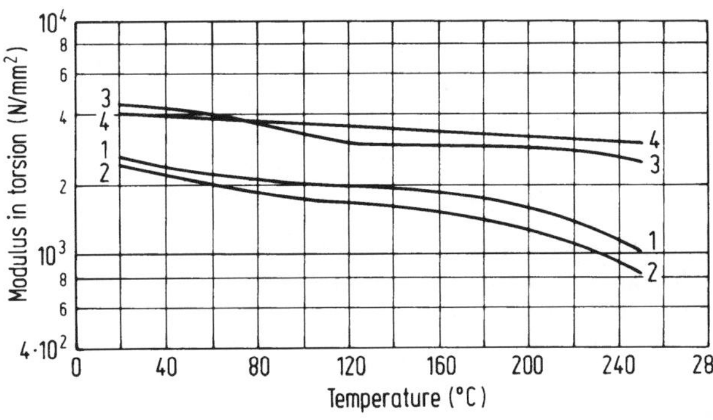

Fig. 10.13. Modulus in torsion in dependence upon the temperature, type of filler and pretreatment fo the molding material[48]. The designation of the curves is indicated in Fig. 10.12

two components. Shrinkage means the difference of the dimensions of the cold tool and the cold molded part (after 24 hours). The cold part, however, is subject to further dimensional changes, if it is, for instance, treated with heat. The following factors influence the shrinkage:

- different coefficients of linear thermal expansion of the molded part and the tool
- reduction in volume due to chemical reaction, cross-linkage and splitting-off of volatile components,

– elastic deformation of the molding after removal due to energy storage,
– processing conditions, for instance cavity pressure

Postmold shrinkage (Table 10.11) is due to the diffusion of volatile components through the surface layer, to postcuring reaction or reorientation of fillers.

The total shrinkage depends to a high degree on the volatiles content of the molding material. The volatiles content of wood flour phenol molding compounds is usually between 3 and 4%.

10.6.3. Thermal Expansion

Due to lower production costs compared to those of shaped metal parts, thermosetting plastic parts get more and more attention in the machine and automotive industry. Since they are mostly used in connection with metal parts, comparable thermal expansion is an essential prerequisite. The expansitivity of polymers is not exactly independent of temperature: however, the gradual increase with rising temperatures is so small that in most cases it is negligible. The volume-temperature function shows

Table 10.11. Typical values for the shrinkage and postshrinkage of phenol molding compounds[1)]

Type DIN 7735	Filler	Shrinkage %	Postshrinkage %
11	Mineral flour	0.3–0.5	0.2–0.3
13	Mica	<0.2	0.05–0.15
16	Asbestos	<0.2	0.1–0.2
31	Wood flour	0.4–0.8	0.15–0.3
51	Cellulose	0.2–0.5	0.2–0.3
71	Textile	0.2–0.5	0.2–0.3

Table 10.12. Linear expansion coefficients between 20–120 °C of some phenol molding compounds[48, 49)] compared to some metals (^{x} preheated for 10 minutes at 110 °C)

Type DIN 7735	Filler	Resin content %	Linear expansion coefficient 10^{-6} K $^{-1}$
31	Wood flour	40–45	35–45
31^x	Wood flour	40–45	30–35
71	Textile fibers	40–50	25–30
12	Asbestos	35–40	20–26
	Carbon	35–38	18–22
	Graphite	30–35	10–13
Phenolic resin, non-filled			60–70
Steel			11–12
Copper			16–17
Aluminium			20–23

a sudden rise at the transition point; the expansitivity is higher above the glass temperature due to the now possible structural motion and rearrangement.

References

1. Schönthaler, W.: Verarbeiten härtbarer Kunststoffe, Düsseldorf: VDI-Verlag 1973
2. Martino, C. F.: SPE Journal, March 1967, P. 96
3. General Electric Co: Genal Phenolic Molding Compounds, Technical Bulletin
4. Houston, A. M.: Materials Engng. *6*, 1975
5. Knop, A., Müller, R., Schönthaler, W.: Kunststoffe *66*, 633 (1976)
6. N. N.: U. S. Plastics Sales, 1977 Modern Plastics Internat., Jan. 1978, 43
7. N. N.: Chem. Ind., August 1977, 437
8. N. N.: The UK Plastics Industry in 1977, Europ. Plastics News, Jan. 1978, 11
9. Rowland, R.: Kunststoffe *67*, 535 (1977)
10. N. N.: Kunststoffe *67*, 542 (1977) according to: Syndicat des Fabricants de Matières Plastiques
11. Japan Chem. Week: 14. 3. 1977, P. 3
12. Bovensmann, W.: Plastifizieren und Homogenisieren von Duromeren mit Schnecken-Plastifizier-Geräten. In: Wirtschaftl. Herst. v. Duromerformteilen. Düsseldorf: VDI-Verlag 1978
13. Schönthaler, W.: Neue Entwicklungen bei Phenolharz-Formmassen. In: Wirtschaftl. Herst. v. Duromerformteilen. Düsseldorf: VDI-Verlag 1978
14. N. N.: Modern Plastics Internat., July 1975, P. 46
15. Wake, W.: Fillers for Plastics. London: Iliffe Books 1971
16. Noll, W.: Asbest. In: Ullmanns Encyclopädie d. techn. Chem., Vol. 8, 4 Ed. Weinheim: Verlag Chemie 1976
17. Yada, K.: Acta Cristallogr. *27*, 659 (1971)
18. Berger, H.: Verarbeitung v. Asbest mit Kunststoffen u. Kautschuk. Stuttgart: Gentner (1961)
19. Fulmer Special Report Nr. 5: Asbestos: Characteristics, Applications and Alternatives. Fulmer Res. Inst. 1976
20. Bassow, N. I., et al: Plaste u. Kautschuk *19*, 507 (1972)
21. Pall, G.: Plaste u. Kautschuk *18*, 665 (1971)
22. Hoechst AG: Härtbare Formmassen Hostaset-Verarbeitung, Technical Bulletin
23. Heinle, P. J., Rodgers, M. A.: SPE-Journal, *25*, *56* (1969)
24. Rothenpieler, A., Hess, R.: Kunststoffe *62*, 215 (1972)
25. Ehrentraut, P.: Kunststoffe *56*, 10 (1966)
26. Rhyner, H. G., Bläuer, B., Leukens, U.: Prakt. Erfahrungen in d. Duromerverarbeitung. In: Wirtschaftl. Herst. v. Duromerformteilen. Düsseldorf: VDI-Verlag 1978
27. Müller, K.: Plaste u. Kautschuk *22*, 225 (1975)
28. Bauer, W.: Kunststoffe *67*, 628 (1977)
29. Bauer, W., Woebecken, W.: Verarbeitung duroplast. Formmassen. München: Carl Hanser 1973
30. Fritzen, A., Bruncken, K., Eisfeld, D.: Pressmassen. In: Kunststoff-Handbuch, Bd. 10–Duroplaste. München: Carl Hanser 1968
31. Todd, W. H.: Modern Plastics Internat., August 1976, P. 46
32. Schönthaler, W., Niemann, K.: Maschinenmarkt *82*, 1638 (1976)
33. Matsuda, S., Tsujita, T.: Japan Plastics, Oct. 1967, P. 36
34. General Electric Co.: Injection Molding Manual, Technical Bulletin
35. Bichler, M., Rothe, J.: Kunststoffberater 8/1976, P. 370
36. Asahi Organic Chemicals Ind.: Japan Plastics, June-July 1976, P. 6
37. N. N.: Modern Plastics Internat., June 1973, P. 21
38. Meiki & Co.: Recycling Machine for Thermosetting Resin Scrap, Technical Bulletin
39. Kistler Instrumente AG: Druckmessung in Spritzgußwerkzeugen, Technical Bulletin
40. Hartmann, E.: Kunststoffe *64*, 106 (1974)

41. N. N.: Modern Plastics Internat., June 1975
42. Sächtling, H. J.: Kunststoff Taschenbuch, 20. Ed. München: Carl Hanser 1977
43. Bachmann, A., Müller, K.: Phenoplaste. Leipzig: Verlag 1973
44. von Meysenbug, C. M.: Kunststoffkunde für Ingenieure. München: Carl Hanser 1968
45. Gilfrich, H. P., Wallhäuser, H.: Kunststoffe *62*, 519 (1972)
46. Verb. Dtsch. Ing./Verein Dtsch. Elektrotechniker: VDI/VDE Richtlinie 2478, Phenoplastformstoffe mit Holzmehl
47. Knop, A., Schönthaler, W.: unpublished
48. Bakelite GmbH: Härtbare Formmassen, Technical Bulletin

11. Heat and Sound Insulation Materials

In Europe today, approximately 45% of the total end energy is used to heat buldings. The ever increasing prices for energy and limited petroleum reserves made it imperative for the consumer and the legislator to improve structural heat insulation. Heat insulation became an economic and political necessity.

Phenolic resins are used as bonding agents for mineral-, glass- and textile fiber mats and, in minor quantities, for the manufacturing of foam for thermal and acoustic insulation.

The effectiveness of construction and insulation materials is shown in Table 11.1.:

Table 11.1. Thermal conductivity of some construction materials compared to phenolic resin foams and mineral wool mats[1, 2]

Material	Density g/cm^3	Thermal conductivity W/m K
Steel	7.84	50
Stone	2.50	2.32
Reinforced concrete	2.40	2.09
Solid brick	1.70	0.75
Glass	2.54	0.81
Particle board	0.65	0.13
Phenolic resin foam	0.40	0.039
Styrene foam	0.25	0.030
Mineral fiber mats	0.15–1.20	0.031–0.039
Air	0.00129	0.0064

Of the three mechanisms of heat transfer, which are heat conduction, convection and radiation, convectional heat transfer is reduced by fiber and foam insulation materials[2, 3]. Air circulation is prevented by dividing the air space into small sections reducing the temperature gradients drastically (Grasshoff-value), i.e. a heat conductivity close to that of resting air is approached.

Fibrous or porous absorbents are used in civil engineering to control sound and to protect from noise and sound, which is not only disturbing, but also detrimental to human health (above 90 dBA). One differentiates between air sound and body sound spreading in solid materials. Protection against air sound is obtained by insulation (transformation of acoustic energy into heat by friction) and/or reflection. The intensity of reflection depends upon the mass and stiffness of the material. Soft construction elements such as mineral fiber mats or foams damp the sound mainly by their insulating properties (ca. 90%) and only partly by reflection.

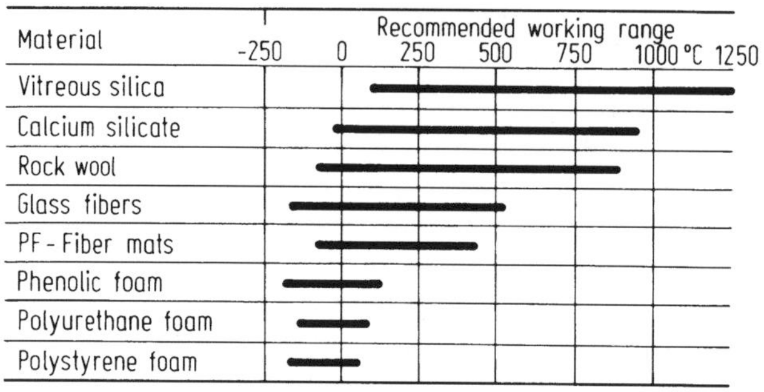

Material	Recommended working range -250 0 250 500 750 1000 °C 1250
Vitreous silica	
Calcium silicate	
Rock wool	
Glass fibers	
PF - Fiber mats	
Phenolic foam	
Polyurethane foam	
Polystyrene foam	

Fig. 11.1. Temperature range of application for some insulating materials[3)]

The flammability of a material is a decisive factor for its application in the building construction area. In general, desirable material properties for minimizing fire hazards are:

– high decomposition temperature and low combustibility of gases generated,
– low heat of combustion,
– low rate and amount of smoke generation,
– low toxicity of the gaseous combustion products.

The Federal Aviation Administration (FAA) did a lot of research work on smoke generation of plastics used in the interior of aircraft[4)]. Experimental tests and aircraft fires clearly demonstrate that smoke and gases generated by a fire may seriously impair escape, cause injuries and limit survival. The low smoke density of phenolic resins should be an essential argument for their use in the fields of construction and transportation.

In West-Germany, DIN 4102 rates phenolic resin bonded mineral wool as "nonflammable" A 1 or A 2, according to resin content, phenolic foam as "flame resistant" B 1 or "normal flammable" B 2, depending on additives and facings. It is further taken into consideration that low flammability of a construction material alone is not sufficient for safety in case of fire. An effective fire preventive construction depends upon the combined effect of low heat conduction and high thermal resistance.

11.1. Inorganic Fiber Insulating Materials

The formation of fibers or wool of silicate melts was first observed by chance on gaps in blast furnace walls. Rock fibers are said to have been produced in Wales (GB) as early as 1840. The first rock wool plant was built in 1870 in Greenfield, USA[5)]. Today, apart from minor quantities of slag wool, mainly glass- and mineral fibers are used. Blankets and felts with a specific weight of 30–80 kg/m^3 may be clad on one side with asphalt paper or quilted on both sides. The blankets can also be clad with plastic or aluminium foils which act as vapor barriers. Water vapor permeation, which increases the conductivity, is a serious problem in the low temperature range. Blankets for insulation of pipe lines and containers can be reinforced by wire mesh. One piece pipe insulation is produced for pipes of 20–900 mm diameter[6)].

The blankets may be placed around the house, on exterior walls, basement masonry, floor, partitions and roof. Apart from heat insulation, sound insulation is also a desired aim.

Fig. 11.2. Heat insulation of buildings with mineral fiber mats. (Photo: Deutsche Rockwool Mineralwoll GmbH, D-4390 Gladbeck)

Fig. 11.3. Heat insulation of pipe lines with mineral fiber shells. (Photo: Grünzweig & Hartmann und Glasfaser AG, D-6700 Ludwigshafen

Prefabricated shells are used to insulate pipe lines for hot water, steam or oil etc., or in the form of blankets for technical furnaces, reactors and containers. The limit of application is 450 °C. Although the binder is slowly degraded at temperatures above 250 °C, the effectiveness of the insulation is not affected. Cold insulation includes refrigeration equipment for the storage of foodstuffs, gas liquefication plants, house-hold appliances, refrigerating cars, ships etc. Glass wool is mostly used in the lower temperature area and for domestic purposes, mineral wool in areas of higher temperatures and for industrial application[7]. As far as the quantity (volume) is concerned, glass fiber insulations take the first place followed by mineral wool. The market share of slag fibers is relatively small and decreasing (< 10%). In Europe, approximately 1.4 million tons of mineral fiber insulating materials were produced in 1976. The leading producing countries are West-Germany, the Scandinavian countries and France.

11.1.1. Inorganic Fibers and Fiber Production

Mineral fibers are multi-component systems with the main components (Table 11.2.) being SiO_2, Al_2O_3, CaO, MgO with mean eutectic points at 1,170, 1,220 and 1,345 °C[5].

The chemical composition of the melt decides the thermal resistance, the devitrification temperature and devitrification rate.

The mineral wool production process is explained in Fig. 11.4. Distinct sediment or eruptive stones e.g. diabase, are melted by addition of lime and foundry coke at approx. 1,500 °C in a cupola furnace. The higher the content of silica, the "longer"

Table 11.2. Chemical composition of mineral fibers[5]. (^{x} = including also 5–12% iron oxide)

	SiO_2	Al_2O_3	CaO	MgO	B_2O_3	$Na_2O + K_2O$
Glass fiber	50–65	3–15	5–15	2–5	1–12	1–18
Slag fiber	30–35	10–20	40–45	2–8	–	–
Stone fiber^x	50–55	6–15	25–35	2–6	–	2–3
Basalt fiber^x	45–50	12–15	9–12	7–10	–	2–4

the melt gets. The upper limit of SiO_2 content is determined by the increased melting and spinning temperature[8].

The molten material then flows to four rotating spinning wheels (3,000–5,000 rpm) and is spun to thin fibers with diameters of between 3–7 μm by centrifugal force. The formulated phenol resin is added as bonding agent in the blowing chamber along with oil (~0.2%) to make the wool dust-free and water-repellent[9].

The melt yield is within the range of 70% calculated on the rock amount. The resulting material may contain at least 40% of non-fibrous material generally in the form of small pellets, called shot, depending upon the method of production and chemical composition. The overall fiber yield may therefore be as low as 40%.

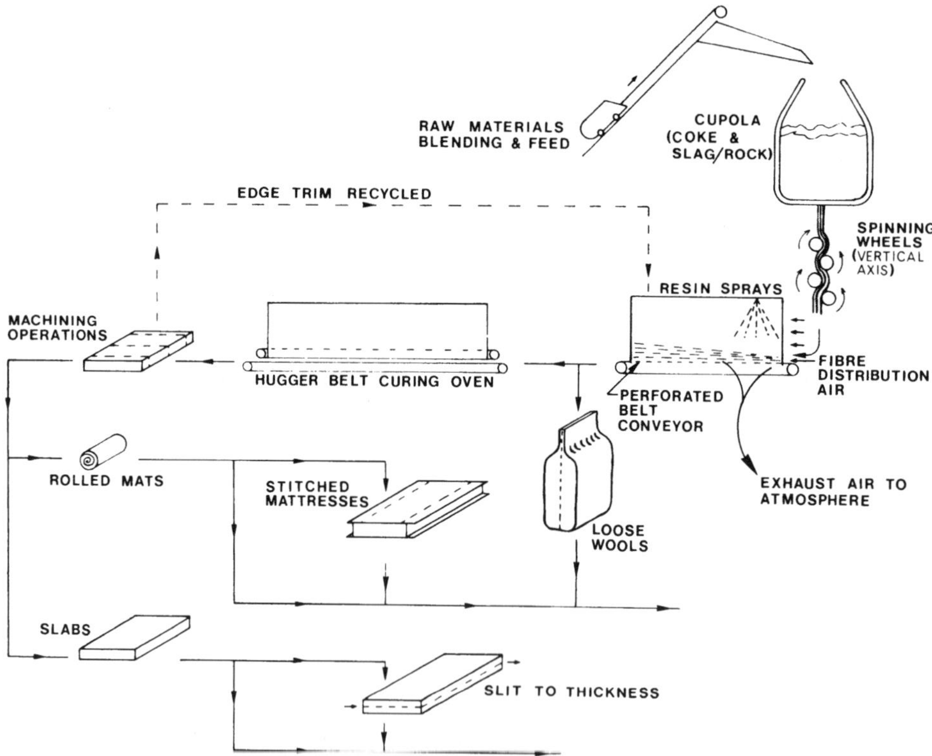

Fig. 11.4. Mineral fiber production process. (Drawing: Newalls Insulation Co. Ltd., Washington, GB)

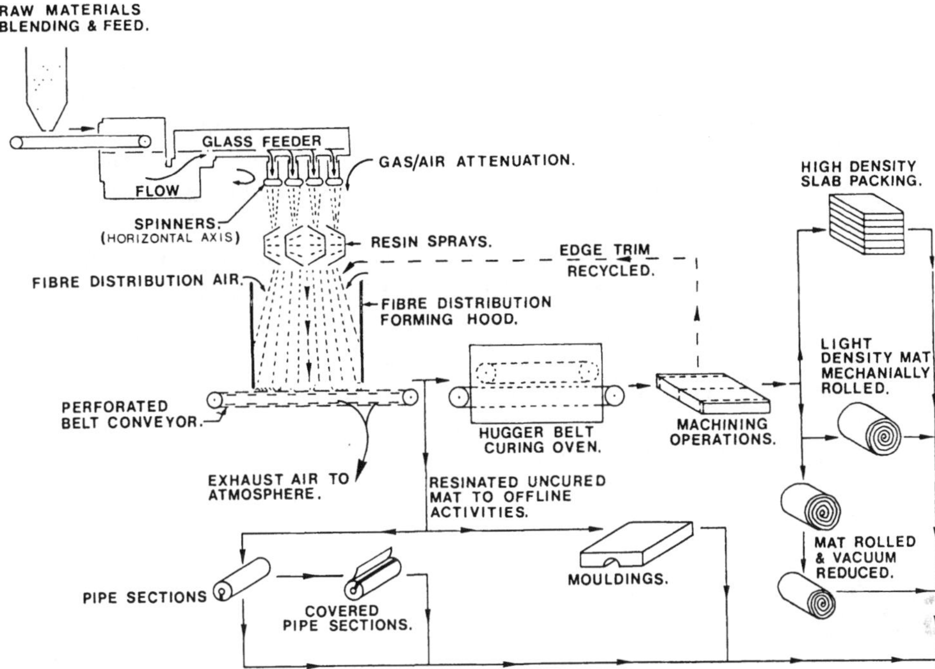

Fig. 11.5. Glass fiber production process. (Drawing: Newalls Insulation Co. Ltd., Washington, GB)

High silica glass fibers[10], on the other hand, are almost shotfree (0.1%). The absence of shots frequently justifies the higher material costs because of the higher effectiveness in relation to weight. The production of noncontinuous glass fibers is shown in Figure 11.5.

In collecting chambers, the fibers are sucked onto a conveyor belt to build up a mat of desired thickness and density. Then the wool is passed through a kiln in which the phenol resin is completely hardened. Several additional processes follow, such as cutting to size, shape forming, controlling and packaging.

11.1.2. Resins and Formulation

Generally, an aqueous resol solution is sprayed onto a mass of hot fibers and then the mass is further heated until the resin is wholly cured. Since the highly diluted resol is spread around the fibers as an extremely thin film and is subjected to relatively high temperatures of the order of 200 °C and higher, a significant amount of low molecular portions of the resol, mainly phenol, formaldehyde and saligenin, are volatilized. Because of this the economy of the production is considerably reduced, and through the emission of phenols and formaldehyde a serious environmental problem is created. Due to the huge amounts of air required in the blowing chamber and therefore low concentration of combustible substances, the treatment of the exhaust air has not yet been satisfactorily solved.

The application efficiency of a resin binder solution becomes a decisive quality mark:

$$E = \frac{W}{G \cdot V \cdot S} \times 100 \, (\%)$$

E stands for the efficiency, W for the increase in weight of the fibrous mass after curing, G for the specific gravity of the binder solution, V for the volume of resin used and S for the percentage of solids in the resin as evaluated by standardized methods. The realized resin efficiency in large technical plants today is within the range of 60–75%.

The water dilutable phenol resols used at present are obtained by reacting phenol with aqueous formaldehyde solution at temperatures below 70 °C. Alkali hydroxides and alkaline earth hydroxides, which can be precipitated with sulfuric acid at the end of the reaction are qualified for condensation catalysts. The formaldehyde ratio used is between 1.5 and 4 mol to 1 mol phenol. In general, the higher the formaldehyde molar ratio the higher the efficiency of the resol. The characteristic properties[11] of an appropriate resol resin very frequently used in Europe are indicated in Table 11.3. The reduction of water dilutability with increasing storage time at 21 °C is shown in Figure 11.6.

Table 11.3. Properties of a phenol resol for the production of mineral- and glass fiber mats[11]

Dry solids content	%	40
pH	–	7.0
Viscosity at 20 °C	mPa · s	8–10
B-time at 130 °C	min	8–10
Dilutability with water	ratio	1:10
Content of free phenol	%	1.0
Content of free formaldeyhde	%	3.5
Storage life at 20 °C	days	14
Storage life at 10 °C	days	35

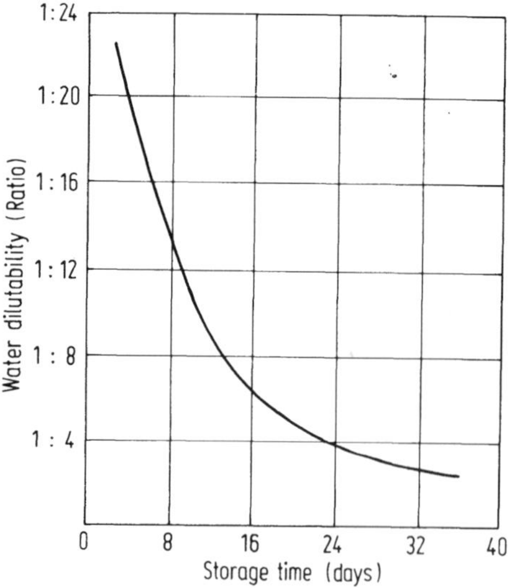

Fig. 11.6. Water solubility of the phenol resol of Table 11.3 in dependence on the duration of storage at 21 °C. (Drawing: Bakelite GmbH, D-5860 Iserlohn-Letmathe)

The following formulation may be used for the production of fiber mats:

100 pbw	phenol resol (40% dry solids)
7 pbw	20% ammonia solution
0.02 pbw	amino silane (γ-aminopropyltriethoxysilane)
0–800 pbw	water

Ammonia serves to bind the free formaldehyde and for pH adjustment to a low alkaline value. Aminosilane acts as coupling agent to improve the humidity resistance and to increase the mechanical strength. In general, the resin coat is adjusted through resin dilution. The resin content in mineral fiber mats is between 1–5%, mostly 3%. For glass fiber mats a higher resin content between 5–14%, mostly 7% is used. In order to reduce the content of free formaldehyde, urea can be added to the phenol resin, either at the end of the reaction or immediately prior to use by the producer. On the other hand, urea addition in larger amounts has the disadvantage of reducing the humidity resistance considerably. Thermal and ageing resistance are also reduced. Yet, urea is often added because of cost advantages. The addition of lignin, calcium- or magnesium lignosulfonate[12, 13)], dicyandiamide[14, 15)] or melamine is recommended to improve PF-resin/urea blends or to reduce binder cost.

11.1.3. Properties of Fiber Mats

The mechanical strength depends upon the fiber length and diameter, the content of bonding agent and the proportion of shots. For the statistical mineral fiber diameter distribution the following indication may be given:

25% < 2 μm; 20% 2–4 μm; 25% 4–6 μm; 30% > 6 μm

It is obvious that the relevant properties[17, 18)] depend to a great extent on material density (Table 11.4.).

Furthermore, these materials are resistant to humidity, do not rot, are chemically neutral and temperature resistant to a great extent. The temperature resistance is at first limited by the organic bonding agent. The thermal decomposition of the bond-

Table 11.4. Physical properties of mineral fiber mats[17)]

Property	Test method	Unit	Test value
Density	DIN 18165	kg/m^3	20–80
Behavior in fire	DIN 4102	–	fire-proof A 2
Thermal conductivity	DIN 4108	W/mK	0.040
Spec. thermal capacity C_p	–	kJ/kg °C	0.84
Noise reduction coefficient	DIN 52212		
125 Hz			0.17
500 Hz			0.76
1,000 Hz			0.86
4,000 Hz			1.00

ing agent starts at approximately 250 °C. The devitrification range of glass fibers is between 650–850 °C depending on composition, of mineral fibers between 750 and 1,000 °C.

11.2. Phenolic Resin Foam

Of all foamed plastics, phenolic resin foam offers remarkable properties, in particular fire resistance, low smoke generation, high temperature resistance and good thermal and acoustical insulation properties which highly qualifies it as insulation material in the construction business[18)]. Inspite of this the market for phenolic foams is developing very slowly. The reasons for this are inadequate official rating and standardization in most countries and the relatively high material costs. A further field of phenolic foam application is flower pin material.

Phenolic foams were employed for the first time at the beginning of the forties as replacement for balsa wood in the German aircraft industry[19)]. The current West European market is within the range of 6–8,000 tons per year (1977), an increase to 20,000 tons per year has been predicted by 1980[20)]. The main producers in Europe are France and West-Germany.

The novolak/HMTA mixtures used at the beginning lost their importance completely. Phenolic foams are now commonly produced by acid curing of resol-type resin by the addition of blowing agents, surfactants and colorants[19, 21)].

11.2.1. Resins and Additives

Aqueous resols made from phenol and formaldehyde at a molar ratio (1 : 1.5–2.5) are generally used. Substituted phenols can be used as additives to improve flexibility. There is a trend towards lower formaldehyde ratios in order to keep the separation of formaldehyde low during the foaming process. Sodium hydroxide or alkaline earth hydroxides are used as catalysts. The following properties are typical for the technical foam resins of today (Table 11.5.):

Table 11.5. Properties of a phenol resin for the room-temperature foaming process[22)]

Characterization		Resol type
Dry resin content	%	80
Viscosity, 20 °C	mPa · s	2,500–3,000
Gel time, 130 °C	min	7–8
Free phenol content	%	3–4
Free formaldehyde content	%	1–2
Storage stability, 20 °C	weeks	4–5

The volatile components mainly consist of water and phenol. Solvents are seldom used. After completion of the reaction, which is performed at temperatures between 60 and 90 °C, the solid resin content and viscosity are adjusted by distillation. The viscosity of the resin is an important criterion for the adjustment of foam density.

Foam density is further regulated by the type of surfactant, amount of blowing agent, temperature and acid-reactivity of the resin. The acid-reactivity is influenced by the storage time. The storage life of highly concentrated resols is limited, in general, 4 months at room temperature; the acid-reactivity decreases with increasing storage time. At the same time, the viscosity increases considerably; maximum viscosity should be 10,000 mPa · s.

The cross-linking of the resol is effected by addition of strong inorganic or organic acids e.g. hydrochloric acid, phosphoric acid, *p*-toluene-sulfonic acid or phenol-sulfonic acid. Of the inorganic acids, especially hydrochloric acid/ethylene glycol mixtures (1 : 1) or phosphoric acid are used. The high H^+ ion activity of the hydrochloric acid is an advantage but is also a corrosion problem. Phosphoric acid mostly used in combination with other stronger acids, for instance sulfuric acid or *p*-toluene-sulfonic acid, results in foams with improved flame resistance. The phenol-sulfonic acid is chemically built into the macromolecular system so that the hazard of corrosion of metal parts is prevented. However, it is considerably more expensive than the inorganic acids. Sulfonated phenol novolaks[23, 24] or resorcinol novolaks[25] have been recommended as curing agents. The acid cured phenolic resins are generally very brittle, have low impact strength and high friability. This problem has not yet been satisfactorily solved.

Most of the flexibilizing additives reduce the acid-reactivity and the flame resistance. Long chain diols, polyols, PVAc, PVA, epoxy resins and diisocyanates[39] are described as modifying agents[19]. Nonreactive polymers, for instance polyamides, PVC, polyesters, acrylonitrile-butadiene copolymers and styrene copolymers have also been recommended. Different fillers like talc, asbestos or glass fibers can be added to improve the homogeneity and increase the compressive strength. Several inorganic fillers (antimony trioxide) improve the flame resistance; however, they increase the specific weight and heat conductivity. Boric acid, boric acid esters, halogenated phosphoric acid esters, urea, thiourea, dicyandiamide, melamine, and others may be used to improve the flame resistance[23, 26].

Blowing Agents

Liquids with a low boiling point and low evaporation heat are mixed with the aqueous phenol resol, and the hardener is then added in a vessel equipped with a fast running agitator or by use of a three-component mixing head. After a short period of time, the temperature rises due to the exothermal curing reaction so that the blowing agent evaporates.

Chemical blowing processes, i.e. formation of gas by chemical reaction, have been recommended occasionally but are not technically applied at the present time. Among others, alkali- and ammonium carbonate or hydrogen carbonate (blowing with CO_2) or metal powders like aluminum, magnesium (blowing with H_2) and a number of organic compounds, which under the influence of acids or temperature separate nitrogen, e.g. benzene-sulfonic acid-hydrazide, N,N′-dinitroso-terephthalic acid-N,N′-dimethyldiamide, were recommended. The air mixing process also did not succed.

The liquid blowing agent must start to evaporate at a moment when the viscosity or plasticity of the mixture and the reaction rate are in a state in which the developing gas can be held and the viscosity of the system permits further formation of gas bubbles. Suitable blowing agents are trichlorofluoromethane (BP 27.3 °C), trichlorotrifluoroethane (BP 47.6 °C), dichloromethane (BP 40.2 °C) and *n*-pentane (BP 36.1 °C). The quantity of the blowing agent depends upon the desired density of the foam. The following quantities are necessary to produce 1 m^3 of foam (Table 11.6.):

Table 11.6. Guide formulation for phenolic foam production[22)]

Formulation		Foam density kg/m^3			
		40	50	60	70
Resin 80% solids, 3,000 mPa · s	kg	54	66	81	95
n-Pentane	kg	3.4	3.0	1.7	1.3
Hydrochloric acid/ ethylene glycol 1 : 1	kg	5.0	5.8	9.3	10.9

n-Pentane is still used for block foaming despite its flammability because of its relatively low price, minimal odor problem and low physiological hazards. If facings of low gas permeability are used, chlorofluorocarbons (Freon, Kaltron) should be preferred.

Surfactants

In order to produce homogeneous, reproducible foam structures, the required surfactant must be soluble in water, non-hydrolyzable and resistant in the acidic pH area. The non-ionic types are the most commonly used materials, anionic and cationic surfactants have lost their importance. The quantities to be added vary up to 3%. Polyoxyethylene sorbitane fatty acid esters, siloxaneoxyalkylene copolymers and castor oil-ethylene oxide adducts are preferred[19)]. Surfactants have a considerable influence on the relation of open and closed cells, on the wettability and water absorption (important for floral pin application) etc., so that the right choice is one of the most important and most difficult tasks.

11.2.2. Foaming Equipment

In the slab stock process, which is the prevailing one at present time, molds up to a capacity of 4 m^3 of simple material like wood are in use. Since the adhesion exceeds the foam tensile strength, the use of a release agent is required. The simplest way is to line the mold with paper or polyethylene foil. After the mixing of the components in a dissolver or agitated vessel, the acid is added last, the homogeneous mixture is fed into the mold and in the closed mold foamed and cured at room temperature or in an air circulating oven at 50–60 °C, depending upon the reactivity of the resin. Under these conditions, foaming starts after a few minutes and is finished after about 20 minutes. The mold is so constructed that very little overpressure can be built up. After a waiting time of 1–1.5 hours the material is removed. The panels or sandwich panels with rigid and/or water impermeable facings are then fabricated in a separate operation.

In the in situ process, the three components – resin, catalyst and blowing agent – are mixed in a controlled ratio and then introduced directly into the space to be foamed out. For both processes, slab stock production or in situ pouring, three component pouring machines have been developed[27]. The conditioning system, proportionating pumps, circuits and mixing head must be designed in order to resist the high aggressiveness of the strong inorganic acids. The output of present units is within the range from 5 up to 65 kg/min. Non-flammable blowing agents should be used with this type of equipment. The in situ process is expected to gain a substantial market in the construction business. Newer continuous production lines with double metal belt equipment incorporate systems for the simultaneous lamination of rigid or flexible facings[28]. Apart from the fact that the process is more economic, those foams have some qualitative advantages. The panels show a slight density gradient and have a higher abrasive resistant surface. The components conveyed by proportionating pumps are mixed in one or two mixing units, acid is the last ingredient to be added, and are sprayed over the total width of the heated steel belt. The reactivity of the resin is adjusted so that the foaming starts after 30–60 seconds. The curing times vary between 4 and 9 minutes.

11.2.3. Foam Properties

The cell structure of phenolic foams is extremely small and spherical and is not matched by any other cellular plastic. The ratio of open to closed cells depends upon the density, the kind of surfactant and the formulation. Within the density range of 35–80 kg/m^3, they are composed of 40% open and 60% closed cells. Lower density foams may contain up to 90% open cells. The intercommunicating cells lead to higher thermal conductivity, higher water absorption and water vapor permeability, but also higher sound absorption.

Tensile stress/strain and similar functions of various foams indicate that phenolic foam has a better resistance to creep under load compared to polystyrene or polyurethane foam. The tensile, flexural and compressive strength, however, are 2–3 times less, if they are compared to thermoplastic foams of the same density[29]. If higher strength[30–32] is required, reinforcement with paper honeycombs, glass fibers, asbestos or hollow microspheres[33] is necessary. The compressive strength is already

Table 11.7. Mechanical properties of phenolic resin foams depending upon the density[22]

Property	Test method	Unit	Density kg/m^3			
			40	50	60	80
Flexural strength	DIN 53423	N/mm^2	0.38	0.46	0.55	0.78
Compression strength	DIN 53421	N/mm^2	0.20	0.26	0.34	0.55
Tensile strength	DIN 53571	N/mm^2	0.19	0.23	0.30	0.46
Modulus of elasticity	DIN 53423	N/mm^2	6.4	9.8	13.0	20.1

increased by facing with papers, cardboard or metal foils. Due to the excellent resistance to chemicals, practically all glues can be used, even those which contain solvents. Hot bitumen in particular can be used for bonding and sealing on the spot. Phenolic foams can be used in a wide temperature range from −195 °C up to +130 °C. Temperature stress up to approx. 200 °C is permissible for a short period of time. The thermal expansion coefficient is within the range of $20–30 \cdot 10^{-6} K^{-1}$. The initially yellow-white foam discolors to brown under temperature stress or long term storage. At 130 °C the phenolic foam shrinks by approximately 1% and a distinct weight loss can be observed. An increase in strength can be achieved by post curing. If phenolic foam is to be used as insulation material[34)], water absorption is prevented by coating with bitumen, lacquers or impermeable facings. The foam absorbs only small amounts of water vapor if stored in humid air. This moisture is relatively quickly given off at drying[35)]. On the other hand, high water absorption is necessary, if foam is to be used as floral pin material. This is achieved by use of special surfactants and low foam density.

11.3. Sound Insulating Textile Fiber Mats

Textile fiber mats are used to control and suppress noise, especially in passenger cars, offices, auditoriums and other rooms requiring sound insulation[36, 37, 38)]. The textile mats are made of cotton, wool or acryl and polyester rags which are finely shredded by special machines and the shreds worked into endless fiber fleeces. The fixing of the fibers is performed by a more or less spot welding of the fiber crossings for which to a large extent powdered novolak/HMTA resins are used. The powder resin is uniformly scattered on the fleece web from a hopper which covers the whole width of the web and is distributed by roll brushes, toothed rolls or other proportionating devices. The bonding agent is also uniformly distributed in the interior of the fleece by special operations. The amount of resin applied is approx. 20–35%, according to the desired strength. The resin loaded fleece is carried by two wire works through a heat channel about 20 m long. The resin melts at first, glues the fibers and bonds them firmly together by the curing process. The final curing is

Fig. 11.7. Examples for application of PF-bonded textile mats for sound insulation in the automotive industry. (Photo: J. Borgers GmbH & Co. KG, D-4290 Bocholt)

performed by hot air (180–200 °C) which is blown vertically on the surface of the fleece or sucked through it. According to a special process, steam is blown onto the fleece prior to the hot air curing in order to fix the powder resin. The temperature must be strictly controlled because high temperatures in the interior of the fleece may easily cause ignition due to the flammability of the material. Flame retardant mineral substances should therefore be added to the powder resin. After curing, the fleece material is quickly cooled to 40–50 °C, cut to size, stacked and packaged.

For certain purposes, the fleece material must be post-formed. This is accomplished by partial cure in the oven and post-forming in appropriate molds. Principally, ground phenol-novolaks with a distinct grain size distribution are used. The resins contain about 10% HMTA as curing agent and mineral fillers as flame retardant additives.

References

1. Buderus: Handb. Heizungs- und Klimatechnik, 32. Ed. Düsseldorf: VDI-Verlag 1975
2. Harrison, M. R., Pelanne, C. M.: Chem. Engng. *19*, 62 (1977)
3. Probert, S. D.: Insulation, Sept. 1968, P. 190
4. Gross, D., Loftus, J. J., Lee, T. G., Gray, V. E.: Smoke and Gases Produced by Burning Aircraft Interior Materials. NBS Building Science Series *18*, P–B 193736, Washington 1969
5. Klingholz, R., Eberle, H.: Mineralfasern. In: Ullmanns Encyclopädie d. techn. Chemie, Bd. 12, P. 537. München: Urban u. Schwarzenberg 1959
6. Deutsche Rockwool Mineralwoll GmbH: Technical Bulletin
7. Harrison, M. R., Pelanne, C. M.: Cost-effective Thermal Insulation. Chem. Engng. Dec. *19*, 62 (1977)
8. Gould, T. R.: Refractory Fibers. In: Encyclop. Chem. Technology, Vol. 17, 285. New York: Wiley 1968
9. Jungers Verkstads AB: Mineral Wool Plants, Technical Bulletin
10. Tooley, F. V.: Handbook of Glass Manufacture, Vol. II, Books for Industry Inc.. New York 1974
11. Bakelite GmbH: Bakelite-Harz M 425, Technical Bulletin
12. West Virginia Pulp and Paper Co.: DE-PS 1226926 (1962)
13. Fiberglass Ltd.: US-PS 1316911 (1971)
14. Owens Corning Fiberglass Corp.: US-PS 2604427
15. Westvaco Corp.: US-PS 3463747
16. Newalls Insulation Co Ltd.: Technical Bulletin
17. Grünzweig & Hartmann AG: Technical Bulletin
18. N. N.: Modern Plastics Int. July 1974, P. 14
19. Papa, A. J., Proops, W. R.: Phenolic Foams. In: K. C. Frish, J. H. Sounders (ed.): Plastic Foams, Vol. II. New York: Marcel Dekker 1973
20. N. N.: Modern Plastics Int. July 1976, P. 8
21. Benning, C. J.: Plastic Foams. New York: Wiley-Interscience
22. Bakelite GmbH: Bakelite-Harz PS 274, Technical Bulletin
23. Valgin, V. D., Novak, W., et al.: DE-AS 1958 186, DE-AS 2133628
24. Valgin, V. D.: Europlastics Monthly *46/7*, 57 (1973)
25. Ciba-Geigy AG: US-PS 1488527
26. Saint-Gobain Industries: DE-AS 1769471
27. Secmer, S. A.: Technical Bulletin
28. Brünig, K.: Kontinuierl. Herst. v. Phenolharzschaumplatten, 5. Int. Fachtagung für Schaumkunststoffe 26./27. Mai 1975, Düsseldorf
29. Phillips, T. L., Lannon, D. A.: British Plastics, May 1961, P. 236
30. CdF Chemie: Phenexpan, Technical Bulletin

31. Deutsche Texaco AG: Resiphen, Technical Bulletin
32. Dynamit AG: Phenobil-D, Technical Bulletin
33. Okuno, K., Woodhams, R. T.: Cellular Plastics, Sept./Oct. 1974, P. 237
34. Jünger, H., Weissenfels, F.: Kunststoffe – Plastics *2*, 47 (1971)
35. Schmidt, H. D.: Kunststoffe im Bau *10*, 8 (1975)
36. J. Borgers-Wattefabrik: Technical Bulletin
37. Eisele, D.: Melliand Textilberichte *56*, 916 (1975)
38. Keller Ges. für chem.-techn. Produkte mbH: Idikell B, Technical Bulletin
39. Schafer, R. J., Baker, L. A., Linden, G. L., Frisch, K. C., Tripathi, V.: J. Cellular Plastics, May/June 1978, 146

12. Industrial Laminates and Paper Impregnation

The hydrophilic structure of non-cured phenolic resins is the decisive factor which qualifies them for the impregnation of paper and cotton fabrics to be used for the manufacturing of electrical and decorative laminates, molded parts, filter papers and battery separators. Low molecular, preferably mononuclear phenol alcohols penetrate into the capillary cavities of the cellulose fibers and, due to the cross-linking reaction, fill the cavities[1], whereas resins of higher molecular weight coat the fibers so that they become water-repellent. During the hardening reaction, at between 150 and 190 °C throughout these applications, a chemical reaction between the cellulose and phenol alcohols may occur which contributes to the increased water- and chemical resistance.

12.1. Electrical Laminates

In printed circuits, the conventional three-dimensional wire network is replaced by a two-dimensional arrangement consisting of a printed network of conductors on an insulating support. The importance of this invention, which is attributed to C. Parolini (France 1926), for electronics is comparable to the invention of printing made by Gutenberg[2, 3].

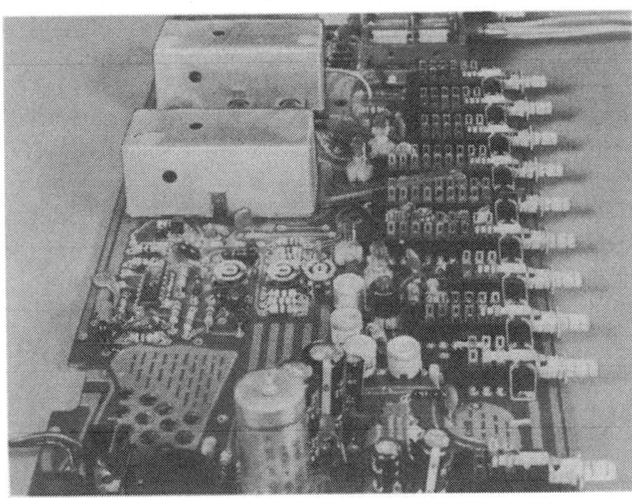

Fig. 12.1. Printed circuit board of a radio equipped with electrical components. (Photo: Isola Werke AG, D-5160 Düren)

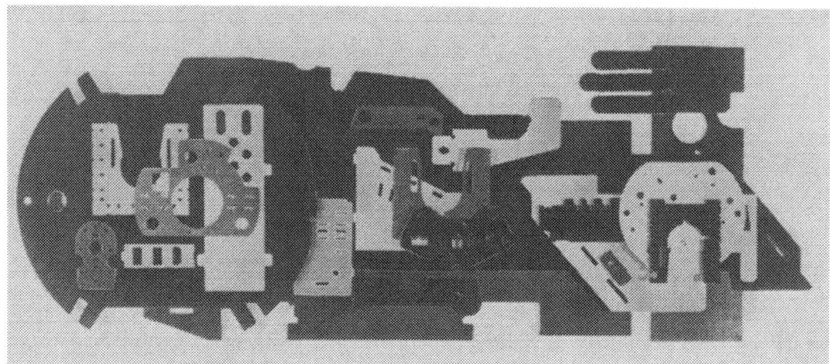

Fig. 12.2. Punched parts made of phenolic resin paper laminates. (Photo: Isola Werke AG, D-5160 Düren)

The supporting insulating material consists of a number of thermosetting resin-impregnated foils of paper or glass fabric (prepreg), which are then pressed and hardened in a heated press. The conducting pattern is made either by abstraction methods – the insulating material is covered completely with a copper foil and the pattern is created by selectively removing the unwanted areas – or by additive methods. In this process, the pattern is made in the form needed by plating[4)].

The standardization for paper reinforced base material in various countries is given in Table 12.1. However, it must be pointed out that the various national standards – DIN, NEMA, BS – do not always exactly correspond. The table therefore only displays comparable values. Due to the extraordinarily fast development of materials and the everchanging requirements of the market, the standards are not always up to date; especially with regard to flame resistance and tracking resistance higher requirements are set forth than stated in NEMA for FR-2. Here, the specification of the Underwriters' Laboratories subject 94 Rate V-1 and V-0 (vertical extinguishing test) is applied. Materials with higher tracking resistance have also been developed. The description of essential quality marks and recommendations for the application according to NEMA specification are compiled in Table 12.2.

Phenolic resin paper laminates (Figs. 12.1 and 12.2) are also used for construction elements in mechanical engineering and automotive electrics. Especially high mechanical strength is required in the low tension area (Hp 2061 in Table 12.1). High dielectric strength is required in the field of high tension, i. e. for application in transformers and switches (Hp 2061.5, Hp 2064).

Important fields of application for higher qualities Hp 2062.8, 2062.9 and 2063 are high-frequency and television engineering and the manufacturing of radio and television sets (DIN 40802).

12.1.1. Materials

The following description of the basic materials, paper, resins and production refers mainly to the more sophisticated copper clad laminates for printed circuit boards. Similar, somewhat less stringent prerequisites are applicable to all other classes.

Table 12.1. Comparison of various national and international standards for industrial laminated thermosetting sheets[5)]

National Standards									
West-Germany	DIN 7735		Hp 2061	Hp 2061.5	Hp 2061.6	Hp 2062.8	Hp 2062.9	Hp 2063	Hp 2064
Switzerland	VSM		S-PF-CP 1	S-PF-CP 2	S-PF-CP 3	S-PF-CP 4	–	S-PF-CP 4	–
France	NF		P, PO	Pa	PP, PPO	–	PPPOO	PPP	–
Great Britain	BS		P 1	P 2	P 3	P 4	P 4	P 4	–
USA	NEMA-LI 1		X, XP	XX	XXP	XXXP	XXXPC	XXXPC/FR-2	–
ISO/R 1642			PF CP1	PF CP 2	PF CP 3	PF CP 4	–	PF CP 4	–
Reinforcement			Paper						
Flexural strength d < 10 mm	N/mm^2	min.	150	130	130	80	60	80	130
Impact strength a_{k10}	kJ/m^2	min.	20	20	15	8	–	7	20
Impact strength, notched	kJ/m^2	min.	5	4	4	2.5	–	2.5	5
Tensile strength	N/mm^2	min.	120	100	100	70	60	70	100
Compression strength	N/mm^2	min.	150	150	100	120	–	–	100
Fission force	N	min.	2000	2000	2000	2000	–	–	2000
Modulus of elasticity	N/mm^2	≈	$7 \cdot 10^3$	$7 \cdot 10^3$	$7 \cdot 10^3$	$7 \cdot 10^3$	$5 \cdot 10^3$	$7 \cdot 10^3$	$7 \cdot 10^3$
Resistance between plugs	Ω	min.	–	–	$5 \cdot 10^7$	10^{10}	10^{10}	10^{10}	–
Breakdown voltage, ∥, 90 ± 2 °C, 1 min, d = 25 mm	kV	min.	15	40	25	25	20	20	–
Breakdown voltage, ⊥, 90 ± 2 °C, 1 min, d = 3 mm	kV	min.	15	40	30	30	25	25	–
Dielectric loss tgδ, 50 Hz, 96^h/105 °C		max.	–	0.05	0.08	0.08	–	–	–
Dielectric loss tgδ, 1 MHz, 24^h H_2O		max.	–	–	–	–	0.06	0.05	–
Dielectric constant		≈	5	5	5	5	5	5	5
Tracking resistance	grade		KC 100	KC 100	KC 100	KC 100	KC 100	KC 100	KC 100
Electrolytical corrosion	grade	max.	–	–	–	–	–	AN 1.4	–
Glow resistance	grade		2 b	2 b	2 a	2 a	2 a	2 b	2 b
Thermal conductivity	W/m · K	≈	0.2	0.2	0.2	0.2	0.2	0.2	0.2
Coefficient of thermal expansion	10^{-6}		20–40	20–40	20–40	20–40	20–40	20–40	20–40
Limiting temperature	°C		120	120	120	120	90	120[9)]	120

Table 12.1. (Continued)

National Standards:						
West-Germany	DIN 7735		Hgw 2082	Hgw 2082.5	Hgw 2083	Hgw 2072
Switzerland	VSM		S-PF-CC 1	S-PF-CC 2	S-PF-CC 3	S-PF-GC 1
France	NF		C	–	CC	–
Great Britain	BS		F2, F 3	–	F 2	PF 1
USA	NEMA-LI 1		C	CE	L	G-3
ISO/R 1642			PF CC 1	PF CC 2	PF CC 3	PF GC 1
Reinforcement			Cotton fabric			Glass fabric
Flexural strength d < 10 mm	N/mm^2	min.	130	115	150	200
Impact strength a_{k10}	kJ/m^2	min.	30	20	35	50
Impact strength, notched	kJ/m^2	min.	10	10	12	40
Tensile strength	N/mm^2	min.	80	60	100	100
Compression strength	N/mm^2	min.	170	150	170	150
Fission force	N	min.	2500	2500	2500	2000
Modulus of elasticity	N/mm^2	≈	$7 \cdot 10^3$	$7 \cdot 10^3$	$7 \cdot 10^3$	$14 \cdot 10^3$
Resistance between plugs	Ω	min.	–	10^7	–	10^8
Breakdown voltage, ∥, 90 ± 2 °C, 1 min, d = 25 mm	kV	min.	8	20	8	20
Breakdown voltage, ⊥, 90 ± 2 °C, 1 min, d = 3mm	kV	min.	5	5	5	25
Dielectric loss tgδ, 50 Hz, $96^h/105$ °C		max.	–	–	–	–
Dielectric loss tgδ, 1 MHz, 24^h H_2O		max.	–	–	–	–
Dielectric constant		≈	5	5	5	5
Tracking resistance	grade		KC 100	KC 100	KC 100	KC 100
Electrolytical corrosion	grade	max.	–	–	–	A/B 2
Glow resistance	grade		2 b	2 b	2 b	2 a
Thermal conductivity	W/m · K	≈	0.2	0.2	0.2	0.3
Coefficient of thermal expansion	10^{-6}		20–40	20–40	20–40	10–20
Limiting temperature	°C		110	110	110	130

Abbreviations: DIN: Deutsche Industrienorm, VSM: Verein Schweizerischer Maschinen-Industrieller, NF: Norme Française, BS: British Standards, NEMA: National Electrical Manufacturers Association, USA

Table 12.2. Characteristics and recommendations for the application of laminated thermosetting sheets according to NEMA-LI 1, 1971 (R 1976)

X: Primarily intended for mechanical applications where electrical properties are of secondary importance. Should be used with discretion when high-humidity conditions are encountered.

XP: Primarily intended for hot punching. More flexible and not as strong as Grade X. Intermediate between Grades X and XX in moisture-resistance and electrical properties.

XPC: Primarily intended for cold punching and shearing. More flexible and higher cold-flow but lower in flexural strength than Grade XP.

XX: Suitable for usual electrical applications. Good machinability.

XXP: Better than Grade XX in electrical and moisture-resisting properties and more suitable for hot punching. Intermediate between Grades XP and XX in punching and cold-flow characteristics.

XXX: Suitable for radio frequency work, for high-humidity applications. Has minimum cold-flow characteristics.

XXXP: Better in electrical properties than Grade XXX and more suitable for hot punching. Intermediate between Grades XXP and XX in punching characteristics. This grade is recommended for applications requiring high insulation resistance and low dielectric losses under severe humidity conditions.

XXXPC: Similar in electrical properties to Grade XXXP and suitable for punching at lower temperatures than Grade XXXP. With good punching practice at ambient temperature. This grade is recommended for applications requiring high insulation resistance and low dielectric losses under severe humidity conditions.

The following materials are required to manufacture one square meter of an 1.5 mm thick, one side copper clad laminate (FR-2 or XXXPC):

295 g copper foil, 35 μm
30 g thermosetting copper adhesive
1,170 g phenolic resin, dry weight
970 g paper

All materials are subject to strict selective criteria and quality control to ensure trouble free production, favorable electrical properties and workability, giving special emphasis to punchability[6, 7)]. In 1978 approximately 10 million m^2 of these laminates were produced in Western Europe.

Paper

Bleached kraft paper as well as cotton linter paper is being used. Cotton linter paper is easier to impregnate, offers qualitative advantages, especially with reference to electrical properties and punchability, but is, however, relatively expensive. Kraft paper is cheaper and somewhat more favorable as far as the mechanical strength is concerned. Since cotton linter paper, which has been impregnated only once, meets the high requirements, while kraft paper has to be impregnated twice, economic advantages are finally outweighed by the condition of the impregnation unit (depreciation, impregnation speed etc.). The paper comes in rolls, at the usual weight per unit area of between 80 and 140 g/m^2, between 2–3,000 m long. Cotton linters are short fibers remaining on the seeds after stoning. Further raw materials for cotton linter paper are rags. The sorted out and cleaned rags are cooked under pressure in rotating cookers in alkaline chemicals, shredded in a hollander and bleached with a chlorine lye.

The raw material for kraft paper is wood, mainly conifer, because of the longer fibers. The wood chips are cooked in a mixture of sodium hydroxide and sodium sulfide to a firm, tough pulp – therefore the name "Kraftzellstoff" (sulfate cellulose or sulfate pulp).

The natural tendency of paper to absorb liquids by capillary action is desirable in technical papers which are to be used for laminates, filters and vulcanized fibers. Therefore, an essential criterion is the absorbency of the paper. To evaluate the suction height, paper strips 15 mm wide and 180 mm long are dipped vertically in water and the height is measured up to which distilled water of 20 °C rises within 10 minutes. The permeability to air is the indication for the porosity of the sheet structure. It can be determined by the Gurley densometer. Further important properties are the conductivity of the aqueous extract, which gives information about the content of inorganic ions, and the wet tensile strength. The orientation of the cellulose fibers during paper manufacturing in Fourdrinier- and cylinder machines leads to anisotropic swelling and anisotropic thermal expansion of the laminate which results in twist and bow. Production factors are also responsible for this occurrence. To place the papers cross-wise would be one way to prevent this, but it is, however, too expensive for mass production. There are a number of process engineering precautions which considerably decrease the bow.

Resins

In the past, mostly cresol and xylenol resols were used to manufacture high quality laminates because they were supposed to have better electrical properties, but this finding is no longer correct. The better compatibility of cresol resins with natural oils results in mechanically tougher and less brittle resin films. In the meantime, phenol resols have been developed which are just as good as cresol- and xylenol resins as far as their general performance is concerned and also offer economic advantages. The price difference between phenol and cresols and xylenols will increase in the future. Phenol based resols gain more and more importance.

Two resins of different MW are necessary regardless whether the impregnation is single or double. The preimpregnating resin, a low molecular pure phenolic, contains only a minor portion of two-nuclear phenols ($< 20\%$). It mainly consists of mono-, di- and trimethylolphenol in addition to free phenol ($< 15\%$).

Hydrophobic, oil-modified resols of medium average MW are used for the second impregnation. The characteristic properties and differences of both resins are indicated in Table 12.3.

Calcium-, magnesium- or barium hydroxide are used as catalysts for the production of pre-impregnating resins which are precipitated at the end of the reaction by the addition of sulfuric acid. A low content of inorganic ions is essential for electrical

Table 12.3. Properties of phenol resols used for the manufacturing of copper-clad laminates according to NEMA FR-2, V-1[8)]

		1st Impregnation	2nd Impregnation
Resin content	%	45	60
Solvents		Water	Acetone/Methanol/Toluene
Viscosity	mPa · s	10–15	800–1,000
Gel time 130 °C	min	15–20	18–25
Gel time 150 °C	min	6–8	8–10
Water solubility	ratio, min.	1:5	1:0.1
Free phenol content	%	10–15	5–7
Free formaldehyde content	%	< 3	< 0.1
Ash	%	< 0.1	< 0.1
Storage stability	weeks	~ 4	~ 12

applications. Only deionized water is used for the production of resins. The most frequently used catalyst for the main impregnating resins is ammonia or HMTA; primary, secondary and tertiary amines are also suitable. Working procedures for the manufacturing of cold punching laminates, corresponding to NEMA FR-2, Grade V-0, are given in Table 12.4. The different manufacturing of core and cover sheets with reference to the paper (weight per unit area) and formulation (resin content, release

Table 12.4. Production of electrical laminates according to NEMA FR-2, V-1: Materials, formulation and impregnation conditions[8)]

Paper: Kraft, 80 and 120 g/m^2
Suction height acc. to Klemm 70–100 mm

1st Impregnation			
Formulation	21.2 kg Bakelite resin IV 2446 9.5 kg Bakelite resin IZ 7266 15.1 kg Water, deionized 54.2 kg Acetone		
Drying plant	Temperature °C, max.	150–180	
Prepreg properties	Resin content %	18 ± 1	
	Volatiles content %	2 max.	
2nd Impregnation			
	Core sheets (6) 120 g/m^2		Cover sheets (2) 80 g/m^2
Formulation	100 kg Bakelite resin IS 7340 20 kg Methanol		100 kg Bakelite resin IS 7340 20 kg Methanol 2 kg Release agent
Drying plant	Temperature °C, max.	150–180	150–180
Prepreg properties	Resin content %	50	60
	Resin flow %	10–15	0.5–1.5
	Volatiles content max.	2–3	2–3

agent) as well as prepreg specification (flow) is indicated. If cotton paper is used, one pass impregnation with mixed resins may be sufficient, provided the way of working is appropriate.

The resins for the second impregnation must be flexibilized in order to meet the requirements for punchability. A multitude of compounds has been recommended to improve the flexibility. The resins used at present time mostly contain phosphoric acid esters (triphenyl- or diphenylkresylphosphate) and tung oil. Tung oil, obtained from the fruits of the tung tree, which superficially resembles an apple tree, is also known as China wood oil since the oil came originally exclusively from China. Plantations are now also to be found in some areas in the U. S., Argentina, Paraguay and New Zealand. Tung oil is largely composed of glycerides of α-elaeostearic acid which has three conjugated double bonds[9]. Type F tung oil (Aleurites fordii) contains 79–82% elaeostearic acid, together with about 12% oleic acid, the residue being saturated fatty acids. The type M oil (Aleurites montana) contains smaller quantities of elaeostearic acid (75–78%).

Mechanisms of the reaction of phenol and phenol resols with olefinic double bonds are described in Chapter 3.3.5 and 17.1. The formation of a chroman ring is very unlikely under conditions customary in the production of resins because of the necessary formation of quinone methides as intermediate compounds. It is more likely that the reaction proceeds like a Friedel-Crafts alkylation. The special advantage of tung oil is that a considerable increase in flexibility and toughness, i. e. improvement of punching quality, is achieved without essentially worsening the resistance to solvents. A disadvantage, however, is the price instability of this natural material, which makes a long-term calculation very difficult. Also the enhanced flammability is a serious disadvantage.

Chemically similar oils, for instance oiticica oil[9, 10], follow the price movements of tung oil very closely. Oiticica oil contains up to 80% licamic acid (4-keto-octadeca -9, 11, 13-trienoic acid). Formulated resins are inferior to some extent in comparison to those based on tung oil, so that they are no real alternatives. For this reason the development of tung oil-free resins using synthetic additives has been intensified and has in some cases been successful. Aromatic phosphoric acid esters improve the punchability as well as the flame resistance. Since they act as external flexibilizers, which are not integrated in the macromolecular system, the solvent resistance is diminished. Aliphatic phosphoric acid esters are not suitable due to their easy hydrolysis. To improve the flame resistance bromine containing additives, e. g. tetrabromobisphenol-A, octabromodiphenyl, pentabromophenol, pentabromodiphenyl ether[11], brominated epoxies, and others may be used. However, the market prefers halogen-free laminates to an increasing extent. This can be achieved by using melamine-modified resins for the preimpregnation and addition of inorganic fillers.

12.1.2. Production of Electrical Laminates

A plant for the production of industrial laminates consists of three units: the impregnation plant, the press plant and the finishing line. In general, horizontal driers are used for paper impregnation (Fig. 12.3), although vertical driers are just so suitable.

Fig. 12.3. Horizontal impregnation machine for the production of industrial and decorative laminates. (Photo: Charatsch AG, CH-5620 Bremgarten)

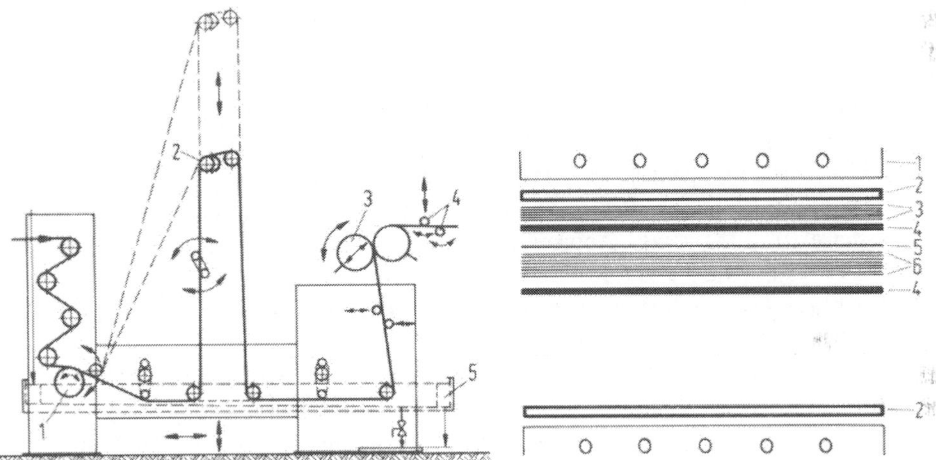

Fig. 12.4. Impregnation with prewetting. (Drawing: Impla, I-20060 Cassina de' Pecchi)
1 prewetting roll; *2* mobile extension rolls;
3 squeeze and doctor roll; *4* smoothing rolls;
5 heated bath

Fig. 12.5. Structure of a press package in an opening.
1 upper heater plate; *2* steel plate;
3 equalizing sheets of kraft paper;
4 press sheet; *5* copper foil;
6 resin impregnated papers

Formerly, their use was restricted by the low wet tensile strength of the paper. Nowadays, this problem is solved technically[12)].

To displace the air from the paper web, prewetting is recommended so that the paper is allowed to "breathe". The following dipping distance should be as long as possible (Fig. 12.4). In the drier, the wet paper web is blown at from above and below by jets, leading to a smooth sinus form of the web which is carried and positioned ("lay on air system", Vits Maschinenbau, Langenfeld) in this way.

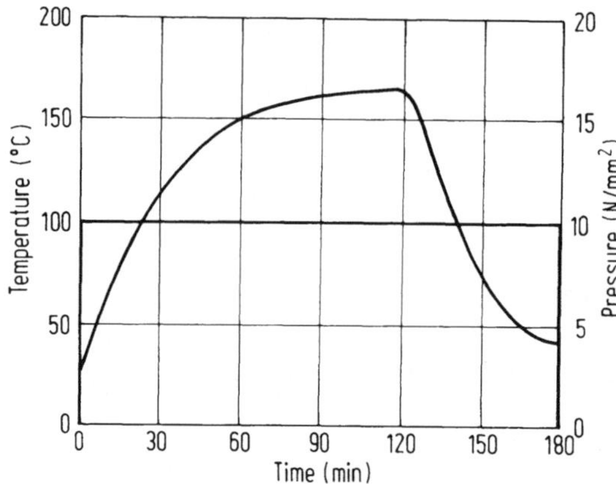

Fig. 12.6. Temperature- and pressure cycle for the production of copper-clad laminates according to NEMA FR-2 and XXXPC

Two operations are hereby taking place at different times. The temperature is raised slowly for some time remaining below the boiling point of the solvents to avoid craters and streaks. After the solvent is almost evaporated, the temperature of the web rises and the condensation reaction of the resin begins. The evaporation and advancement of the resin, tested by the determination of the volatiles content and resin flow, is adjusted by the temperature in the drier and the impregnating speed. At last the web is cooled down in a cooling section or by cooled rolls and rolled up or cut to size. The prepregs are laminated, after the required properties are confirmed (see Table 12.4), either immediately or after intermediate storage in air-conditioned storage rooms to avoid the absorption of water, in a multidaylight press. The newer presses have 20–22 openings. Approximately 8–12 laminates are produced simultaneously per opening (Fig. 12.5).

The heat is supplied by hot oil or steam. The real temperature course in the laminate (Fig. 12.6) depends upon the heating rate, the rate of heat transfer and the number of laminates in an opening. Lower curing temperatures (< 160 °C) improve the punchability, impede, however, other criteria. The cooling rate is important for the extent of bow. Slow and programmed cooling, especially in the upper temperature range, results in considerable improvements. Pressure stages are not customary; the pressure is always between 10 and 12 N/mm^2 (Fig. 12.6). After cooling, a series of finishing operations follows[13].

A compilation of laminate properties is found in the literature[14].

12.2. Laminated Tubes and Rods

Laminated tubes and rods of phenolic resin impregnated papers or cotton fabrics are used as insulation and machine elements in electrical engineering, mechanical engineering and in the textile industry because of their high dielectric and mechanical strength and low specific weight.

Further examples are winding supports in transformers, insulating parts of quenching chambers, switch shafts, compressed gas capacitors, bobbins and guide rolls for paper and textile machines. The use of cotton fabrics results in high mechanical strength, which is necessary for slide bearings and bearing shells.

Quality requirements for tubes are established in national standards (NEMA, DIN; Table 12.5). Natron kraft paper is preferred for the manufacturing of laminated tubes, but with lower absorbency than paper recommended for flat laminates. The papers are coated with phenolic resin on one side, dried and rolled up before they are made into tubes in special machines[14]. The coated papers are wound around a preheated core by means of electrically heated (150–170 °C) counter pressure rolls. During this operation, the resin melts and binds the individual layers. The laminate and core are then put into a drying oven for complete curing. After cure the core is removed and the tube is machined to size. The cure requires thorough control of the curing conditions to avoid blisters and delamination. The phenol resols are not flexibilized. Often *ortho*-cresol mixed with phenol is used to reduce the reaction rate. Thorough control of the volatile components content and resin flow are further prerequisites of a trouble free production.

Fig. 12.7. Laminated tubes and bobbins. (Photo: Ferrozell Ges. Sachs & Co, D-8900 Augsburg)

In the manufacture of bobbins for the textile industry (Fig. 12.7), which are commonly revolved with 4–12,000 rpm, special emphasis must be given to a perfect cylinder shape without differences in density to guarantee satisfactory running. High quality bobbins with a wall thickness of 6 mm or more can be run at speeds up to 20,000 m/min. In special cases balancing of the bobbin by metal pins is necessary[15].

12.3. Cotton Fabric Reinforced Laminates

Higher strength of laminates and shaped parts is obtained by the use of cotton fabrics as reinforcing material instead of paper. Such shaped parts are used as construction material in mechanical engineering and as insulating materials in electrical engineering.

Table 12.5. Minimum mechanical and electrical properties of tubes according to NEMA and DIN depending upon the reinforcement[5)]

National Standards: West-Germany	DIN 7735		Hp 2065	Hp 2066	Hp 2067	Hgw 2085	Hgw 2086	Hgw 2075	Hp 2068	Hgw 2088
USA	NEMA-LI 1		X	XX	XXX	C	L	G-3	–	–
Reinforcement			Paper	Paper	Paper	Cotton fabric	Cotton fabric	Glass fabric	Paper	Cotton fabric
Specific gravity	g/cm^3		> 1.05	> 1.1	> 1.05	1.15–1.4	1.15–1.4	1.4–1.7	1.2–1.4	1.2–1.4
Flexural strength	N/mm^2	min.	100	100	100	80	80	200	80	80
Tensile strength	N/mm^2	min.	50	–	50	50	50	100	–	–
Compression strength	N/mm^2	min.	40	50	50	40	40	100	70	70
Modulus of elasticity	N/mm^2		$0.6 \cdot 10^4$	$0.6 \cdot 10^4$	$0.6 \cdot 10^4$	$0.6 \cdot 10^4$	$0.6 \cdot 10^4$	$1.7 \cdot 10^4$	$0.7 \cdot 10^5$	$0.7 \cdot 10^5$
Resistance between plugs	Ω		–	10^5	$5 \cdot 10^5$	–	–	10^7	10^5	10^5
Breakdown voltage, ‖, 90 ± 2 °C, 1 min, d = 25 mm	kV	min.	25	–	25	10	10	15	10	5
Breakdown voltage, ⊥, 90 ± 2 °C, 1 min, d = 3mm	kV	min.	25	–	–	5	5	15	15	5
Tracking resistance	grade		KC 100	KC 100	KC 100	KC 100	KC 100	KC 100	KC 100	KC 100
Limiting temperature	°C		120	120	120	120	120	130	120	120

They distinguish themselves by high specific strength, resistance to continuous heat up to 110 °C, high wear resistance, very low water absorption and resistance to lubricants, solvents, acids and weak alkalies. These materials possess a very good machining property and are used for cog-wheels, driving rolls, linings for sliding surfaces, guide ledges, switch base plates, wheel bearings and other applications.

Because the laminates are quite thick the reactivity of the resins must be carefully balanced. The high exothermic heat may result in carbonization in the center of the laminate, in especially unfavorable cases even in explosions. Cresol resols with high *ortho-* and *para*-cresol content are preferred because of their slow curing rates. In most cases, these resins are not flexibilized in order to meet the high requirements

Fig. 12.8. Applications for cotton fabric reinforced laminates. (Photo: Isola Werke AG, D-5160 Düren)

concerning the compression strength and resistance to delamination. Cotton fabrics with area weights between 130 and 200 g/m^2 are used. Fabrics for technical application contain considerable proportions (4.5–5.5%) of size mostly based on polyvinyl alcohol, in order to facilitate the manufacturing of the fabric and its handling. Special electrolyte-free fabrics, which are washed with diluted NaOH prior to their use, are used for electrical applications. The phenolic resins contain methanol, ethanol, acetone or toluene as solvents. The impregnation is accomplished with solutions of approximately 50%. After a preimpregnation, the fabric is squeezed out to ensure better penetration of the resin; then the desired amount of resin is applied by a second impregnation. The resin content of the prepreg is, in most cases, adjusted to 50% and the resin flow to 5%. In no case shall the proportion of volatile components exceed 3%, especially if thicker laminates are concerned. Explosions may occur due to the gas-steam pressure which can not be controlled. The prepregs are pressed at a pressure of 10 N/mm^2 without pressure stages and at temperatures of 160–165 °C. The curing time is approximately 4 min/mm; a curing time of approximately 17 hours for instance and additionally 8–10 hours for cooling for laminates 250 mm thick must be considered.

12.4. Decorative Laminates

The main application for decorative laminates, also called high-pressure laminates (HPL), is the furniture industry. Here, they as well as low-pressure laminates, decorative papers, wood veneers, coatings and vinyl facings are used to laminate particle boards and other wood materials used mainly for the manufacturing of kitchen cabinets. The hard and easy to clean surface is an important prerequisite. Further fields of application are furniture for children, laboratory furniture, wall elements in buildings, ships, boats and rail cars. The high surface hardness, scratch resistance, light colors and fastness to light of decorative laminates are attributed to the hard surface layer of melamine resin impregnated cellulose. The core sheets layers are of kraft paper impregnated with phenolic resins. Sodium hydroxide catalyzed aqueous resols (60% dry solids) are almost exclusively used.

Conventional sizes range between 3.6 x 2.0 m, the thickness of the laminates is mostly 1.3 mm[18]. In general, the definition for electrical laminates is also applicable to HPL. The individual layers consisting of different kinds of paper are also impregnated with different kinds of thermosetting resins. The components of an HPL, 0.3–1.5 mm thick, are indicated in Table 12.6.

The production of decorative laminates is very similar to the production of electrical laminates. The impregnation is accomplished on the machines shown in Figure 12.3. The speeds of impregnation are generally higher; they are 10–16 m/min for decorative papers and 30–50 m/min for phenol resin impregnated kraft papers. The volatiles content, resin content and flow are the most essential criteria of prepreg quality. They are pressed in a multi-daylight press (Fig. 9, Chapter 9) with 6–12 laminates per daylight. The pressure is between 8 and 12 N/mm^2, the temperature between 135–160 °C.

The main problem of the HPL production is indicated in Table 12.6. The asymmetrical structure due to different qualities of paper and resin may lead to internal stress, bow and cracks because of the different thermal expansion and swelling when exposed to heat and moisture. In addition, the thermal expansion (approximately $45 \cdot 10^{-6} K^{-1}$) is different in length and cross direction due to the fiber orientation.

Table 12.6. Structure of an HPL-laminate[18]

Number	Definition of paper	Type of Paper	Resin	Resin content %
1	Overlap paper	α-Cellulose 20–50 g/m^2	Melamine	67
1	Decorative paper	α-Cellulose 80–160 g/m^2	Melamine	50
1	Barrier sheet	Kraft paper, bleached 80–160 g/m^2	Melamine	35
X	Core sheet	Kraft paper, unbleached 80–160 g/m^2	Phenol	35
1	Back paper	Kraft paper, unbleached 80–160 g/m^2	Melamine	50

The materials – plywood and particle boards or inorganic materials – also have different coefficients of thermal expansion.

Polyvinyl acetate dispersions, urea- and urea-melamine resins, phenol- and phenol-resorcinol resins are used as adhesives according to the requirements and equipment. The adhesive coat is approximately 80–150 g/m^2. The adhesive bonding can be made at ambient or elevated temperatures up to 60 °C and pressure of 0.015 to 0.5 N/mm^2. The curing time is guided by the formulation, hardener type and temperature.

The properties specified in national standards (Table 12.7), despite of mechanical properties, are e. g. abrasion resistance, scratch resistance, light fastness and resistance to cigarette glow, hot pot bottoms and boiling water[14, 19].

In Europe, the post-forming of decorative laminates which was introduced in the USA 20 years ago, is being increasingly employed due to the tendency towards round edges in furniture manufacturing (Fig. 12.9).

Post forming is a process in which almost completely cured phenolic laminates are heated up to a degree just below the blistering point until pliable and then

Table 12.7. Some national standards covering decorative laminates

USA	NEMA	LD 1
West Germany	DIN	53 799
Great Britain	BS	3794
France	NF T	54-001
Japan	JIS-K	6903

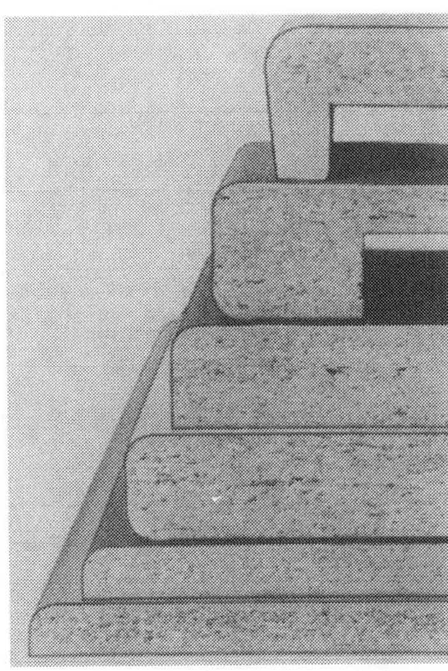

Fig. 12.9. Postformed profiles of particle boards. (Photo: Resopal Werk H. Römmler GmbH, D-6114 Groß-Umstadt)

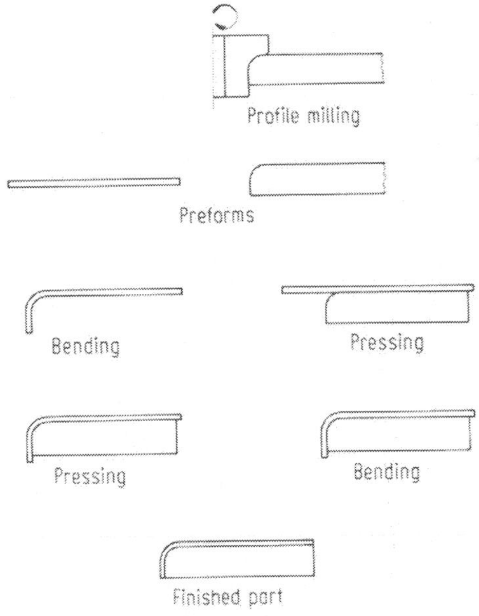

Fig. 12.10. Postforming working process[20]

quickly shaped in a suitable mold. For convex forming the exterior plies (pattern sheet) are strained to tension, the back plies, however, to compression. The center remains free of strain. Kraft paper can be stretched only to a certain extent.
In the USA, therefore, it is still customary to produce core sheets of crepe paper. In Europe, this process is not practicable due to production problems. Here, flexibilized phenolic resins, which show a certain thermoplasticity are used. The moisture content of the laminate plays an important role in the deformation ability[20]. Water facilitates the elongation of the cellulose fiber, affects the sliding of the interfaces and acts as internal lubricant in the resin. The laminates are heated up to temperatures of between 140 and 220 °C by profiles or pipes heated by fuel oil or electricity, sometimes also by IR-radiators. Additional heat can be applied by hot air. PVAc and UF-resins are used as glues.

Phenolic resin impregnated kraft papers (40–80 g/m^2 area weight) are used for overlayed plywood and fiber boards for outdoor applications. The main goal is to improve weather and abrasion resistance and to reduce water absorption. More important examples are concrete form plywood (Fig. 9.14) and construction elements for railway cars and other transportation vehicles[16]. The resin take up is relatively high, in general between 120–200% depending on paper weight. Release agents (stearates, about 1%) can be added to the impregnating resin. The prepregs, which may be pigmented, generally have a storage life between 4–6 months if stored at proper conditions, i. e. not exceeding 25 °C and 65% humidity. Plywood covering can be made at plywood manufacture or in a separate operation at the following conditions: pressure 1.5–3 N/mm^2, temperature 100–120 °C, molding time 5–10 minutes.

12.5. Filters

High porosity and filler-free papers, impregnated with phenolic resins, are used to make oil, fuel and air filters for the automotive industry among others. This impregnation gives the paper the required strength and swelling resistance without reducing the porosity considerably.

Generally, phenol novolak solutions in low boiling solvents like methanol and acetone, which already contain the required amount of HMTA, are used. Such solutions have a viscosity of 400–500 mPa · s at 65% solids content and a gel time of 7–10 minutes at 130 °C. Prior to impregnation, they are diluted to approximately 10–18%. An impregnation machine is shown in Figure 12.3. The resin

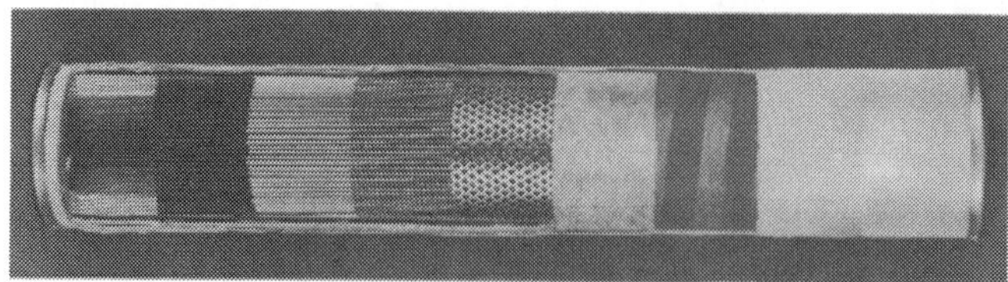

Fig. 12.11. Filter element of a water separator. (Photo: Faudi Feinbau GmbH, D-6370 Oberursel)

Fig. 12.12. Automobile battery and battery separators made of phenolic resin impregnated paper.
(Photo: Moll Akkumulatorenfabrik, D-8623 Staffelstein)

content of the paper should be 20–30%. The temperature and impregnation speed are adjusted so that the resin is only partially cured. The webs are then cut to size, folded and heated up to 180 °C in a tunnel oven, and the resin then completely cured. Those filters must have high porosity, resistance to gasoline and oil, low swelling tendency and mechanical strength as well as sufficient aging resistance. The bursting pressure must be at least 0.2–0.3 N/mm^2. It should not drop below 0.1 N/mm^2 after 24 hours of service at 160 °C.

12.6. Battery Separators

Separators in batteries serve to keep the distance between electrode plates and as electrical insulation between electrodes placed at narrow intervals (Fig. 12.12). They should however, not impair the ion migration and thereby the current flow. Materials for separators are mainly paper and sintered PVC, rarely polyethylene fiber sheets.

Paper separators are made by impregnating paper of controlled porosity with PF-resins, curing, corrugating to form the rib and cutting to size. The resins useful in practice have an average MW ranging between 130 to 300 with a phenol to aldehyde ratio preferably between 1 : 1.8 and 1 : 2.3. Basic catalysts, for example alkali hydroxides, are used for the reaction[22].

Special kraft/cotton linter mixtures are suitable for the manufacturing of separators. The fibers are protected from acid attack by coating them with phenolic resin. The separator must be resistant to sulfuric acid and must not give off substances which affect the electrode reaction. The resin content is generally between 25–50%. The oxidation resistance is a very important criterion in view of the very high potential of the PbO_2 electrode besides high ion permeability. A test device and test performance for this property are indicated in the literature[22]. Paper quality and resin properties as well have a decisive influence on the resistance to oxidation. Separators for automobile batteries, approximately 1.8 mm thick, or separators for heavy duty and stand-by batteries up to 6 mm thick, must also have sufficient mechanical strength. Sometimes, to stabilize the separators, they are covered with perforated plastics sheets or glass fleeces.

References

1. Schwaner, K., Komaromi, C.: Plaste u. Kautschuk *15,* 413 (1968)
2. Knewstubb, N. W.: Plastics *27,* 49 (1962)
3. Eisler, P.: The Technology of Printed Circuits, Chapman & Hall Ltd. 1956
4. Coombs, C. F.: Printed Circuits Handbook. New York: McGraw Hill 1967
5. Isola Werke AG: Herstellverfahren u. Anwend.-möglichkeiten gedruckter Schaltungen. Technical Bulletin
6. Herbert, J.: Electronic Production, October 1974, 35
7. Tillessen, R.: Feinwerktechnik & Meßtechnik *85,* 8 (1977)
8. Bakelite GmbH: Bakelite Phenolharze-Bindemittel für Elektroschichtpreßstoffe. Technical Bulletin
9. Nylen, P., Sunderland, E.: Modern Surface Coatings. New York: Interscience 1965
10. Thomas, A.: Fette und Öle. In: Ullmanns Encyclopädie d. techn. Chem., Vol. *11,* 4. Ed. Weinheim: Verlag Chemie 1976
11. Dynamit Nobel AG: DE-OS 2 142 890
12. Caratsch AG: Technical Bulletin
13. G. Siempelkamp & Co: Press Lines and Finishing Lines for the Production of Technical Laminates. Technical Bulletin
14. Franz, A., Jüngerich, W., Schröder, W.: Schichtstoffe, Hartpapier u. Hartgewebe. In: R. Vieweg, E. Becker (ed.): Kunststoff-Handbuch Bd. 10 – Duroplaste. München: Carl Hanser 1968
15. Isola Werke AG: Spinnspulen aus Hartpapier CARTA für d. Chemiefaser-Ind., Technical Bulletin
16. Enzesberger, W.: Holz als Roh- u. Werkstoff *27,* 31 (1968)
17. Böhme, P., Karger, S.: Holztechnologie *10,* 130 (1969)
18. Steinmetz, O. H.: Dekorative Schichtstoff-Platten. Stuttgart: DRW-Verlag
19. Resopal Werk H. Römmler GmbH: Techn. Daten Resopal-dekorative Schichtpresstoffplatten, Technical Bulletin
20. Menge, W.: Verfahrenstechnik des Verarbeitens dekorativer Schichtstoffe im Post-forming-Verfahren zum Fertigen von Möbelelementen. In: Verbund v. Holzwerkstoff u. Kunststoff in d. Möbelindustrie. Düsseldorf: VDI-Verlag 1977
21. Berndt, D.: Galvanische Elemente. In: Ullmanns Encyclopädie d. techn. Chem., 4. Ed. Vol. 12. Weinheim: Verlag Chemie 1976
22. Monsanto Co.: US-PS 3, *926,* 679 (1975)

13. Coatings

All structural metals corrode to some extent in their natural environment. Especially critical is the corrosion of iron and steel, which in the presence of oxygen and water are destroyed very fast. Usually corrosion does not take place in the absence of either one[1]. High-performance organic coatings are used to provide the passivation, galvanic protection and barrier properties, which are necessary for corrosion inhibition. High adhesion and low water vapor and oxygen transmission (the reasons are explained in Chapter 18) are the decisive reasons for the high effectiveness of phenolic coatings. Further advantages are the excellent resistance to chemicals and temperature. Since non-modified phenolic resins result in very brittle coatings because of structural characteristics of cured resins, a lot of research work was performed in order to flexibilize the resins without limiting the advantages of phenolic structures.

Therefore, phenolic resins are never used alone in coatings but always in combination with more flexible, hydrophobic resins, e. g. epoxy-, alkyd- or natural resins, maleinized oils and polyvinylbutyral. Because of their individual coloration and tendency towards discoloration, they are used mainly as primers and undercoats. The solubility in hydrocarbons and aromatics, if required, is achieved by use of alkyl phenols. The cure is mostly performed by baking at temperatures between 160–200 °C, the cross-linking at room temperature by the addition of acids or by prereaction with drying oils. Comprehensive publications concerning the chemical modifications of phenolic resins for coating applications have been issued by Hultzsch[2]. The most important fields of application for phenolic coating resins are automotive primers, coatings for metal containers, anticorrosive marine paints and printing inks. The highest rates of growth are observed for electrodeposition primers for the automotive industry.

13.1. Automotive Coatings

Automotive coatings consist of an undercoat and a topcoat, sealers as midcoats between two layers may be used as a barrier against resin migration to avoid discoloration and to improve intercoat adhesion[4]. The main material to be coated is cold rolled steel, the surface of which is pretreated in several operations as alkali cleaning, acid etching and chromate rinse. The formation of a 2–3 μm thick crystalline surface layer consisting of hydrated zinc iron phosphate was found to be most useful. The automotive undercoat is applied by electrodeposition or dipping.

The electrocoating process is gaining more, and more importance. In principle, an electrodeposition paint consists primarily of a pigmented water-thinnable amino- or, in rare cases, alkali-solubilized polycarboxylic acid resin, which in many cases is modi-

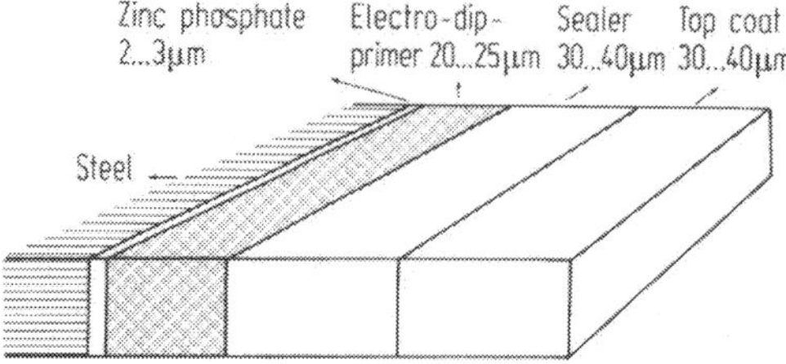

Fig. 13.1. Metal pretreatment and coating systems for automotive body corrosion protection

Fig. 13.2. Electrophoretic dip priming. (Photo: Volkswagenwerk AG, D-3180 Wolfsburg)

fied with phenolic resins. The fundamental property which such vehicles must have is the capability of being converted readily from a highly hydrophilic (water soluble or at least water dispersible) state to a highly hydrophobic state after deposition and curing[5].

The modern automotive enamel topcoat is in most cases an acrylic resin[6] containing hydroxyl groups (2-hydroxyethylacrylate), which is cross-linked with MF-resins at moderate conditions, or an alkyd-melamine resin combination which is often used in Europe[7].

13.1.1. Water-Borne Paints and Electrodeposition

Water-borne or water-reducible coatings[8] gain ever more importance due to ecological and safety reasons as maintenance and industrial paints on interior and exterior metal substrates. The most frequently used method to obtain water solubility of macromolecules is the introduction of carboxy-, sulfone- and sulfomethyl (see also Section 19.9) groups and following ion formation by reaction with ammonia, amines or inorganic bases. It is important that the carboxyl group is not fixed over a saponifiable ester link but over a non-hydrolizable carbon-carbon bond on the polymer backbone.

Mononuclear hydroxymethylphenols are per se water soluble, higher molecular weight resols only by addition of higher amounts of inorganic bases, however, such resins are not suitable for coatings. The first water soluble phenolic coating resins, called Residrines, were developed by Hönel. Alkyd resin was used as carboxyl group carrier. Many water-reducible phenolic resins containing coatings are described in the literature and patents[9-14]. Also carboxyl group containing monomers (13.1)[13] and (13.2) can be used.

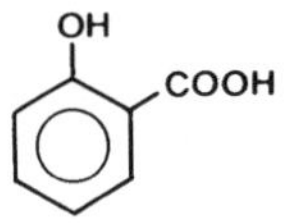

OH
COOH

(13.1)

Salicylic acid

CH_3
HO — C — OH
CH_2
CH_2
COOH

(13.2)

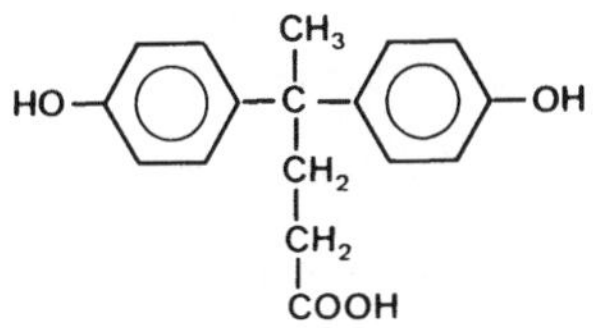

4,4-Bis(4-hydroxyphenyl)
valerianic acid
("Diphenolic acid")

A steadily increasing portion of current automotive body priming is done by the electrodeposition process using water-borne paints. Natural drying oils, expoxy esters, polybutadiene-, alkyd- and acrylic resins are used as vehicles. Maleinized oils, which are obtained by the addition of MSA to natural oils, mainly linseed oil, in proportions between 5 and 25%, exhibit high impact strength and flexibility but unsatisfactory corrosion resistance because of the hydrolizable ester link[14]. The completely non-saponifiable liquid diene polymers including cis-1,4- and 1,2-polybutadiene[15] are superior in this respect. However, the impact strength of these films is minor, so that combination with linseed oil is sometimes a good compromise[16]. To meet the high requirements for corrosion resistance and throwing power, such vehicles must be cross-linked with PF-resins[17, 18]. The phenolic resins to be used in those combinations can be of novolak as well as of a resol type. Up to now, BPA-resols etherified with methanol or butanol have proven to be the most efficient. This applies especially to the corrosion resistance[19]. To increase the flexibility of the film, an addition of *p*-alkylphenols is favorable. Normally, the phenolic resin is prereacted with the carboxylic polymer, which improves the paint stability and avoids the bleeding problem. Novolaks are prereacted with the non-hydrolized maleinized polymer under half-ester formation.

Of the anticorrosive pigments lead silico chromate may be used. Lead- and barium chromates result in stable paints but show only low anticorrosive action. The use of aqueous anionic vehicles and the application method restrict the choice and level of pigmentation. Suitable pigments are titanium dioxide, iron oxide, carbon black and most organic pigments which are stable under alkaline conditions[5, 20]. Extenders used include calcium carbonate, talc, silica, barites and China clay. The most frequently used solvents are glycol ethers, alcohols or cyclohexanone; in general, coupling solvents which are partly compatible with both phases. Water immiscible solvents as defoamers yield smoother films; pigment wetting agents and surfactants may be used as additives.

The preparation of the vehicle and the paint formulation can be performed in the following way: The polybutadiene resin is heated in the presence of acetylacetone (~0.04%) as catalyst and copper naphthenate (~0.6%), to prevent gelation, with agitation in a nitrogen atmosphere up to 100 °C, and then 20 parts of maleic anhydride are added. The temperature is raised to 190 °C and held for approximately 3 hours. The α-allyl carbon addition is believed to be the main reaction mechanism, rather than 1.4 dipolar (Diels-Alder) addition.

After the complete or only partial hydrolysis of the anhydride group, the prereaction with the phenol resol resin is performed at 70–80 °C. Melamine formaldehyde resins and extenders (e. g. hydrocarbon resin) may be added at this stage. Then, the amine (60 °C, 30 min) and the pigment paste, which was prepared in a separate process, are added. The paint, as applied in the tank, has a solids content of 12–15%, a pH of 6.7–6.8 and a conductivity of 3,000–3,500 μS.

The vehicle should have a high ratio of MW to ionic charge, and all the components of the paint must be transported and deposited in the ratio in which they appear in the formula. The amine-neutralized and thereby negatively charged polymer particles move, if current is passed through the system, to the article to be coated which has been made the anode, and the now insoluble polymer forms a semipermeable film from which almost all water is removed by electro-osmosis. Initially, deposition takes place at areas of high current density, and as the film resistance increases, the deposition will continue in shielded sections until the whole surface is covered by a uniform film. This ability of the paint to coat completely recessed and shielded areas is designated as throwing power. The deposite film contains approximately 90% of solids and is sufficiently strong and adherent in the noncured state to permit rinsing with water to remove the surplus paint called dragout[21]. During the polymer deposition at the anode, the positive ammonia ions move to the cathode where, after the discharge, the amine is reformed. A major improvement during the past years has been the use of ultrafiltration to avoid waste problems and to reuse the dragout paint. Anodic electrodeposition causes some dissolution of iron zinc phosphate and iron leading to film staining and reduced salt spray resistance[22]. This can be avoided by cationic deposition which should offer also better corrosion resistance[23, 24].

13.2. Coatings for Metal Containers

The choice of a container coating is influenced by requirements dictated by the product to be packed, by the container construction and by the manufacturing methods used in making or filling the container[25]. Metal containers are made of steel (cold-rolled steel, blackplate), chromium plated steel, tinplate, aluminium or aluminium alloys. The sheet stock is generally coated by coil coating in highly automated units (Fig. 13.3). The following fabrication into the final product partly involves extreme deformations. Therefore, one of the major prerequisites of a coating is the flexibility of the film. As far as food containers are concerned, it is essential that the taste or smell of the content is not affected by the coating. The use of coatings for food packaging is subject of official regulations, for instance in

Fig. 13.3. Coil coating line, above, pretreatment section, below, coating and curing section. (Photo: Eisenmann KG, D-7030 Böblingen)

the USA issued by the Food and Drug Administration (FDA) and the US Department of Agriculture (USDA).

The coating must be resistant to sterilization, since not only the content, but also the container must be sterilized. Beer cans are sterilized at 63 °C/45 min, meal products and cans usually at 120 °C/90 min. The problems of food packaging are extremely versatile. Most of the food stuffs with few exceptions are of acidic character. Proteins containing sulfur cause black stains on the metal because of the formation of iron- and tin sulfide, as in fish or pea cans. Tomatoes and spinach also cause discoloration of the coating. Zinc oxide can be added to bind the sulfur under formation of insoluble white zinc sulfide.

Phenolic coatings offer excellent resistance to corrosion, to chemicals and to sulfur staining compounds. Therefore, they are used for pails, drums, collapsible tubes, aerosol cans and for the interior and exterior of food containers. If direct contact with the content must be avoided, a thin phenolic prime coat and a vinyl lining may be used. Mainly, resols based on cresols, sometimes also based on phenol, are used. The yellow individual coloration (gold lacquers) of ammonia catalyzed resols and the low flexibility are disadvantages. Lighter or almost colorless films can be obtained by the use of etherified phenol- or BPA-resols and phenol/cresol

Table 13.1. Epoxy/phenolic can coating formulation, solids content 40%[26]

25.6 pbw	Epoxy resin (Rütapox 0197)
14.4 pbw	Phenolic resin (Bakelite 100)
0.3 pbw	Silicone oil (leveling agent)
20.0 pbw	Ethylglycol
39.5 pbw	Ethylglycol acetate
0.2 pbw	Phosphoric acid

blends. Bisphenol-A is employed when taste is a critical factor. Alkylphenol resins are occasionally used to improve the flexibility. Curing is effected at 180–220 °C for 15–20 min, temperatures up to 300 °C have been employed. In general, the phenolic resin must be flexibilized with other polymers, for instance epoxy-, polyvinylbutyral- or alkyd resins. A standard can coating formulation is described in Table 13.1.

The phosphoric acid promotes a very fast cure. Dodecylbenzenesulfonic acid is also approved by the FDA as catalyst. Microvoids caused by traces of inorganic salts and hydroxides in the resins or by undercuring which can also occur by traces of basic catalysts which neutralize the phosphoric acid, often are the cause of lacking sterilization resistance. Prereaction of the resin at 60–80 °C results in most cases in a considerable improvement of the coating performance.

13.3. Marine Paints

Marine coatings must prevent corrosion, which is always electrochemical in character, prevent fouling and minimize surface roughness. Impact strength is also an important property. They have to prevent current conduction and must be resistant to cathodic protection. The coating of underwater sections consists of three basic coats: the primer, an anticorrosive- and an antifouling paint. Hot rolled steel as used for ships' plating is first cleaned by shotblasting to remove the iron oxide layer. Then a quick drying shop primer – the most successful ones are based on phenolic resin/polyvinylbutyral combination, zinc dust or aluminum/epoxide resin combination or zinc silicate, 10–15 μm thick – is applied[27]. When ships are repaired, the painting operation starts with the cleaning of the steel surface by brushing, flame or sand blasting. Then a wash primer[28] based on polyvinylbutyral/phenolic resin/phosphoric acid with a dry film thickness of 10–15 μm is applied. The anticorrosive paint for the second coat may consist of oil- and rosin-modified phenolic resin, chlorinated rubber, vinyl chloride/vinyl acetate copolymers, coal tar/epoxide resin combinations or zinc silicate. These paints are normally applied by rollers, seldom by spraying, either in the dry dock or at sea in the open air. The coat is 100–120 μm thick. The antifouling paint consists of a biocidal material, in most cases copper(I) oxide, sometimes organic tin- and lead compounds which are even more effective, a bonding agent, fillers and solvents. Copper(I) oxide is normally placed in a partially soluble matrix composed of rosin and an oil based polymer[29]. Insoluble matrix coatings, based on vinyl resins, often called contact coatings, require higher loadings of biocides. The use of phenolic resins has been recommended[30].

13.3.1. Shop Primers

A 10–15 μm thick coat of shop primer should protect steel plates from corrosion during the building operations in the shop or outside for a maximum of 12 months. At the same time, it should improve the adhesion of the following coats. The formulation is restricted by the requirement, that it must not affect the quality of the welds. Some active pigments[20, 31, 32], for instance chromates, should be omitted for ecological reasons. The separation of gases and vapors detrimental to human

Table 13.2. Formulation of a wash primer[26)]

		pbw
Resin	Phenol resol resin (Bakelite LG 721)	12.7
	Polyvinylbutyral (Mowital B 30 H)	12.1
	Epoxide resin (Rütapox 0191)	0.9
	Zinctetraoxychromate	8.6
	Talcum, micronized	5.7
	Aerosil	0.3
	Wetting agent	0.4
	Isopropanol	42.9
	n-Butanol	16.4
Hardener	Phosphoric acid, 85%	6.0
	Isopropanol	29.0

health during plate cutting and welding must be prevented. No slag inclusions, gas bubbles or ash must be found in the welding seams. A shop primer's composition corresponds to a great extent to the one of a wash primer which is described in the next section (Table 13.2). The essential components are polyvinylbutyral, phenolic resin and micronized iron oxide.

13.3.2. Wash Primers

This name is somewhat misleading. A wash primer is applied by brush or by spraying. Up to World War II, the pretreatment of the cleaned steel surface with diluted phosphoric acid and red lead was the only, although unsatisfactory primer for ship bottoms. Modern primers for steel surfaces are built on polyvinylbutyral, phenolic resins and alcoholic phosphoric acid solutions[33)]. An example of a two component primer formulation is given in Table 13.2.

Phenol, sometimes mixed with cresols, is mostly used as monomer. The flexibility can be improved by the addition of *p*-alkylphenols.

13.3.3. Oil-modified Phenolic Resin Paints

The most important applications for oil-modified phenolics are anticorrosive undercoats for ship painting and boat paints. Similar multicoat systems are customary for other transportation vehicles. Paints for railway cars may consist of an epoxy primer, an urethane oil/alkyd-modified phenolic undercoat and an urethane oil/alkyd based finish[34)]. Alkyl- and arylphenolic resins can be cooked with drying oils due to their reduced self-reactivity[2)]. Tung oil is preferred, sometimes linseed oil or castor oil are used. The proportion of phenolic resin is between 25 (resols) and 100% (novolaks) in relation to the oil, depending on self-condensation ability. The reaction of novolak resin made of *p*-tert.-butylphenol, *p*-octylphenol or *p*-phenylphenol is performed by the following procedure in order to prevent gelation. One half of the resin is dissolved in the oil and heated up to 190 °C for 60 min. Then, the rest of the resin is added and cooked at 230–240 °C until foaming ends and then for an additional 30 min, till the end of the reaction. After cooling, the modified resin is diluted with white spirit and aromatic solvents. In order to accelerate air drying, cobalt- or lead siccatives are added, supplemented by additives to prevent wrinkling and skin formation. Such coatings are tack-free between 6 and 16 hours at ambient temperature, depending on the content of tung oil.

13.4. Printing Inks

Up to the beginning of the 20th century, linseed oil in combination with carbon black and other pigments was exclusively used to make printing inks. New printing procedures and new materials to be printed made it necessary to supplement the linseed oil by the more effective synthetic polymers. The different printing procedures are divided into four main types: relief, planografic, intaglio and screen printing[35]. There is a clear tendency towards roller offset and rotation intaglio printing at the present time. Relief printing is of no importance nowadays.

The most important application for phenolic resins as binder for printing inks are the publication gravure and offset inks which are used to a large extent for the printing of magazines and catalogues. Because of the extremely high speeds of the paper web (up to 800 m/min), the drying rate becomes the decisive factor of the printing ink performance. Phenolic resin-modified colophonium resins give off solvents very fast and completely. Mostly they are mixed with other resins due to cost. Publication gravure inks may therefore contain 50–60% toluene and approximately 15% of a colophonium-modified phenolic resin, 15% calcium resinate, changing proportions of hydrocarbon resin and approximately 5–10% pigments. The resins should effect a fine and uniform distribution of the pigments during the ink preparation on a three-roll mill and also during the printing process. The solvent must be released very fast and a hard film formed. The modified colophonium resins (type B inks in the USA) are superior to the resinates (type A) and hydrocarbon resins as far as rate of solvent loss and film hardness is concerned. Cresols, xylenols or alkylphenols[36] are used as phenolic components. The compatibility with solvents and mineral oils is important in the choice of these components. In France, the use of toluene is not allowed, white spirit is used instead. Apart from solvent release, melting point, viscosity, yellowing and smell are further criteria.

13.4.1. Rosin-Modified Phenolic Resins

Rosin-modified phenolic resins are used in printing inks, in oil lacquers and as additives to alkyd paints because of their good compatibility with natural oils ("art copals") in which they improve the drying and shine. However, the designation "phenolic resin-modified rosin" would be more accurate because of the low phenolic resin content, generally at about 20%. Colophonium consists mostly of rosin acids, which belong to the class of the diterpenes (abietic-, laevo-pimaric-, neo-abietic acid etc.), and changing amounts of neutral components (rosin acid esters and several diterpenes[37, 38]). The reaction mechanism with phenolic resins and the structure of the formed products, however, cannot be exactly defined (13.3). The molecular structure and the MP (between 100–140 °C) decide on their solubility in aromatics and white spirit[2, 3]. Low molecular, "oil-reactive" alkylphenol resols are most frequently used to modify colophonium. They transform the monofunctional

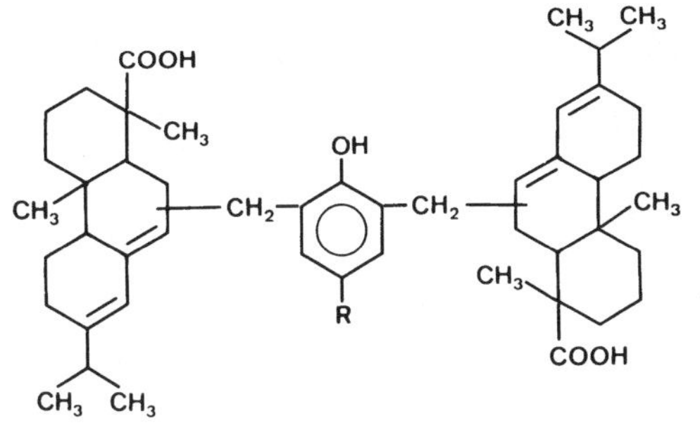

(13.3)

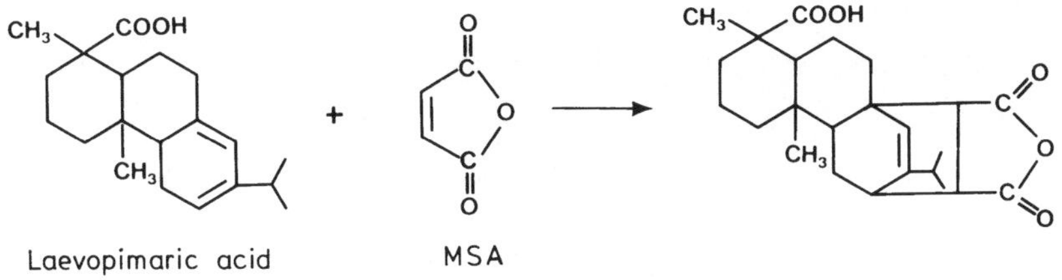

Laevopimaric acid + MSA → (13.4)

rosin acids into a polycarboxylic acid polymer (polyester) either by chroman ring formation (see 3.3.5 and 17.1) or more probably by alkylation favored by the acidity of the carboxyl group. The phenolic resin is added to the molten colophonium at 110–140 °C. Under these conditions the resin must be easily soluble, because otherwise the self-condensation of the resol would first occur. Then, the temperature is raised to approximately 250 °C, whereby polyalcohols like glycerol or pentaerythrite can be added last for esterification and to increase the molecular weight. The decarboxylation starts above 250 °C. In individual cases, cooking with novolaks at relatively high temperatures is performed. Apart from the joint condensation of all components, rosin acids can also be prereacted with formaldehyde (Prins reaction mechanism, 2.17), yielding hydroxyl group containing compounds. The softening range of colophonium is raised by approximately 45 °C to 105 °C by such reactions. Maleic acid anhydride (MSA) may also be slowly added to rosin acids at temperatures above 125 °C (1,4-dipolar addition mechanism, 13.4). The maleinate resins which are formed by this process have considerable economic importance. Phenolic resin esters can be prepared by reaction with anhydrides.

13.5. Other Applications

Chemical equipments and plants are not only endangered by environmental influences and high temperatures, but in addition, by aggressive chemical compounds. Such components are containers, reactors, pipe lines and heat exchangers. After the usual pretreatment of the steel to a surface roughness of 30–60 μm, differently pigmented heat curing phenolic coatings are applied in several operations either by dipping or pneumatic or electrostatic spraying. The thickness of one coat is 30–40 μm. After predrying for approximately one hour at room temperature, the coat is heated slowly up to 120 °C with a holding time of 15–30 minutes. This process

Fig. 13.4. Phenolic resin coated heat exchangers before cure, drying oven in behind. (Photo: Secova Ges. für Korrosionsschutz, D-4100 Duisburg)

is repeated until the desired thickness of the coat (up to 250 μm) is reached. It is then baked at 180–200 °C for two hours[39].

Non-plasticized phenol resols can be used as sole bonding agents as well as in combination with polyvinylbutyral, epoxy resins or furane resins. The alkali resistance can be considerably improved by alkylation of the phenolic hydroxyl group with allyl alcohol, allyl chloride, dichloropropanol or epoxide compounds.

Cresol-, xylenol-, and phenol resols have a certain importance in electrical engineering, either alone or in combination with flexibilizing resins (PVB, PVF, alkyds) as impregnating-, dynamo sheet- and wire varnishes. The requirements for this application are strong adhesion on copper, brass and aluminum, high electrical insulating property, temperature resistance and resistance to humidity and transformer oil[15].

Protective coatings are used extensively in nuclear power plants for corrosion protection, maintenance and to facilitate the removal of radionuclide soils (decontamination). The high resistance of phenolic resins to high energy radiation was pointed out in Chapter 7.3. The decontamination of concrete is difficult to perform. Epoxy, epoxy-phenolic or phenolic surfacings are used instead of vinyl paints in critical areas[40].

References

1. Berger, D. M.: Chem. Engng., August 1977, p. 77
2. Hultzsch, K.: Phenol-Lackharze u. abgewandelte Phenolharze. In: Vieweg, R., Becker, E. (ed.): Kunststoff-Handbuch Bd. 10, Duroplaste. München: Carl Hanser 1968
3. Kittel, H. (ed.): Lehrbuch d. Lacke u. Beschichtungen Bd. 1, Teil 1, Stuttgart, Berlin: W. A. Columb.
4. McBane, B. N.: Automotive Coatings, In: R. R. Myers, J. S. Long (ed.): Treatise on Coatings, Vol. 4, Formulations. New York: Marcel Dekker 1975

5. Rousse, R. E.: J. Oil & Col. Chem. Assoc. *55*, 373 (1972)
6. Kline & Co. C. H. Fairfiled: The Kline Guide to the Paint Industry (1975)
7. Volkswagenwerk AG: Technical Bulletin
8. McEvan, J. H.: The Role of Water in Water-Reducible Paint Formulations. J. Paint Technology *45*, 33 (1973)
9. Vianova Kunstharz AG: ÖS-PS 180 407; ÖS-PS 299543
10. Reichhold-Albert-Chemie AG: US-PS 1 254 528; US-PS 1 254 529; DE-OS 1 669 286; DE-OS 2 046 458
11. Dtsch. Farben-Z. *27*, 370 (1973)
12. Kansai Paint Co. Ltd.: DE-OS 2 237 830
13. Vianova Kunstharz AG: DE-PS 113 775
14. Daimer, W.: Dtsch. Farben-Z. *27*, 358 (1973)
15. Chemische Werke Hüls: US-PS 3 546 184
16. Kita, R., Kimi, A.: J. Coatings Technology *48*, 53 (1976)
17. Ford Motor Co: DE-OS 1 794 354
18. BASF AG: DE-OS 1 949 294
19. Rifi, M. R.: J. Paint Technology *45*, 73 (1973)
20. Anisfeld, J.: Farbe u. Lack *81*, 1024 (1975)
21. Levinson, S. B.: J. Paint Technology *44*, 41 (1972)
22. May, C. A.: J. Paint Technology *43*, 43 (1971)
23. Verdino, H.: J. Oil & Col. Chem. Assoc. *59*, 81 (1976)
24. Hatch, M. J.: Progr. Organic Coatings *1*, 61 (1976)
25. Gerhardt, G. W., Seagren, G. W.: Coatings for Metal Containers. In: R. R. Myers, J. S. Long (ed.): Treatise on Coatings, Vol. 4, Formulations. New York: Marcel Dekker 1975
26. Bakelite GmbH: Phenolharze für d. Lackindustrie. Technical Bulletin
27. James, D. H.: Marine Paints. In: R. R. Myers, J. S. Long (ed.): Treatise on Coatings, Vol. 4, Formulation. New York: Marcel Dekker 1975
28. van Oeteren, K. A.: Korrosion u. Korrosionsschutz. Hannover: C. R. Vincentz 1967
29. Bufkin, G., Bounds, R., Thames, S. H.: Paint and Varnish Production, Feb. 1974, P. 25
30. Pitre, A. S., Saroyan, J. R.: US-PS 2 579 610
31. Hare, C. H.: Paint and Varnish Productions July 1974, P. 19
32. Ruff, J.: Dtsch. Farben-Z. *25*, 199 (1971)
33. Lampe, W.: Die Funktion der Phosphorsäure in Wash-Primern. Kunstharz-Nachrichten Hoechst AG
34. Timmins, F. D.: J. Oil & Col. Chem. Assoc. *60*, 291 (1977)
35. Sixtus, H.: Druckfarben. In: Ullmanns Encyclopädie d. techn. Chem. 4. Aufl. Bd. 10, Weinheim: Verlag Chemie 1976
36. Hoechst AG: US-PS 1.428 285
37. Barendrecht, W., Less, L. J.: Harze, natürliche. In: Ullmanns Encyclopädie d. techn. Chem. Vol. 12, 4. Ed., Weinheim: Verlag Chemie 1976
38. Nylen, P., Sunderland, E.: Modern Surface Coatings. New York: Interscience 1965
39. Ernst, H., Mushak, J.: Plaste u. Kautschuk *24*, 284 (1977)
40. Gad, W.: Kunststoffe im Bau *11*, 5 (1976)

14. Foundry Resins

The production of shaped parts of metal melts can be performed either in permanent molds or in lost molds[1, 2]. Permanent molds made of metal, graphite or ceramics are used to cast low melting non-ferrous metals. Lost molds consisting of fire-proof sands, an inorganic or organic bonding agent and sometimes various additives, are used for iron and other metals. Molten metal is poured into the mold cavity and solidifies to the desired shape. The mold becomes brittle under the influence of the temperature of the molten metal and can be readily removed from the casting. One mold (and core) must be used for each casting.

The foundry industry is a market of increasing importance for phenolic resins. During the last five years, it was not the quantitative increase of the foundry business which attributed to the increase in use of phenolic resins, but various changes in the production processes and the type of the bonding agents used. The structure of the foundry industry is different in different countries. As far as the turnover is concerned, West-Germany has the greatest in West-Europe followed by Great Britain, France and Italy.

Table 14.1. Production of cast iron and steel in West-Germany in million tons per year[3)]

Production (mio. to/a)	1977
Gray cast iron	2.93
Spheroidal graphite cast iron	0.59
Cast steel	0.28
Malleable cast iron	0.21
total	4.01

Table 14.2. Use of foundry products in particular industries. West-Germany, 1977[3)]

	%
Automotive industry	33
Machine construction	31
Construction business	13
Steel industry	11
Others	12

Fig. 14.1. Metal casting. (Photo: Halbergerhütte, D-6700 Ludwigshafen)

Fig. 14.2. Phenol resin bonded sand mold and core by shell process. Cylinder block for an automobile motor.
(Photo: Halbergerhütte, D-6700 Ludwigshafen)

In West-Germany 494 foundries with 103,300 workers (1977) and in the USA approximately 4,500 foundries with 375,000 workers (1975) were in operation. A concentration process is taking place at the present time in all industrial countries and is still continuing. The most important markets for foundry products are listed in Table 14.2.

For some time, the automotive industry has been a very important market in all western countries due to the high rates of growth. The machine construction industry is currently stagnant. The steel industry has suffered the greatest losses.

14.1. Mold- and Core-Making Processes

Molds for casting operations and cores, separate shapes which are placed in the mold to provide castings with contours, cavities and passages, are composed mainly of sand and a suitable binder. The binder may be of organic as well as inorganic nature. Natural inorganic binders are clays (montmorillonite, glauconite, kaolinite); synthetic ones are, for instance, sodium silicate, cement or gypsum. Mainly phenol-, furane- and urea resins are used as organic binders. Various oils and carbohydrate binders are of minor importance. However, there also are physical forming processes for sand molds (magnetic molding process and vacuum process).

14.1.1. Inorganic Binders

Green sand molds are made of sand, clay, water and sometimes organic additives and are used without further conditioning or drying (green). They are widely used as the least expensive method of producing a mold, but their dimensional accuracy and stability is low. Ferrous and non-ferrous castings can be produced in rapid production cycles[4)]. Higher strength and dimensional stability can be obtained by drying in dry climates (skindried molds) or by oven drying (dry sand molds).

Sodium silicate is a further inorganic binder (CO_2 process) with the attraction of low cost, cleanliness and low toxicity. The disadvantages of this process are relatively low strength, limited storability and low breakdown capability of the mold. Many organic additives have been recommended to accelerate hardening, enhance the breakdown capability and reduce water absorption. Cement bonded sands can be used for the production of very large castings, but the low hardening rate has limited their wider use. Quick setting cements and accelerating additives have been recommended recently[5].

14.1.2. Organic Binders

Although Baekeland recommended sand as filler for phenol-formaldehyde resins in one of his patents in 1907, it took more than 30 years until Croning in Hamburg succeeded in developing a shell molding process after a lot of experiments, in which synthetic thermosetting resins were used for the first time instead of oils, starch or waste materials of the cellulose production. Phenolic resin bonded cores and molds possess a number of favorable properties which are necessary for economic mass production of high quality castings. These are:

- short hardening time,
- high dimensional stability and hardness of cores and shells,
- good storability of cores and shells,
- high castings surface quality,
- good breakdown capability of cores and shells,
- low expenditures for cleaning.

The most important thermosetting resins used today are furane-type resins based on furfuryl alcohol (including FA/UF- and FA/PF/UF-resin blends), phenolic resins (including PF/UF- and PF/FA/UF-resin blends) and, by a considerable margin, polyurethanes. The most important foundry processes using synthetic polymer materials are the no-bake, shell molding, hot-box and cold-box processes. Newer publications indicate, that one of the most important considerations in the selection of a process is its energy consumption. Therefore, today there is a marked trend towards the cold-set processes. Considering the binders, phenol resins and furane resins gain further importance to the disadvantage of the inorganic binders, but the organic materials should satisfy the requirements of work safety and environmental protection to the greatest extent. Further factors to be considered are:

- type of metal to be cast,
- number of pieces and production rate,
- volume of the casting,
- dimensional accuracy,
- availability of skilled personnel,
- extent of capital investment,
- types of sand available.

14.1.3. Requirements of Foundry Sands

Silica sands of high purity are used for most sand molding operations. While some countries like West-Germany, Belgium, The Netherlands and France have very good qualities of sand (99,8% SiO_2), other countries, however, like Italy or Japan must import sand for the production of high-quality iron and steel castings. The chemical purity as well as the morphology of the grains affects the economy and strength considerably. Sands with poor surface quality need much larger amounts of binder to obtain adequate strength, thus increasing costs and the amount of gases evolved at the casting process causing casting defects and higher environmental pollution[10, 11].

It is obvious that considerably larger amounts of acidic hardeners are needed when sands with higher portions of basic oxides (Na_2O, K_2O, CaO, MgO) are used in the cold-set process in order to obtain comparable hardening rates. This is also the case with a higher humidity content. Advantages of silica sand as mold and core material are low costs, ready availability in different gradings, consistency in composition, compatibility with all types of chemical binders and adequate thermal and chemical resistance in connection with metal melts.

On the other hand, there also are some disadvantages, for instance the relatively high thermal expansion which frequently results in expansion defects, especially if large castings are involved. Silica sand can react with certain molten metal oxides. Frequently, problems arise at the casting of austenitic manganese steels. The high and abrupt expansion of quartz (silica) is caused by inversions of the crystalline structure depending on temperature and pressure.

$$\alpha\text{-quartz} \underset{}{\overset{575\,^\circ C}{\rightleftharpoons}} \beta\text{-quartz} \overset{870\,^\circ C}{\rightleftharpoons} \beta\text{-tridymit} \overset{1470\,^\circ C}{\rightleftharpoons} \beta\text{-cristobalit} \overset{1705\,^\circ C}{\rightleftharpoons} \text{melt}$$

trigonal — hexagonal — hexagonal — cubic

These expansion-type defects can be avoided by other materials like chromite, zircon or olivine. These sands, of course, are considerably more expensive and some of them are only available to a limited extent, so that their use requires a thorough economic analysis. The thermal expansion of chromite, zircon and olivine (Table 14.3) is uniform and clearly smaller compared to that of quartz. They are used in just about the same order as listed below.

A further, very important factor is grain shape and size distribution. When liquid metal comes into contact with the mold, steam and different gases of decomposed organic binders are generated. The mold must be porous enough to allow the escape of gases and so prevent an excessive pressure build up in the cavity which possibly might result in casting defects and damage of the mold. On the other hand, the generation of gases has some advantages, for instance preventing metal penetration and giving a good casting surface and excellent delineation to the casting. However, a balance between gas evolution and permeability must be maintained[1].

Table 14.3. Properties of different sand materials

Characterization	Sand material			
	Silica	Chromite	Zircon	Olivine
Chemical composition	SiO_2	$Cr_2O_3 \cdot FeO$	$ZrSiO_4$	$(Mg, Fe)_2\,SiO_4$
Specific gravity g/cm^3	2.64	4.5	4.6	3.2
Expansion at 900 °C %	1.56	0.65	0.25	1.02
Volume price relation, silica = 1	1	9–11	60–70	6–7

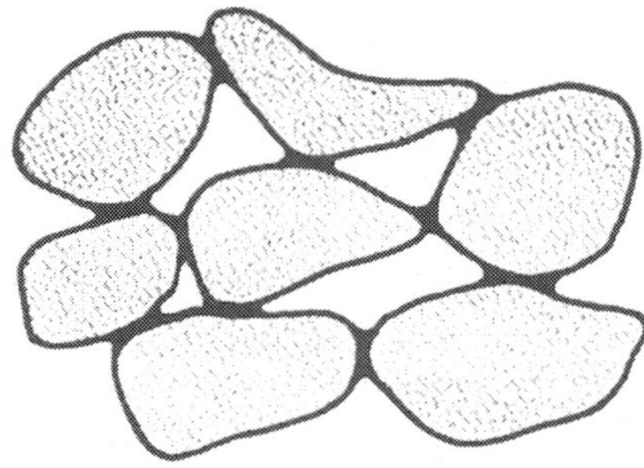

Fig. 14.3.
Distribution of resin in a bonded sand[22]

If the sand grains have a perfectly round shape and uniform diameter, the total volume of pores per unit volume is independent of the diameter of the grains. The pressure increase due to friction dependent upon the void diameter must be considered. Although large pores are desirable to permit the escape of gases, they can permit the penetration of the liquid metal into the voids. Fine sands and multimesh sands offer better resistance to erosion and better surface quality of the casting, but require more bonding material. All of these factors must be considered and invariably a compromise must be made.

Due to the highly increased costs for sand, transport and disposal, sand reclamation becomes increasingly important. Plants are available on the market for the following regeneration processes:

- mechanical regeneration (breaking, grinding and dust removal)
- pneumatic regeneration (impact screening)
- wet regeneration (washing process for inorganic binders)
- thermal regeneration (incineration process)

Almost 200 plants are already in operation in West Germany. Occasionally, reclaimed sand needs a lower amount of binder because the sand pores are sealed.

Recently, the problems which may be caused by deposits of used synthetic resin bonded sands have been thoroughly studied. Phenol levels found at used sand elution with water, are definitely lower than those determined in the sewage of household[12].

14.2. Shell Molding Process

In the shell molding process a mixture of sand and phenol resin, today only resin coated sand, is placed against a heated metal pattern (250–280 °C) by means of a dump-box or blowing with special machines into hot, closed molds. Prior to this, the mold is sprayed with a release agent, usually based on silicone. The resin melts on the hot surface and binds the sand grains. A solid shell is formed, the thickness of which depends upon the time of direct contact, the temperature of the mold and the hardening rate of the resin. As soon as the shaped molding has reached a thickness of 4–7 mm, usually after 20–30 seconds, the unhardened surplus mixture is removed and fed back into the chute. The shell is only hardened on one side. The reverse side is then completely cured by infrared radiation or in a tunnel oven. This cured shell constitutes one half of the mold. The mold is assembled by clamping both shell halves together or gluing them together with thermosetting resins (bonding

Fig. 14.4. Shell process core blowing machine. (Photo: Röperwerk, D-4060 Viersen-Dülken)

Fig. 14.5. Core box. (Photo: Röperwerk, D-4060 Viersen-Dülken)

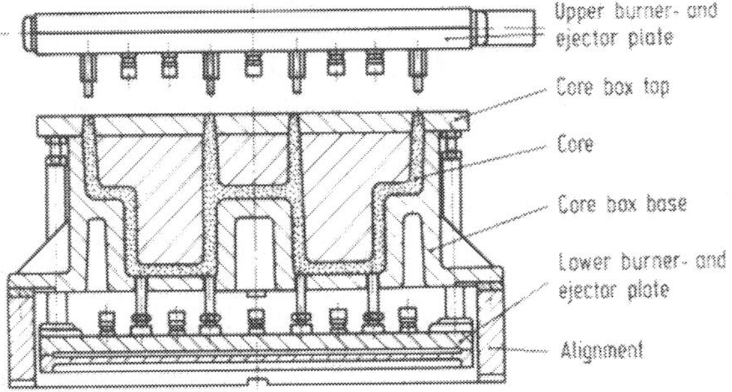

Fig. 14.6. Construction of a core box with heater plates. (Drawing: Röperwerk, D-4060 Viersen-Dülken)

time 20–40 seconds), after the required cores are set. Shell molds can often be poured flat without further support depending on the casting volume and metalostatic pressure.

This process was invented by J. Croning around 1944[13]. It first aroused interest in the U.S. after he introduced it there himself in 1948/49. Since the shell process distinguishes itself by a very accurate reproduction of all the contours of the model, very narrow tolerances can be maintained at the casting operation[14, 15]. Furthermore, very clean and dense casting surfaces can be obtained. Hollow cores can be made easily and economically and thus considerable amounts of mold material can be saved. This process, known in West-Germany as the Croning process, is of particular importance to the automotive industry because of its high accuracy in comparison to other processes.

Precoated Resin "Shell" Sand

Economic series production of complicated cores with automatic core blowing machines (Fig. 14.4) necessitates a free flowing resin coated sand with a controlled grain size distribution and uniform resin coat. While a number of foundries carry out their own precoating operation, many others purchase precoated sand from outside suppliers. The "warm" and "hot" coating processes differ from each other by the physical state of the resin i. e. whether dissolved or granulated novolak is used.

The warm coating process, the resins and formulation of the mixture to be used are described in Table 14.4. The coating time is 7–14 minutes. Mixing times which are longer than necessary for drying must be avoided. The thin resin coat can be rubbed off causing a remarkable strength reduction. The pneumatic conveying of the sand in pipelines more than 8 m long is therefore not advisable.

In the hot coating process, the emission of solvents is prevented by the use of granulated phenol resins and an aqueous HMTA-solution (Table 14.5). The novolak resin with a MP of 70–75 °C must have a low melt viscosity so that a quick and uniform coat can be obtained. The aqueous HMTA-solution should only be added after the hot mixture is cooled with water to about 100 °C in order to prevent excessive resin advancement. Lumps are formed which fall apart during the cooling operation. When this happens, the release agent is added. The sand is emptied at 60–65 °C, sieved and cooled to 30 °C in a fluidized bed in order to prevent lumps, and then conveyed into a silo.

In the USA, low condensed water-borne resins which are stabilized with small amounts of alcohol are used for sand coating. In order to obtain a free flowing sand with these low-melting novolaks, the resin must be advanced at coating. Higher temperatures are required due to the high heat of evaporation of water. For these reasons, the process is more complicated and did not have any success in Europe. The high proportion of free phenol (~10%) also constitutes a considerable environmental problem.

Shell Sand Properties

To check the quality of the coated sand, the melting point, cold tensile strength and hot tensile strength are tested. Since the price for the resin coated sand is decisively influenced by the resin costs, only as much bonding agent as necessary is added. A larger amount of resin is required for the production of shells (3–4%) which are subjected to higher stress during casting than cores. In general, a resin content of 1.8–2.5% is sufficient for cores (Fig. 14.7). Further properties must be tested if complicated cores and shells are involved, for instance gas volume, shell permeability,

Table 14.4. Precoated resin shell sand: warm coating process[16)]

Resin properties:	
resin content	65–67%
viscosity at 25 °C	2,000–2,500 mPa · s
flow distance of novolak resin	50–60 mm
softening range of novolak resin	72–75 °C

Formulation:
100.00 pbw silica sand H 32 (dry)
0.36 pbw hexamethylenetetramine (HMTA)
4.50 pbw resin solution (3 kg solid resin)
0.20 pbw calcium stearate

Process description:
weigh cold sand into the mixer;
add HMTA, mix for 15–20 seconds;
add resin solution, mix for 1–2 minutes;
blow hot air (150–180 °C) into the mixer and continue to mix until the lumps which are formed during the evaporation of the solvent fall to pieces and the coated sand can still be cumulated by hand under pressure;
add mold release agent and continue to mix without supply of air until the sand is dry and flows easily;
empty the mixer and sieve the sand through a shaking sieve with a mesh size of 0.5 mm;
transport the coated sand over an air bed conveyor in which the rest of the solvent is driven out and the sand cooled down, and then by a belt conveyor into the silo

Table 14.5. Precoated resin shell sand: hot coating process

Formulation:	400.0 pbw silica sand		
	12.0 pbw granulated phenolic resin		
	3.0 pbw water		
	4.3 pbw aqueous HMTA solution of 40%		
	0.5 pbw calcium stearate		
Process description:	heat the sand up to 120–130 °C		10–12 s
	feed hot sand into the mixer		10–12 s
	add resin		8 s
	mixing time		45 s
	add water and HMTA solution		5 s
	cool with cold air		140 s
	add calcium stearate		5 s
	mix		10 s
	empty		18 s
	total cycle time	approx.	240 s
	sand flow out temperature 60–65 °C.		

surface hardness, peel back and thermal shock resistance. The stick point of the coated sand, meaning the softening temperature of the resin, depends on the degree of preadvancement and reactivity of the resin. A deviating melting point may indicate slow cure, low tensile strength and tendency to peel back. The cold-tensile strength reflects the ability of the shell and core to be handled without damage. The hot-

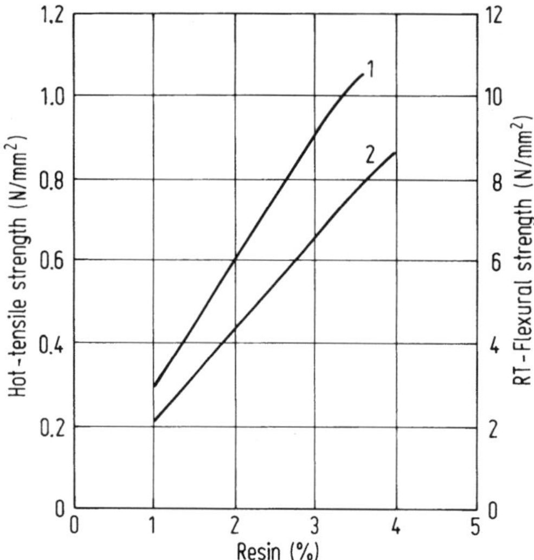

Fig. 14.7. "Hot coated" shell sand, hot-tensile strength (1) and RT-flexural strength (2) depending on resin content. HMTA 12%, calcium stearate 6%, based on resin weight. Cure at 280 °C for 3 min.

tensile strength is an important indication of the dimensional stability of the shell at casting, depending strongly on resin quality. Strength and performance are improved significantly by modifying novolak resins with salicylic acid[17)] or resol resins. Strength and temperature resistance are also controlled by the proportion of HMTA. High proportion of HMTA, but not more than 18%, effects a higher cross-link density and thereby higher thermal resistance. But the cores and shells become more brittle. The best results are obtained at a 10–13% HMTA level.

To prevent some frequently occurring mold defects in the shell casting process, such as mold cracking, peel back, soft mold or low hot-tensile strength, various additives can be added[17–19)]. The thermal expansion of the sand (Table 14.3) at casting may lead to mold cracking. Apart from the use of sands with lower thermal expansion, cracking can be prevented by the use of thermoplastic additives. The most common additive is a modified naturally occurring wood resin, called Vinsol, which is a complex mixture of substituted phenols, resin derivatives and hydrocarbon resins[19)]. Vinsol, handled in pulverized or flaked form, has a softening point of 112 °C (ring and ball). Due to the phenolic structures contained therein it is also able to react with HMTA yielding a higher melting, but still thermoplastic resin. Therefore, the HMTA proportion should be slightly increased.

By the addition of 0.25–0.5% Vinsol, in relation to the sand, thermal shock resistance can be increased and the metal penetration reduced. Excessive addition, however, leads to reduced hot-tensile strength.

Peel back, a serious mold defect frequently occurring during the drain cycle in the dump-box molding, has several causes. On one hand, this peeling from excessive weight may result by a too high resin content. Another cause can be a very wide melting range of the resin due to an unfavorable MWD, excess of release agent, insufficient solvent removal or presence of oily contaminants. But although pattern design, pattern temperature and contact time can affect the tendency towards peel back.

14.3. Hot-Box Process

In principle, the hot-box process is very similar to the shell molding technique. A wet resin/sand mixture is blown into the heated core box. A non-free-flowing sand is obtained from clay-free sand, resin and catalyst. It is more difficult to blow the wet mixture into a confined space than the dry, precoated shell sand, and thus, more difficult to produce accurately and consistently cores of complicated configuration and thin sections[20]. An important advantage to shell molding is the essentially shorter curing time. By the use of a catalyst, a short heat shock is sufficient to start the curing while the final curing is performed outside the mold. After a few seconds, a hard surface layer is formed at the hot walls (180–280 °C) so that the core can be stripped out of the core box without any danger. The heat absorbed by the core is enough to cure the core completely outside of the mold.

The surface quality can be improved by coating with a refractory wash. The hot-box process is applicable to mass production of small and medium volume castings. Examples in the automotive industry are transmission-, case-, crankcase-, body-, cylinder head- and gear box casing cores[21]. Phenol resols modified with UF-resins, FA-resins and FA/UF-resin blends are used as bonding agents together with catalysts. Aqueous solutions of ammonia salts of strong organic or inorganic acids are used as catalysts. Ammonia reacts with formaldehyde contained in the resin whereby the acid is liberated and the reaction started. The hardening process starts while hardener and resin are being mixed, so that the mixture must be processed within a limited time which depends upon the temperature.

The hardening rate of resols is not sufficient to reach the short cycle times which are usually obtained with UF/FA-resins. The reactivity can be increased by the addition of UF-resins. However, higher urea levels would be necessary for phenolics in comparison to FA-resins in order to obtain comparable hardening rates at low temperatures. UF-resins are less thermally stable than phenolic resins and decompose considerably faster at the high casting temperatures. The gas shock caused thereby often leads to pinhole formation in the casting. The PF hot-box resins available on the market today contain approximately 7% of nitrogen. In order to avoid pinholes, ferric oxide is added in the foundries. However, higher quantities of ferric oxide can lead to efflorences adhering to the mold[20a]. Therefore, efforts are made to reduce the nitrogen content to at least 3% maintaining the reactivity.

Appropriate PF-resins, made at a P/F molar ratio of between 1 : 1.4–1.8 and sodium hydroxyde as catalyst have in general the following key properties (Table 14.6). The proportion of urea or UF-resin is indicated by the nitrogen content. Recently, also PF-resins modified with furfuryl alcohol (FA) or furane resins have found increased market acceptance. Because of their low viscosity and better wetting, higher core strength can be obtained. In order to improve the flowability of the mixture and to facilitate the removal of the cores from the mold, lubricants and release agents like linseed oil or stearates can be added. The preparation of the mixture (Table 14.7) is performed discontinuously in a mixer. The sand is first mixed with the hardener for one minute and then with the liquid resin two to three minutes. The release agent is added last and mixed until an even distribution is obtained. The result is a wet, non-flowing mixture which can be processed on automatically operating transfer and spider units at a blowing pressure of 6–8 bar.

Table 14.6. Properties of urea-modified phenol "hot-box" resins[16)]

Dry solids content	70 ± 1%
Viscosity at 20 °C	1,600 ± 200 mPa · s
Nitrogen content	7–8%
Water content	15–17%
Free phenol content	3–4%
Free formaldehyde content	< 0.5%

Table 14.7. Hot-box core sand formulation[16)]

Silica sand H 32, dry	100 parts
Hardener (ammonium chloride)	0.4 parts
Phenolic resin (urea modified)	2.0 parts
Release agent	0.2 parts
Iron oxide	0–1.5

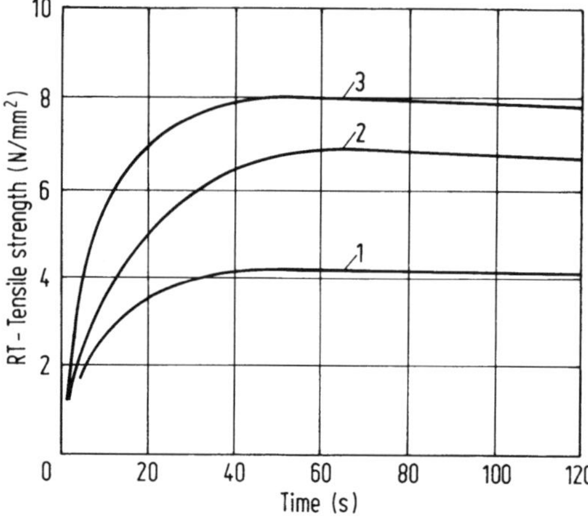

Fig. 14.8. RT-tensile strength of hot-box sands depending on curing time and resin content. Formulation: 100 parts silica sand H 33, 2.0 parts phenolic resin, 0.46 parts catalysts. Testing rods made at 220 °C and a pressure of 6 bar.
1 = 1.2% resin share; *2* = 1.6% resin share; *3* = 2.0% resin share

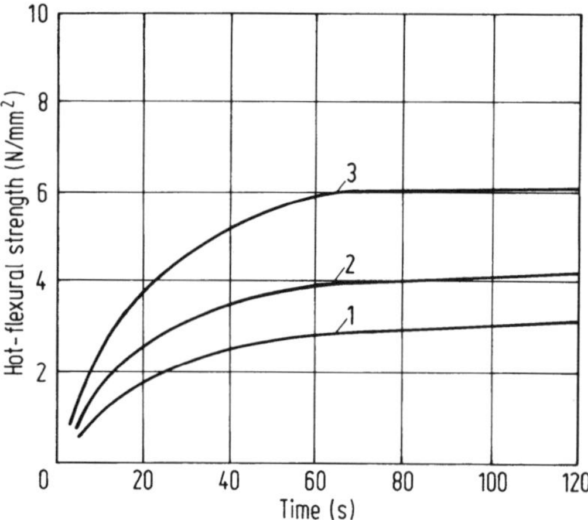

Fig. 14.9. Hot-tensile strength of hot-box sands depending on curing time and resin content. For conditions and designation see Fig. 14.8

14.4. No-Bake Process

The cold-set processes have been known for approximately 20 years. The no-bake process does not require gaseous catalyst addition, contrary to the cold-box process.

The no-bake technique is very economic because the application of heat is not necessary. It is especially suitable for the casting of large pieces of gray-cast iron and cast steel for machine and vehicle construction[22]. Tool costs are considerably lower than those for the hot processes.

The most frequently used binders are those of the furane type: FA/UF-resin blends, UF/PF-resin blends and FA/PF/UF copolymers. Polyurethanes are also used in the USA. UF-resins offer high reactivity and through-cure, low cost and adequate bond strength. The high formaldehyde evolution of UF-resins during processing is of disadvantage. The high nitrogen content (20–25%) and the low temperature resistance lead to the formation of pinholes and metal penetration. Furane polymers are less dependent upon temperature and sand quality variations, and the fume evolution is low. The bond strength developed is high. However, they are very expensive.

Phenol resins offer advantageous prices, acceptable strength and through-cure. Fume evolution is often criticized. In order to reduce it, resins with an especially low content of free phenol and formaldehyde were developed. The high-temperature decomposition reaction and the nature of the gaseous decomposition products are described in Chapter 7.1. Nitrogen free, fast setting phenol resins were introduced recently. A compromise between removal time, casting performance and price is achieved by the combination of these three ingredients (PF, UF, FA).

The use of organic silane coupling agents in no-bake formulations has become common practice[23, 24] to improve the bond between sand and resin. An increase in strength of up to 40% can be obtained by the addition of γ-aminopropyltriethoxysilane or similar silanes between 0.1 and 1% in relation to the resin.

The phenol resins are characterized according to the content of FA, nitrogen, water and free formaldehyde. The content of free phenol and formaldehyde is considerably reduced in the newer resins. At present, they contain approximately 3–4% phenol and less than 0.5% formaldehyde. The resin addition is generally between 1.2 and 2%.

For the cold-setting of furane and resol resins[25] inorganic acids (e.g. phosphoric acid) may be used; however, strong organic acids (phenol sulfonic acid, *p*-toluene sulfonic acid) which are enhanced in acidity by their (small) content of sulfuric acid, are preferred. The most important criteria to differentiate the organic acids available on the market (65–85% acid content) are the content of sulfuric acid and water, viscosity and, last but not least, the price. A high water content delays the curing considerably.

A great advantage of the no-bake technique is the possibility of varying the setting and curing time by varying the catalyst addition rates (40–100% related to the resin). The cure rate depends strongly upon the temperature of the sand. When the temperature of the sand drops below 8–10 °C, only very low strength of the bonded sand is obtained. Further addition of acid in an attempt to speed up the reaction rate reduces the strength significantly[26].

The preparation of the mixture (Table 14.8) can be performed in static mixers or in continuously operating high-speed mixers which permit the use of extremely fast curing resins. These machines, such as Fascold[27], Pacemaster or Gisag[28] have integrated metering and mixing components (Fig. 14.10). Molding rates approaching those of green sand are possible at appropriate automation (carousel).

Table 14.8. No-bake sand formulation[16)]

Silica sand, dry	100.0 pbw
p-Toluene sulfonic acid 65%	0.6 pbw
Phenol/FA/urea-resin	1.2 pbw
Iron oxide	0–1 pbw

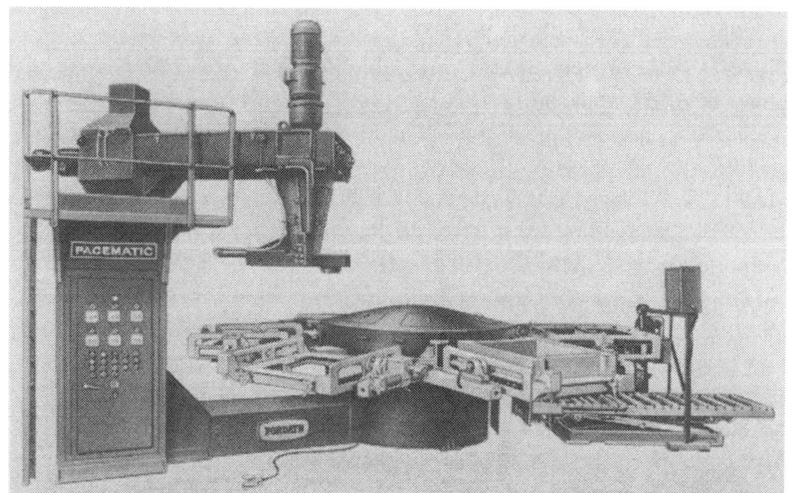

Fig. 14.10. 6-Station high-speed molding machine for the no-bake process. (Photo: Fodarth Ltd, West Bromwich, GB)

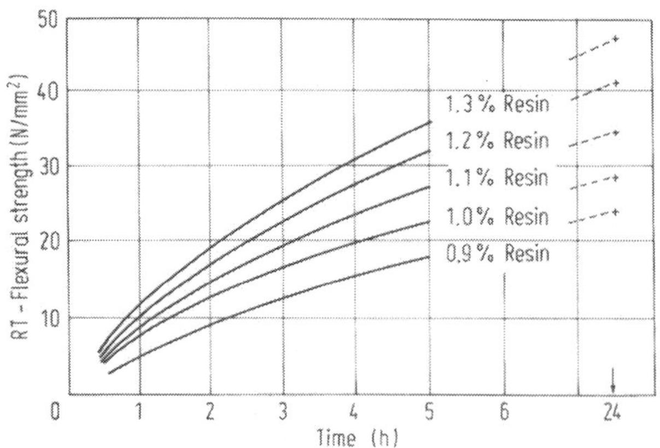

Fig. 14.11. No-bake core sand. Flexural strength at 20 °C depending on temperature and curing time at 20 °C

The acid catalyst is added first to prevent local "burn-red" of the resin under high local concentration of acid. Then, the bonding agent is added and further mixed until the material is uniform. At high nitrogen content iron oxide may be added to prevent the formation of pinholes. The maintenance of a certain temperature of the mixture (between 15–30 °C) is important for a faultless production due to the strong dependence of the reaction rates upon the temperature. A flowable, wet sand is obtained which is immediately filled into the molds and is compressed by vibration or manually. The core boxes can be of nonmetal materials. Since the curing starts as soon as the acid is mixed with the bonding agent, the molding material has only a limited working time and must be processed right away. Deformation of the preform at a certain resin advancement results in reduced strength.

The strip time is the time from discharging the mixed sand into the core box until the time when the core can be safely stripped. The strength increases as curing progresses until casting (Fig. 14.11).

Naturally, it is very desirable that the strength is as high as possible with regard to work time/strip time. Conventional single-trough continuous mixers may be used with sands having strip times of 20–40 minutes. The work time with most resins is then at least 5–8 min.

14.5. Cold-Box Process

The cold-box process[29, 30] is based on the fast reaction of poly-funtional isocyanates with polyols. The reaction, which proceeds at a considerable rate even in the cold, leads to the formation of polyurethanes (See Section 3.4.6). The important characteristic of this process is the use of gaseous organic amines as catalysts[31]. Their application requires specially designed gassing and exhaust systems. "High-ortho" novolak resins (See Section 3.4.3) are used as the hydroxyl component[31, 32]. They may be prepared in a water-free reaction medium in the presence of salts of divalent metals as catalysts. The phenolic resin/isocyanate combination – usually a diisocyanate, for instance 4,4′-isocyanatodiphenylmethane – is used in water-free solution. Chiefly higher boiling esters or ketones are used as solvents. The amount of binder normally used in this process is between 1 and 2 percent, for iron castings between 1.2 and 1.4%. Phenolic resin and isocyanate are used in the ratio 1 : 1. Since the bonding agent system is very sensitive towards moisture, only washed and dry sands can be used. The humidity content should not be higher than 0.1%.

The processing of a mixture of, for example, 100 kg silica sand H 31, 0.8 kg phenolic resin (65% dry resin) and 0.8 kg of diisocyanate (87–88%) is as follows: At first, the sand is mixed with the phenolic resin solution for one minute in a continuous or batch mixer. Then the diisocyanate solution is added and mixed for another minute. A good flowing sand is obtained which has a work time of 1 1/2 to 2 1/2 hours. Special core shooting machines have been developed which are additionally equipped with a gassing device. After the core is shot, the catalysts triethylamine or dimethylethylamine are blown in the form of a fine mist into the tightly closed core box in an air or CO_2 stream. These amine vapors are highly flammable and, mixed with air within certain limits, tend to explode. Therefore, special safety precautions must be taken. The use of CO_2 instead of air as carrier gas is preferred. An extremely fast cure is obtained by the sudden high amine concentration. The calculated consumption of amine is about 0.05% of the sand weight[30].

The cold-box process has some advantages, i.e. high core quality and cure rate, which renders them equally suitable for either mass production or short runs and jobbing work. Prototype tools can be made from cheap material like wood. However, stringent health and safety precautions have to be taken because isocyanates and amines are highly toxic. The amines cause a strong cauterizing effect on the skin and the eyes. After the curing process, the amines must be blown out of the core and destroyed by absorption in phosphoric acid or incineration. Nevertheless, it was shown that these problems can be solved technically. Phenolic resin-isocyanate binders are more widely used in the USA, but, increasing application is also expected in Europe.

14.6. Ingot Mold Hot Tops

To produce steel ingots, the melt is poured into big ingot molds consisting of three sections made of refractories (bottom) and steel. In order to obtain uniform solidification, the cap sides of the ingot mold are insulated with phenolic resin bonded insulation boards called hot tops.

The hot top formulation consists of the fillers sand, asbestos, mineral wool and paper which are mixed to a fibrous pulp with water and bonding resin in a mixer like those used by the paper industry (Table 14.9).

Table 14.9. Materials for the production of hot tops for ingot molds

Sand (fine)	35–40%
Asbestos and mineral wool	15–20%
Paper	5– 6%
Phenolic resin	4– 5%
Water	25–35%

The mixture is pumped into storage bins which are installed over molding tables. The molds consist of wooden frames with a sieve plate at the bottom. The material is poured into the mold, spread out evenly with a trowel and dehydrated by application of vacuum. The boards are removed and cured on a sheet metal or wire support in the oven. The oven temperature is approximately 180–190 °C. After cooling, the boards are sawn or ground to the proper measurements. Powder resins of the novolak type, blended with HMTA or paraformaldehyde, are used as bonding agents.

Recently, phenolic resins have been partly replaced by urea resins or starch due to costs. This, however, reduces the flexural strength and the resistance to humidity.

References

1. Metals Handbook, 8th Ed., Vol. 5, Forging and Casting. Amer. Soc. Metals, Metals Park (1970)
2. Höhner, K. E.: Gießereiwesen und Gußeisen, Temperguß und Stahlguß. In: Ullmanns Encyclopädie d. techn. Chem., 4. Ed., Vol. 12, Weinheim: Verlag Chemie 1976
3. Verein Deutscher Giessereifachleute, VDG, Düsseldorf
4. Middleton, J. M., Met, A.: Foundry Trade, April, 1976, 463
5. Kleinheyer, U.: Gießerei *63*, 487 (1976)
6. Hatton, W. L. S., Cross, S.: Chemical Binders- the Supplier's Point of View. In: Chemical Binders in Foundries, BICRA, Birmingham (1976)
7. Howell, R. C.: Can Chemically-Bonded Sand Replace Green Sand? . In: Chemical Binders in Foundries, BICRA, Birmingham (1976)
8. Kestler, J.: Modern Plastics Intern., Oct. 1976, P. 20
9. Smith, J. R., Rowley, S., Young, J. S.: Can Hot Process as survive? . In: Chemical Binders in Foundries, BICRA, Birmingham (1976)
10. Goettmann, F. P.: Production of Sands for the Foundry Industry. AFS-Transactions *10,* 15 (1975)
11. Middleton, J. M.: Alternatives to Silica for Mould Production. In: Chemical Binders in Foundries, BICRA, Birmingham (1976)
12. Bradke, H. J., Klein, T.: Deponieverhalten und Verwertung von Gießereisanden, Teil I und II: Laborauslaugungen von Gießereisanden zur Ermittlung der eluierbaren Stoffe und des voraussichtlichen Deponieverhaltens. IWL, 5 Köln 51 (1975)
13. Croning, J.: DE-PS 810174, 83293, 832936, 832937
14. Pölzguter: Gießerei *43*, 270 (1956)
15. Meyers, W.: Gießerei 56, 101 (1969)
16. Bakelite GmbH: Bakelite-Resins for the Foundry Industry, Technical Bulletin
17. Hooker Chemical Corp.: DE-PS 1095516 (1960)
18. The Borden Chemical Corp.: DE-OS 1808673
19. Hercules Inc.: Vinsol Resin, Technical Bulletin
20. Berndt, H.: Gießerei *53,* 96 (1966)

20a Berndt, H., Unger, D., Räde, D.: Gießerei *59,* 61 (1972)

21. Awbery, M.: Automobile Castings by the Hot-box Process. In: Chemical Binders in Foundries, BICRA, Birmingham (1976)
22. Beale, D. W.: The Practical Application of Cold-set Processes to Larger Castings. In: Chemical Binders in Foundries, BICRA, Birmingham (1976)
23. Union Carbide Corp.: Technical Bulletin
24. F. Raschig GmbH: DE-PS 2511165
25. Lebach, H.: Angew. Chem. 22, 1598 (1909)
26. Tilch, W.: Prüfung kaltselbsthärtender Formstoffe, Gießereitechnik *22*, 113 (1976)
27. Dhonau, H.: FASCOLD Cold-set Process. In: Chemical Binders in Foundries, BICRA, Birmingham (1976)
28. Christel, M.: Gießereitechnik 22, 158 (1976)
29. Kögler, H.: Gießerei *64,* 95 (1977)
30. Truchelut, J.: Ashland Cold-box Process. In: Chemical Binders in Foundries, BICRA, Birmingham (1976)
31. Ashland Oil Inc.: DE-PS 1583521 (1967)
32. Ashland Oil Inc.: DE-AS 2011365: DE-AS 2349200

15. Abrasive Materials

On October 26, 1907, L. H. Baekeland requested the registration of a patent covering the manufacturing of grinding devices in which, for the first time, phenolic resins were described as bonding agents for abrasive materials like corundum, sand, glass powder, iron oxide and pumice. In the patent description it was pointed out that these abrasive materials possess new properties and remarkable advantages compared to the inorganic bonded abrasive materials known at that time. Apart from good chemical and physical properties, there were simple production methods and low costs. Only one and a half years later, on February 5, 1909, on the occasion of a speech which Baekeland gave at a meeting of the New York Section of the American Chemical Society, he presented a grinding stone which had been manufactured with Bakelite to demonstrate the versatile applications of phenolic resins. This was the hour of birth of plastic bonded abrasive materials. Since then, grinding has become the most important process for the post treatment of metals[1,2]. A second family of abrasive materials is made by bonding a layer of abrasive grains on a flexible support by means of an adhesive. They are called coated abrasives, and sandpaper is a well known example.

15.1. Grinding Wheels

The following bonds are normally used to manufacture grinding wheels[3]:

Inorganic: vitrified bond
silicate bond
magnesite bond

Organic: resinoid (phenolic) bond
rubber bond
shellac bond

The vitrified bond is formed from finely powdered clays, feldspar, fluxes and glassy frits. Dextrine may be used as temporary binder during the forming process. A "green" wheel is made in a hydraulic press and dried in an oven. The firing process in which the temporary binder is burnt out and the ceramic bond is formed, lasts up to 10 days including cooling. The maximum temperature is within the range between 1,100 to 1,400 °C.

Ceramic wheels which constitute more than 50% of all wheels, exhibit high porosity and allow a cool grinding in dry- and wet operation. However, they are very brittle and follow the movements of the machine very exactly. Therefore, they are used for profile and precision grinding, normally up to a circumferential speed of 45 m/s up to max. 60 m/s.

Fig. 15.1. Different kinds and shapes of phenol resin bonded grinding wheels. (Photo: Krebs & Riedel KG Schleifscheibenfabrik, D-3522 Karlshafen)

The resinoid bond is stronger, more resilient and thermal shock resistant than the vitrous bond. The essential advantage of phenol resins in comparison to rubber or shellac is the higher temperature resistance. The actual temperatures developed during grinding vary considerably with the kind of application. For grinding steel by use of high-power operated machines, temperatures of at least 1000 °C (approaching 2000 °C was suggested) in the surface layer are common. Therefore, it is easily understandable that laboratory performance evaluations in those cases can only serve as a guideline. Actual thermal shock and mechanical stress is difficult to simulate.

Resinoid wheels are much less sensitive to impact, push and side pressure than ceramic wheels. The higher stability permits higher rotation speeds and thereby better grinding performances. The simpler manufacturing contributed to their high market share which is over 40%. Further growth can be expected.

15.1.1. Composition of Grinding Wheels

A grinding wheel consists of three basic elements[1, 3]: the abrasive grain, fillers and a bonding agent. Wheels differ by hardness (hard and soft) and grade. The wheel grade indicates the ratio of abrasive grain, bond volume and porosity, which is influenced by formulation and manufacturing. The bond volume is decisive for the hardness of the wheel, which designates the resistance to the breaking out of the abrasive grain and not the hardness of the grain. In the Norton scale the increasing hardness is indicated by the letters E to Z. A difference is also made between dry and wet grinding. Wet grinding is applied to heat sensitive materials. The grinding heat is removed by cooling liquids consisting of an oil in water emulsion and an anticorrosion additive. Mineral oils also are adequate.

Abrasive Materials

Corundum, the naturally occurring crystalline form of aluminum oxide (alumina, Al_2O_3) is very seldom used in industrial abrasives today. Synthetic, fused alumina is the most frequently used abrasive material. High-grade bauxite (hydrated aluminum oxide), purer than the grades for the production of aluminum metal, is the basic material. Water is first removed in a rotary furnace at approximately 1,100 °C. Fused alumina is then made by fusing calcinated alumina together with coke (to reduce metal oxides), iron (to remove silica) and titania (as additive contributing to toughness) in an electrical furnace. Maintaining 2,000 °C is sufficient to keep alumina molten. The cooling rate determines the fineness of the crystalline material. After cooling, the large pieces (2–3 t) are broken up and crushed to abrasive grains. There are different types of fused alumina which differ from each other by their composition, mechanical properties and toughness: regular, high-titania, fine crystal and white alumina[1, 3].

Silicon carbide (carborundum) which is obtained by reacting silica sand with coke in an electric furnace at approximately 2,000 °C, is harder but not as tough as aluminum oxide. Two types are produced, which differ in purity and color: "regular" (black) and "green" material. Pure silicon carbide forms colorless crystals. Fused alumina is preferred for the grinding of high tensile materials like steel. Silicon carbide is more suitable for grinding of hard and brittle materials such as cast iron, stone, glass, ceramics and "hard metals". Boron carbide is extremely hard (~ 2,800 Knoop hardness) but also very brittle. It is used, like diamond, for special purposes only, for instance for the treatment of tungsten carbide. Broken glass and emery are used to a large extent for the production of coated abrasives. In general, the abrasive grain size may vary between 20 μm to 3 mm. In addition, there are different geometric forms available, e.g. sticks, hollow spheres[4] etc.

Fillers and Reinforcements

The stability of phenol resin bonded grinding wheels and their flexural strength, the heat resistance and the toughness in particular are improved by the addition of fine grained fillers. Suitable fillers are aluminum oxide, iron oxide, silicates, chalk and asbestos. The additions of cryolite, pyrite, zinc sulfide, lithopone, potassium fluoroborate, antimony trisulfide and lead chloride have also proven to be of great

Table 15.1. Physical properties of grinding materials

Properties		Alumina α-Al_2O_3	Silicon carbide α-SiC
Crystal structure	–	hexagonal	hexagonal
Density	g/cm^3	3.99	3.2
Hardness, Mohs	–	9	(9.6)
Hardness, Knoop	(100 g load)	~ 2,000	~ 2,800
Melting point	°C	2,050	> 2,200 °C decomposition
Thermal expansion	K^{-1}	6.10^{-6}	$3.4 \cdot 10^{-6}$

advantage. The presence of the two products mentioned last must be clearly indicated because of their toxicity. The hardening is accelerated by the addition of basic oxides like calcium oxide and magnesium oxide. The action of sulfur containing additives (pyrite, zinc sulfide, lithopone) is attributed to the oxidation to sulfur oxides at higher temperatures whereby the formation of strongly adhering metal oxide layers is prevented, the oxidative degradation of the phenol resin bond delayed and thereby the service life of the wheel prolonged. The specific effect of cryolite is probably due to the relatively low melting point. The cryolite melts at high surface temperatures and the developing cavities increase the effectiveness of the grinding. At the same time, the melt can serve as lubricant and facilitate the grinding operation. Different reinforcements are used to improve the strength of the wheels, for instance glass cloth, textile cloth, non-wovens or kraft paper.

Resins

Originally only liquid phenolic resins were used, but today a combination of liquid and pulverized phenol resin are used because of better performance[5]. The liquid resin serves as wetting agent for the abrasive grain, powder resin and fillers. In addition, furfural, furfuryl alcohol, cresols and anthracene oil are also used in combination with liquid phenol resins. Liquid phenol resins and furfural possess very good wetting properties and good compatibility with pulverized phenol resins during all phases of the hardening reaction. The qualitative requirements of phenol resins are high, especially as far as the uniformity of the batches is concerned. Sodium hydroxide or sodium carbonate are preferred as catalysts for the production of liquid resols (Table 15.2) of phenol and formaldehyde. Efforts are being made to produce low viscous resins with the highest possible content of dry solids. The viscosity must be kept within narrow limits so that the temperature of the resin as well as the storage time must be taken into consideration (Tables 15.4 and 15.5).

The storage of liquid resol resins in refrigerated rooms is recommended due to their limited shelf life. At temperatures of 5–10 °C, the increase in viscosity is so small over several months, that no problems will arise. If the storage temperature is higher, supplies should be kept for a maximum of 4 weeks. The importance of the conditioning of the resins, i.e. a constant temperature and humidity as well as observance of the required storage conditions must not be underestimated.

Powder resins are high-condensed novolaks with a low free phenol content, high softening point, medium to short flow and high melt viscosity (Table 15.3).

Table 15.2. Properties of a liquid phenol wetting resin[5]

Dry solids content	70–80%
Free phenol content	15–20%
Water content	ca. 5%
Viscosity at 20 °C	2,500–3,500 mPa · s

Table 15.3. Phenol powder resins for grinding wheels[5]

Melting range	95/98–98/100 °C
Flow distance (125 °C)	25–30 mm
Size distribution (0.06 mm screen)	below 1%
HMTA content	ca. 9%
Humidity content	below 1%

Table 15.4. Viscosity of a phenol resol in dependence on temperature[3)]

Temperature °C	Viscosity mPa · s
0	49,000
10	11,000
20	3,100
30	1,100

Table 15.5. Viscosity change of a phenol resol as influenced by storage conditions[5)]

Storage time days	Viscosity mPa · s	
	Storage at 20 °C	Storage at 40 °C
1	2,600	2,600
10	2,750	5,800
18	3,200	20,000

Oxalic acid is mostly used as catalyst. The resins are mixed with HMTA as hardener and then finely ground. The HMTA portion may be between 6 and 14%, preferably 9%. Lower HMTA content leads to less cross-linking and to softer bonds. High HMTA content results in higher cross-link density and thereby higher hardness and heat resistance of the wheel.

The elasticity and toughness of the phenolic bond can be improved by the addition of 10–20% epoxide resin or polyvinylbutyral. The blast resistance of those wheels is improved significantly.

After a storage time of 3 to 6 months the white pulverized novolaks may turn yellow but other properties will not remarkably change. In general, pulverized resins should not be stored longer than 1 to 2 months. The storage room must be kept cool and dry to avoid humidity absorbtion and lump formation. Prior to use all ingredients should be conditioned to room temperature. Moisture condensation on the resin because of differences in temperature must be avoided.

Phenolic resins with higher resistance to cooling liquids have been developed for wet grinding.

15.1.2. Manufacturing of Grinding Wheels

In the cold molding process, green wheels are made first, then cured in an oven. In the compression molding process, the wheels are formed and cured in one step. Two mixers are used in most cases to prepare the mixture. The abrasive grain is covered with the wetting resin in the first mixer, the powder resin, additives and fillers are mixed in the second. Then, the wetted abrasive grain is mixed with the mixture from the second mixer.

Preparation and processing of the mixture should be performed in an air-conditioned room at a constant temperature and humidity to obtain reproducible plasticity.

The compression molded wheels possess a denser structure, higher strength and longer service life. Cold molded wheels have a more open structure, allow cooler grinding and show, in general, higher cutting efficiency. From the production economy the compression molding process is somewhat less favorable than the cold molding process because more molds are necessary due to the long molding times. Therefore, the compression molding process will only be applied to manufacture special wheels, for instance high-density wheels.

Cold Molding Procedure for Non-reinforced Wheels

A roughing wheel can be manufactured according to the following formulation:

Table 15.6. Formulation for the manufacturing of roughing wheels

74–77 pbw	Abrasive grain NK 24, 30, 36
4– 7 pbw	Phenol liquid resin
10–13 pbw	Phenol powder resin
2– 5 pbw	Pyrite
3– 5 pbw	Cryolite
0.1–0.4 pbw	Calcium oxide
0.2–0.8 pbw	Antimony trisulfide

The final mixture shall be homogeneous, free flowing but still plastic enough, so that it can be kneaded by hand with a slight pressure. During the mixing process, the mixture heats up so that the mixing time shall be kept constant for a given batch volume. The weight ratio of liquid to pulverized phenol resins is generally within the range of 1 : 2 to 1 : 4. It depends upon the size distribution of the abrasive grain, upon the kind and the quantity of the filler and the viscosity of the liquid resin used. The weighed mixture is fed into the mold on a rotating table and evenly distributed. The surface is levelled by a model. Wheels of uneven density and hardness and undesirable unbalance are obtained by uneven distribution and bad flow of the material. The molding pressure is within the range of 10–40 N/mm^2, preferably 15–25 N/mm^2, according to the plasticity of the mixture and the desired hardness of the wheel. Normally, it is molded to volume. The molding time depends upon the wheel thickness, the size of the grain used and the plasticity of the mixture. In general, it is 5 to 30 seconds and must be separately determined for each type of wheel. These "green" wheels are then cured in an electrically heated drying chamber or tunnel dryer with fresh air supply according to the predetermined program. The preforms must be placed at such distances that adequate air circulation is guaranteed. They are placed on porous ceramic or steel plates. Big thickwalled wheels and high pot wheels are often surrounded by a metal ring, and the intermediate space is filled up with sand to prevent deformation of the wheels during the curing process. Deformations happen if they are heated up too fast and the viscosity of the bonding agent decreases faster than the simultaneous condensation reaction can compensate. Blisters may occur if the surface layer is curing too fast and the volatile components cannot escape.

The green bond changes into a fused mass before the temperature reaches 80 °C; the water contained in the resol and formed during the condensation evaporates approaching 100 °C, resol cross-linking begins in this range. At about 115 °C HMTA decomposition starts and ammonia formed will escape to a great extent. Oven temperature is controlled in a way so that 80–85 °C are quickly reached. Between 80–100 °C a longer holding time is maintained so that a homogeneous melt can form and the volatile components can escape. In case the temperature of 100 °C is reached too fast or exceeded, gaseous inclusions result in a fine porous structure of the bond with reduced strength. A further holding time is placed in the range of 120–130 °C to enable the ammonia to escape. The hardness and toughness of the bond can be influenced by the end temperature which shall not exceed 180 °C. After complete cure, the wheels are slowly cooled down to 50–60 °C in the furnace by circulating air.

In this way deformation of the wheels and the occurrence of cracks due to thermal stress are prevented.

High-Speed, Reinforced Grinding and Separating Wheels

In order to increase wheel rotational speed from 45 m/sec to 60, 80 and 100 m/sec, the impact and flexural strength and bursting resistance of the wheels must be considerably raised[6, 7]. The increase to 60 m/sec is reached by use of epoxy-modified powder resins without changing the manufacturing process. A further improvement in bursting resistance is obtained by reinforcement with glass cloth impregnated with phenolic resin. Resins used for impregnation exhibit strong adhesion to glass fibers, high strength and a certain flexibility. Their curing rate must correspond to the reactivity of the other resins used in wheel construction. Glass cloths with appropriate mesh width and differing in area weight and type of weave (plain, leno and basket weave) are used according to wheel type and size of the abrasive grain. Leno weave is preferred for roughing wheels and exterior layers of cut-off wheels, basket weave for big cut-off wheels. Loomstate cloth is used in general, however, higher strength is obtained from cloth made from silane direct sized yarns. A separating wheel and the reinforcing glass cloth is shown in Figure 15.2.

For wheel manufacture the mixture is divided according to the number of prepreg insertions and the mold is charged alternately with the mixture and prepreg layers. The preform is molded at RT and 15–35 N/mm^2 pressure. The cure is finished within 24 hours in the oven at a maximum temperature of between 170–180 °C under slight pressure built up by metal plates.

Fig. 15.2. Phenol resin bonded cutt-off wheel and glass cloth reinforcement. (Photo: Krebs & Riedel KG Schleifscheibenfabrik, D-3522 Karlshafen)

Compression Molding Process

Water-free wetting agents like furfural, perhaps a mixture containing furfuryl alcohol, cresols and minor amounts of anthracene oil, are used to wet the grain in order to reduce the proportion of volatile components. After the preheating of a preform, which may contain glass cloth inlays, at 75–85 °C within 30 min, the cure is performed in hot molds. The molding time depends upon wheel thickness, in general, it takes 2 minutes with an additional 1/2 minute per mm thickness of the wheel. Without preheating the time must be extended. During the molding process breathing of the mold will allow the volatile components to escape. The cured wheel is pushed out of the mold and clamped in a cooling device in order to prevent warping. It can then be postcured for 6 to 10 hours at 140 °C in order to improve wheel service life.

To prevent excessive flow during molding, which may cause irregularities in wheel structure, high molecular weight powder resins are used with lower flow of 15–20 mm.

Snagging Wheels

For heavy surface grinding, for instance of cast billets of steel, snagging wheels are used which are pressed onto the working piece with automatic pendulum grinding machines at a pressure of 20–80 N/mm^2. The billet moves automatically back and forth under the wheel. Snagging wheels with a diameter of up to 800 mm must have an exceptional strength which is obtained by using high performance components and high compression. While normal grinding wheels have a density of 2.4–2.7 g/cm^3, max. 2.9 g/cm^3, it is 3.1–3.5 g/cm^3 for high-density wheels. This high density is only obtained by compression molding. The void volume should not exceed 1%. Heavy-duty abrasives with considerably improved toughness, based on zirconium oxide containing alumina or granular alumina obtained in a sintering technique, are used. The filler portion is relatively high and can be up to 50% of the bond. Basic fillers like calcium oxide and magnesium oxide serve as curing accelerators and catalysts for the reaction with furfural as well as to bind the water developed during the curing process. Only water-free wetting agents, mainly furfural in combination with cresol or neutral oils, are used. Since the mixture contains high amounts of fillers, the production of a free-flowing and dust-free mixture requires a lot of experience. Higher portions of furfural render the mixture unstable and lead to clogging. This can be avoided by adding small amounts of a neutral oil, in proportions between 0.6–1.0% in relation to the total mixture. At first, a preform is made at 15–25 N/mm^2 pressure which is then preheated to 90–130 °C. After 40–60 minutes, the mass is plastic but still shape retaining and so far degassed that the preform is then cured immediately in the hot mold at 20–40 N/mm^2 pressure and 150–170 °C molding temperature within 30 to 60 minutes. The curing in the mold must be performed to such an extent that blisters are avoided at removal and during the following post-curing process. The hot wheels are post-cured in the furnace at 160 °C for 8–12 hours. The subsequent cooling must be done very carefully. The efficiency of a hot molded, high-density wheel is remarkable. A wheel of 600 mm diameter has a service life of 20 hours at a running speed of 60 m/s. During this time, about 400 kg of steel are ground-off.

Fibrous Laminated Wheels

High-speed grinding wheels with high impact, flexural and bursting strength are obtained by reinforcement with kraft paper, non-wovens or fabrics. The webs are coated with liquid phenol resin by roll coating, powdered with abrasive grain and dried in a tunnel dryer at 86–95 °C. As many rondels as necessary are punched out according to the thickness of the wheel and molded between chrome plated sheets in a mold at a pressure of 10–20 N/mm^2 and at 155–165 °C. The cooling is performed either under pressure in the mold or in clamping devices to prevent warping of the wheels.

15.2. Coated Abrasives

Abrasive belts, discs, sheets and drums as shown in Fig. 15.3 are mainly used for polishing actions where high dimensional accuracy is not required rather than to remove thick sections. The efficiency is relatively low due to the single grain layer. However, by use of synthetic resins instead of animal glue, abrasive belts with higher flexibility and joint strength[8, 9] can be made and this material can be used for contour grinding.

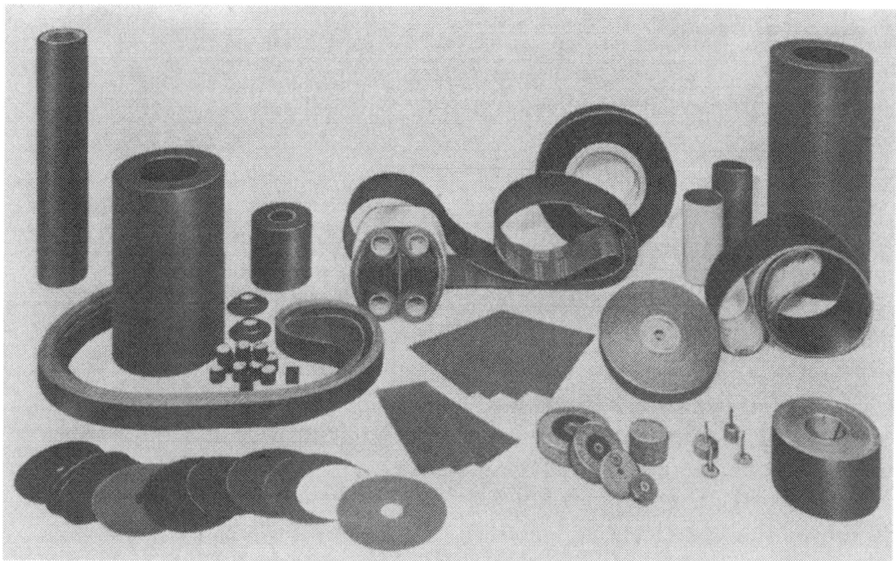

Fig. 15.3. Various kinds of coated abrasives. (Photo: Norddeutsche Schleifmittel-Industrie, D-2000 Hamburg)

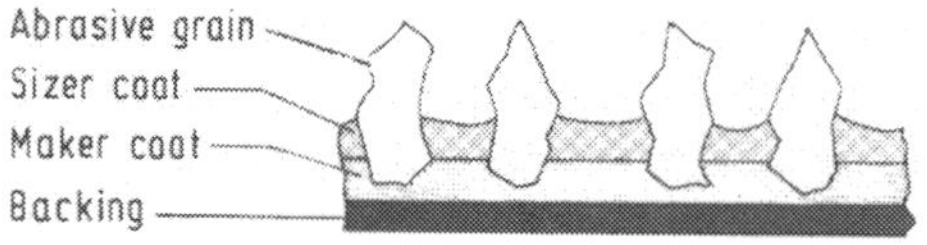

Abrasive grain
Sizer coat
Maker coat
Backing

Fig. 15.4. Structure of a coated abrasive[8)]

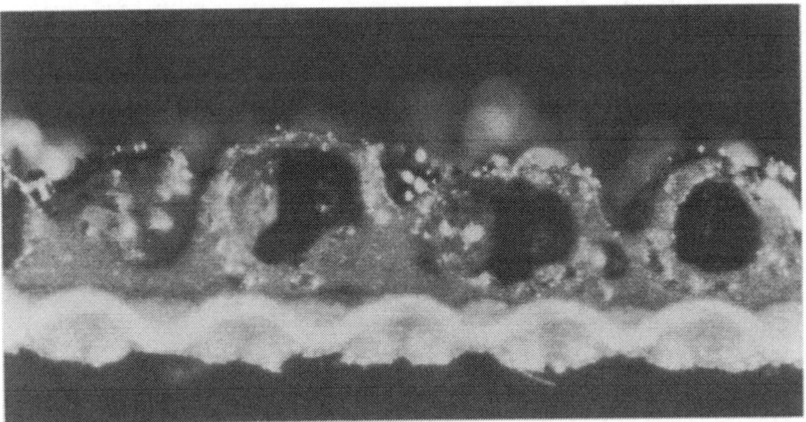

Fig. 15.5. Cross section of a coated abrasive of holow spheres coated with abrasive grains. (Photo: Norddeutsche Schleifmittel-Industrie, D-2000 Hamburg)

In general, the abrasive grain is bonded to the support (backing) by two adhesive coats, the maker coat and the sizer coat (Fig. 15.4 and 15.5).

15.2.1. Composition of Coated Abrasives

Abrasive Materials

Glass (broken and pulverized), flint, emery (impure corundum mixed with iron oxide), garnet and mainly aluminum oxide and silicon carbide are used for coated abrasives[1, 3]. The quantity applied depends upon the type and size of the abrasive grain used and how densely it is scattered over the backing material, designated as open or closed coat. For open coat, spaces are left between the grains so that wet removal of the material ground-off is possible. Backings for coated abrasives are paper, fabrics, vulcanized fibers or combinations thereof.

Kraft paper can, in principle, be used, but in most cases, special papers (70–220 g/m^2) which have already been modified during their production with rubber, acrylic resin and other polymers, are used to improve the flexibility, strength and water resistance. Anisotropic strength is preferred for belts. This can be achieved during the production of the paper by fiber orientation. Specially woven cotton or linen fabrics with high uniformity are used if materials with high flexibility and tensile strength are demanded. Vulcanized fiber produced from pure cotton paper made by the sulfuric acid- or zinc chloride process, is used in particular for abrasive discs because it is very strong and tough.

Adhesives and Coatings

Animal glue, vegetable oils and natural resins are being substituted more and more by synthetic resins, mainly in fields where water and temperature resistance is required. However, animal glue is still widely used today to bond flint, glass and garnet for dry grinding and when low pressures are applied. Compared to phenolic resins, aqueous animal glue requires only short drying times at low temperatures of 30–50 °C and no curing. Therefore, only relatively simple installations with low energy consumption are necessary. Since animal glue is a natural product, there may be differences in quality. Finally, its availability is limited. Phenolic resins are superior to all other bonding agents as regards water- and temperature resistance; they lead to higher grinding efficiency and accuracy. Heavy-duty industrial belts are manufactured with synthetic resins only. Alkyd-, polyurethane- or epoxy resins are also used in some fields. In addition, animal glue and phenolic resin may be used in combination.

15.2.2. Coating Process

The backing material, available in rolls up to 2m wide, is coated with the liquid resin in a roll coating machine and dried as shown schematically in Figure 15.6. The quantity for the first coat, which is also called the maker coat, is approximately 100–400 g/m^2 and depends mainly upon the size of the abrasive grain.

Two processes are commonly used to apply the abrasive grain: gravity coating and electrostatic coating. With the older scattering process, the grain falls out of a hopper onto the resin loaded web or is thrown against the web by a rotating brush as shown in Figure 15.6. In the newer electrostatic deposition process, the grains are flung up to the web from below against gravity by an electrical field. The grain should be fixed in the resin loaded backing vertical to the longitudinal axis. This is achieved more evenly by electrostatic deposition, thus, such belts also show higher grinding efficiency.

The loaded web is then transported into the festoon dryer where it is dried and only precured in the first drying section. The webs are placed in long loops over bars

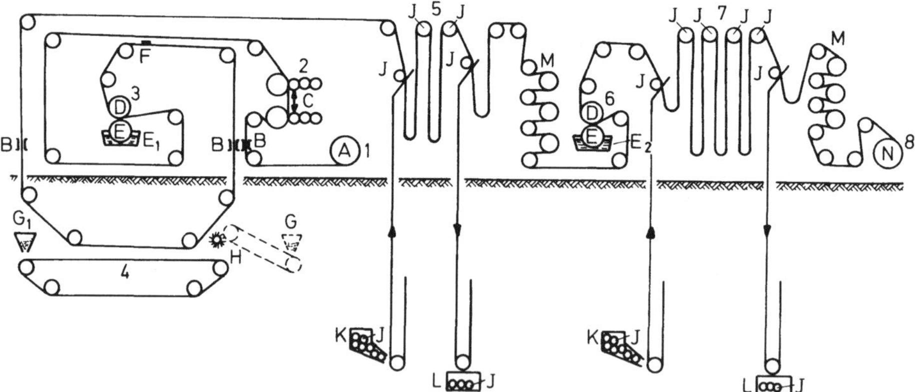

Fig. 15.6. The coating process. (Drawing: Norddeutsche Schleifmittel-Industrie, D-2000 Hamburg) *1* Take-off roll; *2* Printer; *3* Maker coater; *4* Grain coating machine; *5* First dryer; *6* Sizer; *7* Main festoon dryer; *8* Roll-up machine.
A Backing roll; *B* Thickness measuring instrument; *C* Printing roll; *D* Applicator roll; *E* Pick-up roll; *E1* Resin for the 1st (maker) coat; *E2* Resin for the 2nd (sizer) coat; *F* Smoothing brush; *G* Grain hopper, mechanical deposition; *G1* Grain hopper, electrostatic deposition; *H* Throwing brush; *J* Rotating poles; *K* Rotating poles supply; *L* Rotating poles deposit; *M* Tension rolls; *N* Product jumbo roll

which are moved through the different temperature zones by a chain conveyor. By this arrangement it is easier to make adjustments to suit distinct conditions than in horizontal dryers.

The web is then stretched and either fed to the next roll coater or back to the first coater. Here, a second binder coat is applied which is called the sizer coat, and the abrasive grain finally imbedded and fixed. The final drying and curing is performed in the main festoon dryer according to an exact program; above 80 °C the temperature is increased very slowly in order to prevent too fast curing of the surface and the formation of blisters. The reduction of viscosity at higher temperature should be compensated by the increasing cross-linking of the resin so that the abrasive grains do not change their positions. The maximum temperature should not exceed 120–130 °C; higher temperatures would lead to embrittlement of the web. In order to adjust the normal moisture content of the web, reconditioning is performed in the last drying zone. Prior to rolling up, the abrasive belt is subjected to a flexing process, which yields the necessary flexibility by a multitude of fine cross and diagonal cracks without damaging the backing. The webs are then made into rolls, blades, sheets, rondels or endless ribbons.

15.2.3. Abrasive Papers

Abrasive papers for dry grinding are bonded with natural glues and synthetic resins. The grain is fixed to the backing with an aqueous animal glue as maker coat. After grain deposition, a great deal of water is evaporated in about 30 minutes at 40–50 °C. The second coat is made with phenolic resins. They are liquid resins for which the curing rate can be adjusted according to the efficiency of the plant. Resins of medium reactivity and appropriate curing conditions are described in Table 15.7.

The resin for the sizer coat can contain up to 50% inorganic fillers such as chalk flour ($CaCO_3$). This results in high grinding efficiency, increased bond strength and diminished shrinkage. The fillers thicken the bonding agent so that flow-off is prevented. The water resistance and thermal resistance of a pure animal glue bond can be improved by addition of phenolic resins. In order to increase the shelf life of

Table 15.7. Phenol resin characterization and curing conditions

Characterization of the resin:	phenol resol	
	solids content	78–79%
	viscosity at 20 °C	2,800–3,200 mPa · s
	gel time at 130 °C (plate)	7–9 min
	gel time at 100 °C	58–68 min
Curing conditions:	heating up time to 75–80 °C	1/2 hour
	holding time at 75–80 °C	1 hour
	holding time at 90–95 °C	2 hours
	holding time at 105 °C	1 hour
	holding time at 120 °C	1 hour
	total time	5 1/2 hours

such mixtures, solvents and stabilizers can be added. The curing temperature is between 90 and 120 °C, depending upon the amount of phenolic resin added.

In high-performance grinding papers, the maker coat as well as the sizer coat consists of liquid, aqueous phenol resols. Since the liquid resins may penetrate very deeply into the paper web and it may get brittle, latex containing papers are used, to which a barrier coat may be applied. The coat can be made of plasticized thermoset resin solutions or thermoplastic polymer dispersions to prevent excessive impregnation. The same or different phenolic resins may be used for the maker and the sizer coat. In general, the viscosity of the sizer coat is lower than that of the maker coat.

Abrasive papers to be used for wet grinding of, for example, car body coatings must have a high flexibility and elasticity apart from good water resistance. Latex papers or papers modified with acrylic resins or PVC are often used instead of kraft papers. Alkyd resins are preferred for both coatings which, however, are not always satisfactory as far as strength and grinding efficiency are concerned. Phenolic resins give high hardness and good grinding efficiency, but the papers are too brittle. Better results are obtained by the use of epoxy resins and polyurethane resins which result in a very hard, elastic and resilient coat. More work remains to be done on the development of wet grinding materials and it appears that alkylphenol resins are yet another alternative.

15.2.4. Abrasive Tissues

Cotton fabrics with an area weight of 350 g/m^2 are used as backing for coarse granulation with high requirements for strength. Lower area weights of between 230–290 g/m^2 are adequate for lower requirements and fine granulation. The grain is fixed with animal glue for dry grinding and after drying coated with liquid phenolic resins. Fast curing phenol resols are used to shorten the curing cycle and to reduce the thermal stress of the coated web.

A water resistant size is required for fabrics which should be used for wet grinding of glass, stone, synthetic materials, etc. The finish may consist of a PVA-dispersion and a low molecular weight, water dilutable phenol resol, for instance. The finished fabric must be provided with a barrier coat prior to the application of the maker coat, in order to prevent resin penetration and embrittlement of the web. Oil-modified alkyd resins or epoxy resins are suitable for the maker and sizer coat.

15.2.5. Vulcanized Fiber Abrasives

Vulcanized fiber wheels are mainly used for grinding of machine and automobile bodies due to their high grinding efficiency. The wheels usually have a size of 180–230 mm and run with a circumferential speed of 40–50 m/s. Vulcanized fiber sheets must have a high tensile strength, a high peel resistance and sufficient elasticity. Liquid phenol resols of different reactivity are almost exclusively used as bonding agents. High reactive resins are used for short drying equipment with little heat supply or for high coating speeds. However, they are considerably more sensitive in processing. A faultless wetting of the vulcanized fiber web[10)] and a good flow of the resin are important prerequisites. To prevent unrestricted flow, chalk flour is added to the resin at a ratio of 1 : 1. The addition of fillers to the sizer coat contributes favorably to abrasive performance.

References

1. Edler, E.: Schleifen und Schleifmittel. In: Ullmanns Encyclopädie d. Techn. Chem., Vol. 15, 3. Ed. München, Urban und Schwarzenberg, 1964
2. Grinding, Honing and Lapping. In: Metals Handbook, 8th Ed., Vol. 5, American Society for Metals, Metals Park, Ohio (1970)
3. Abrasives. In: Materials and Technology, Vol. 2, Longmann, de Bussy: London 1972
4. Dziobek, K.: Z. ind. Fertig. *65,* 463 (1975)
5. Bakelite GmbH: Bakelite-Harze für die Schleifscheiben-Industrie. Technical Bulletin
6. Bogusch, E., Heiß, W.: Metalloberfläche *29,* 18 (1975)
7. Krains, G.: Berufsgenossenschaft/Betriebssicherheit, Mai 1961, 190
8. Klingspor, C. GmbH: Bandschleifen, Theorie und Praxis, Technical Bulletin
9. Norddeutsche Schleifmittel-Industrie: Schleifmittel. Technical Bulletin
10. Neuschäfer, K. H.: Vulkanfiberschichtstoffe. In: Vieweg, R., Becker, E. (ed.): Kunststoffhandbuch, Vol. 10. München: Carl Hanser 1968

16. Friction Materials

As long as the wheel exists there has been the problem of how to stop it successfully. The first very primitive devices were replaced by others directly attached to the vehicle and had brake surfaces of wood, leather or cork. However, with the development of motor vehicles, new methods had to be found to control the ever-increasing speeds. In the automotive industry, this was accomplished by brake linings of asbestos, thermosetting phenolic resins, and rubber. Linings similar to those shown in Figure 16.1, are installed on brakes of rail vehicles, aircraft, conveying facilities and elevators. Similar materials are also used for clutch facings and bearing elements.

16.1. Friction and Wear of Thermosets

The two basic laws of friction, that the frictional force between sliding surfaces is proportional to the applied load and that the frictional force is independent of the apparent area of contact, are generally valid for metals but do not always hold for polymer/metal combinations[1]. It is further accepted that the frictional force is composed of the rupture of small junctions produced by adhesive force over the regions of true contact and displacement of material resulting from interpenetration of the surface irregularities. The coefficient of friction μ, the proportionality factor between friction force (F) and applied load (L),

$$F = \mu \cdot L$$

however, is not constant if phenolics rub steel. A maximum of friction can be observed at different speeds, loads and temperatures. Polymers exhibit viscoelastic behavior, and therefore, the deformation properties are rate dependent. The deformation, respectively displacement, plays an important role in the friction process. The rate of wear is difficult to predict. In general, it depends upon the material, load, rubbing velocity and time[2]. Since the kinetic energy is transformed into heat during the braking operation, the temperature in the interior of the lining rises to 800 °C. At the boundary of contact far higher temperatures have been measured. The most important requirements for effective friction linings[3] are as follows:

- The coefficient of friction should be high and independent of temperature
- high thermal resistance
- high strength and distinct elasticity
- low wear and abrasiveness
- resistance to hydraulic fluids, gasoline and water
- no noise development

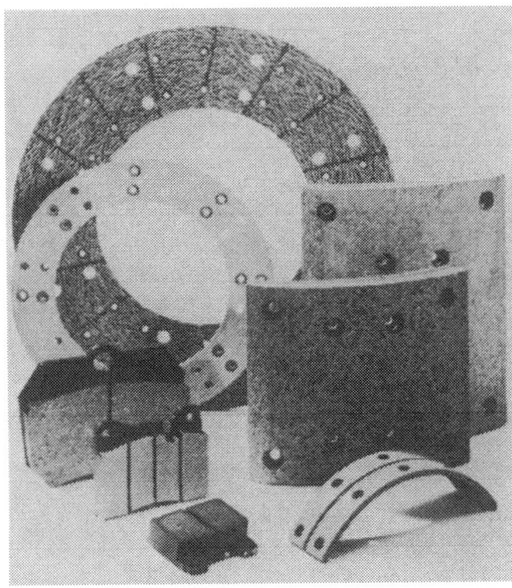

Fig. 16.1. Different types and shapes of friction materials. (Photo: Textar GmbH, D-5090 Leverkusen)

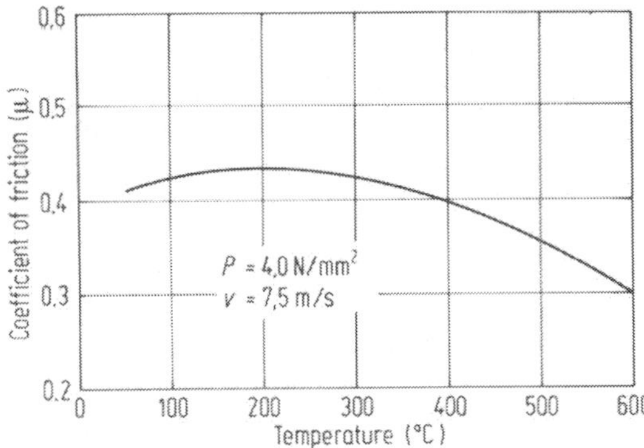

Fig. 16.2. Coefficient of friction μ in dependence on temperature[4)]

The optimum of all these requirements can only be achieved by a compromise. The thermal stability and thermal degradation of phenolics is described in Chapter 7, special resins with increased thermal stability in Chapter 8.

The dependence of the coefficient of friction upon the temperature (Fig. 16.2) and area pressure (Fig. 16.3) of a rubber-modified phenolic bonded disc pad are shown in the following figures. The reduction of the coefficient of friction at rising temperature is called fading.

The abrasion resistance determines the service life of the lining. If abrasion is high, the brake drum may be affected by the developing dust. This leads to erratic braking and often to drum scoring. The abrasion resistance is evaluated by determining the loss of weight and thickness of the lining. Another important factor is the water absorption and the influence of moisture on friction performance.

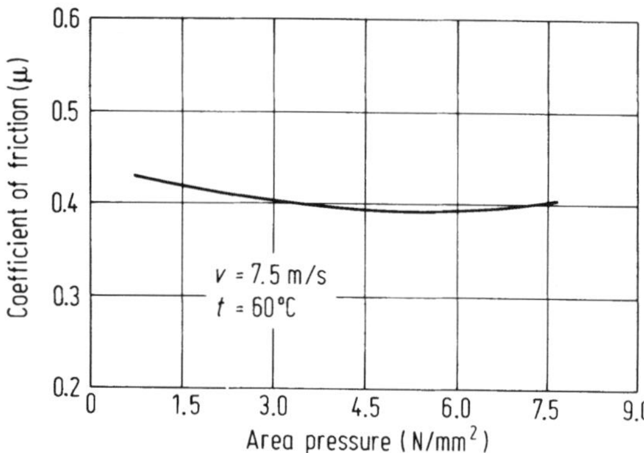

Fig. 16.3. Coefficient of friction μ in dependence on area pressure[4]

16.2. Formulation of Friction Materials

The composition of friction materials is very versatile[5]. In most cases they contain up to 20 components (Table 16.1). The exact composition is generally a well guarded secret.

The requirements and composition vary considerably due to the national speed limits and different dimensions of the lining. In the USA, because of the speed limit of 55 miles per hour, the requirements for temperature resistance are not so high as for the heavy-duty linings in Europe. In general, in the USA the brake linings contain more phenolic resin and rubber than elsewhere.

Table 16.1. Composition of mixtures for friction linings[6]

Asbestos fibers	30–50%
Bonding agent	
phenol- and/or cresol resin	10–15%
rubber	2–10%
Organic fillers	
friction dust	0–10%
Inorganic fillers	
barite	10–20%
calcium oxide/magnesium oxide	0– 5%
China clay/slate flour	0–10%
aluminum oxide	0– 2%
Metals	
powder or shavings	5–25%
Others	
graphite	1– 5%
antimony sulfide	0– 5%
molybdenum disulfide	0– 5%

Fibers

Chrysotile asbestos[7] is the main component of friction linings. Fibers of different lengths, yarns and fabrics are used. The chemical and physical properties of asbestos and physiological hazards at handling[8] are to be found in Section 10.2.2. Asbestos imparts strength and thermal resistance to the lining. Its abrasiveness is relatively low. In addition, cotton, organic and metal fibers may also be added. Carbon fibers in a carbon matrix (see Section 19.1) are recommended for the manufacturing of aircraft brake linings. The low wear rate of carbon, coupled with the appropriate thermal conductivity and high fiber strength, should result in a material of outstanding performance.

Fillers

A series of very different inorganic and organic materials are used to improve the friction performance, particularly thermal conductivity, abrasion resistance and strength[5,9]. Indications are given in Table 16.1. Calcium- and magnesium oxide also serve as curing accelerators for phenolic resins. Copper oxide, iron oxide and zinc oxide and sulfides, e.g. molybdenum disulfide, iron sulfide and zinc sulfide, are often applied. Metals like iron, nickel, magnesium, copper, brass and zinc in the form of powder or shavings are added to improve the thermal conductivity. Graphite and molybdenum sulfide are used as lubricants. Friction dust, a cured and pulverized reaction product of cashew nut shell liquid and formaldehyde, is an important and very often used additive[10]. It is assumed that this material forms a film on the surface of the lining during the braking operation compensating irregularities in friction behaviour and reducing wear.

Resins

In the beginning of the automotive industry, asphalt, natural resins and drying oils in combination with cotton fabrics were used to manufacture friction elements. Today, phenol- and cresol resins are used almost exclusively. In some formulations, cresol resins are preferred, which yield more flexible and tough lining materials. Modification of phenol resins with cashew nut shell liquid, tung oil, furfural and epoxy resin may add to flexibility and friction performance. NBR and less frequently CR are efficient modification components. Non-modified phenolics tend to fading. They became flexibilized and the reduction of the coefficient of friction at rising temperatures is at least partly compensated by modification with rubber. Rubber shows an increasing coefficient of friction with increasing temperature. In most cases the content of rubber in friction linings ranges between 2–8%. At higher amounts the brake tends to grab.

Resins of different consistency and reactivity are applied as required by each individual manufacturing process. Liquid phenol- and cresol resols, catalyzed with sodium hydroxide or ammonia, may contain small quantities of alcoholic solvents to reduce the viscosity. Phenol novolaks, which may be modified with NBR are dissolved in a mixture of acetone, spirit and toluene. This solution also contains HMTA.

Table 16.2. Phenol powder resins for friction linings[6]

Melting range	80–90 °C
Gel time at 150 °C (plate)	1– 2 ½ min
Flow length	45–55 mm

Solid resins in fine, free flowing powder form, blended with HMTA, are used in most instances (Table 16.2). The HMTA level is generally between 8–13%.

Solid cresol resols have in general a lower melting range of 60–70 °C and a gel time of 5–8 minutes at 130 °C. The production of higher melting resols is not possible because of their high reactivity. Pulverized resol resins may clog after some time, therefore, these resins should be ground by the customer shortly before use in order to ensure free flowing mixtures.

16.3. Manufacturing of Brake- and Clutch Linings

The following processes are the most important amongst those used for manufacturing of resin bonded friction materials[5, 10, 11, 12]

1. Impregnation process for asbestos fabric- or yarn linings for drum brakes and clutch facings
2. Wet mix "dough" process for drum brake linings
3. Dry mix process for disc pads and drum linings

These processes also differ from each other by the nature of the bonding agent used, which may be added as a solution or in solid form. The curing conditions are shown in Table 16.3, provided the cure takes place by pressing.

Impregnation Process

Clutch facings are almost exclusively made by impregnation. Asbestos yarns or asbestos fabrics to be used are impregnated by dipping with liquid or dissolved phenolic resins at normal pressure. A faster and more complete impregnation is obtained if vacuum and pressure are applied. The fabric is placed in a pressure vessel which is evacuated for 15–30 minutes and covered with the resin solution. Pressure is applied for approximately 30 minutes. The resin content is adjusted by the concentration of the impregnating solution which may be a rubber solution or a rubber/phenol resin mixture. During the drying operation in an oven the solvent is completely re-

Table 16.3. Molding conditions for production of friction elements

Process	Impregnation	Wet mix	Dry mix
Temperature °C	150–175	150–175	160–180
Molding time, min/mm thickness	0.5–1	0.5–1	0.4–0.6
Molding pressure, N/mm^2	15–20	15–30	15–50

moved and a certain advancement of the resin adjusted. The dried fabric is then consolidated on a calender and cut to the desired size. The curing can either occur in clamped formers in an oven, or by hot pressing. The maximum oven temperature of 180 °C is held for several hours. The total curing time including post-stoving is 24 to 35 hours, depending upon the thickness of the lining. This method results in very dense, high-quality linings. For compression molding (Table 16.3), the flow of the prepreg must be adjusted very accurately in order to obtain an evenly dense texture.

Wet Mix "Dough" Process

A tough dough-like mixture is made of asbestos, fillers, liquid bonding agents and solvents in a heavy-duty mixer. Rubber has to be masticated on a roll mill and dissolved before it is added to the mixture. The homogeneous mixture is forced through an extruder die, which is in accordance with the dimensions of the lining. A dense, continuous ribbon is obtained which is cut into strips of the required length, and the solvent is removed by drying in the oven. The drying has to be done carefully at temperatures of 50–90 °C for several hours. The resin is only cross-linked as far as it is necessary for the following molding process. This method can be varied by compacting the mixture on rolls forming a tape which is dried by the same process as above. Curing can be effected in clamped formers in the oven with a temperature rise of 10 °C/h, starting at 80 °C up to 180 °C, which is then held for 6–10 hours, or by compression molding (Table 16.3) within a short time.

Dry Mix Process

Rubber is masticated first with asbestos (to open out the fibers) in a heavy-duty mixer. The mixture is then prepared in an ordinary mixer by adding the powdered resin and other fillers.

Disk pad linings are molded either directly onto the metal plate or, an underlayer of an asbestos-phenolic resin molding compound is molded into it to improve shear strength. To obtain perfect adhesion, the metal plates are first cleaned by sandblasting, degreased and coated by dipping them in or spraying them with an adhesive solution based on rubber or polyvinylbutyral-modified phenolic resins and then dried. A preform is prepared by cold pressing a weighed quantity of the dry mix in a single mold at a pressure of 7–15 N/mm^2. The preform is pressed on the metal plate in a hot mold (Table 16.3). The same procedure is carried out, if a molding compound is applied as midlayer. The formation of blisters is avoided by repeated breathing. In most cases, multicavity molds are used. The cured linings are then stoved in an oven at about 160–180 °C for 12 to 14 hours.

On the other hand, the mixture may be also cold molded to the metal plate and then cured in the oven in a clamping mold.

Drum linings can be manufactured in the same way. Frequently a post-forming process is applied in which the mixture is first pressed to sheets at about 15 N/mm^2 and at 110–120 °C in a press. After cooling, the sheets are cut into strips. The strips are then heated to soften the resin and bent to the desired radius. Finally they are cured in clamped jigs. For cost reasons, larger shells may be molded and then divided to desired size.

After stoving, the linings are treated further. The resin rich surface area is ground off and the surface subjected to a thermal shock treatment to carbonize it slightly and thus the first sharp braking operation simulated. The linings are then ground to measure and upon visual control made up. The extremely high expenditures for quality control and strict supervision of the manufacturing process has to be pointed out.

References

1. Lancaster, J. K.: Plastics and Polymers, Dec. 1973, 297
2. Hodge, J. C.: Plastics and Polymers, Feb. 1974, 27
3. Abex Corp.: The Eight Critical Properties of Brake Lining Performance. Technical Bulletin
4. Textar GmbH: Technical Bulletin
5. Bohmhammel, H.: Gummi, Asbest-Kunststoffe: *11*, 924 (1973); *12*, 1063 (1973); *1*, 34 (1974); *11*, 926 (1974)
6. Bakelite GmbH: Technical Bulletin
7. Noll, W.: Asbest. In: Ullmanns Encyclopädie d. techn. Chem., Vol. 8, 4th Ed. Weinheim: Verlag Chemie 1976
8. Heidermanns, G., Kühnen, G., Schütz, A., Prochaska, R.: Staub-Reinhaltung Luft *35*, 433 (1975)
9. Manhattan Rubber, US-PS 2025951; Ferodo Ltd. FR-PS 722209; Union Carbide Corp. US-PS 1416492
10. BP Chemicals: Cellobond Phenolic Resins. Technical Bulletin
11. Carrol, W. G.: British Plastics *35/8*, 414 (1962)
12. Batchelor, C. S.: Friction Material. In: Kirk-Othmer: Encyclopedia of Chemical Technology, Vol. 10, 2. Ed. Interscience 1967

17. Phenolic Resins in Rubbers and Adhesives

In general, phenolic resins exhibit adhesive functions in all kinds of application. They are always used in thin layers in connection with fillers or reinforcing fibers. Their adhesion to most materials is very good due to the marked polarity of the phenolic structures. Disadvantageous, however, is their brittleness and the high curing temperatures and pressures required.

17.1. Mechanisms of Rubber Vulcanization with Phenolic Resins

The remarkable improvement in chemical and physical properties of rubber after vulcanization with sulfur[1] was discovered jointly by Goodyear and Hancock in the early 1840's. About a hundred years later, in 1942, Wildschut found that natural and synthetic rubbers can be vulcanized with certain phenolic resins without any additives[2]. The advantage of phenolics in comparison to sulfur vulcanization is the higher strength, heat aging resistance, low compression set and improved chemical resistance. However, the economic significance of phenolic resin vulcanization is rather low. Butyl rubber, which is difficult to vulcanize conventionally because of its low unsaturation and easy reversion, can be favorably cross-linked with phenolic resins.

Resol resins according to the general formula 17.1 are required for cross-linking. The *para* position should be substituted (R = tert-butyl-, octyl- or nonyl group), otherwise the resin condensation reaction would be much faster than the resin-rubber reaction. Appropriate resins with MP's between 55–65 °C, added in the range of between 5–12% depending on rubber type and cross-link density desired, are mostly based on *p*-octylphenol. Activators are used to enhance the vulcanization rate, e. g. pyromellitic anhydride, phthalic anhydride or fumaric acid for NBR, toluene sulfonic acid or chloroacetic acid for SBR, $SnCl_2 \cdot 2H_2O$ or ZnO in combination with chloride ion spending compounds like Hypalon or Neoprene for BR. A second variety of resins is halomethylated ($-CH_2Br$ instead of $-CH_2OH$; 1–2% chlorine or bromine content) and requires therefore no additional activation.

$$HO-CH_2-\left[C_6H_2(OH)(R)-CH_2\right]_n-\left[C_6H_2(OH)(R)-CH_2-O-CH_2\right]_m-C_6H_2(OH)(R)-CH_2-OH \quad (17.1)$$

There are different opinions concerning the reaction mechanism of the crosslinking of rubber with phenolics. Hultzsch[3] pointed out that the reaction should take place (analogous to drying oils) through quinone methides as intermediates followed by

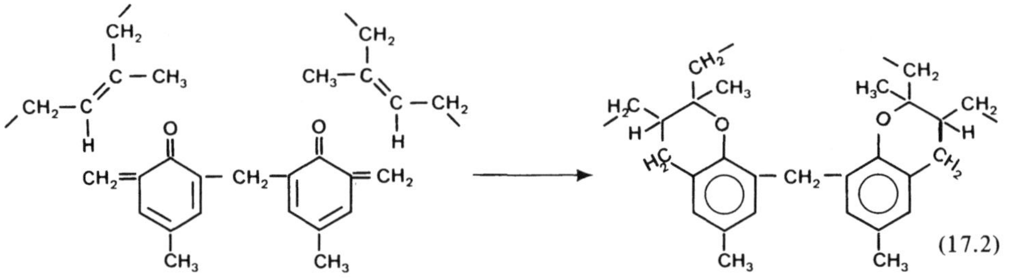

(17.2)

chroman ring formation (17.2) (See section 3.3.5). At least two-nuclear resols are therefore required. However, the fact that cross-linking can be accomplished with mononuclear phenol compounds contradicts this reaction mechanism.

Van der Meer[4] also presumed quinone methides as necessary intermediates. Cross-linking shall, however, occur through hydrogen abstraction from the methylene

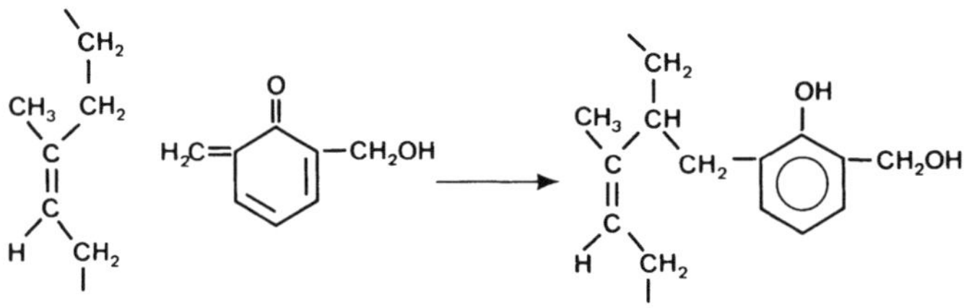

(17.3)

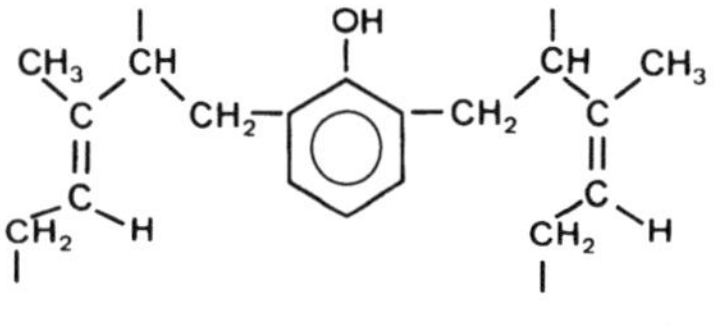

(17.4)

group in the allyl-position of the rubber (17.3). After rearrangement to the aromatic structure a second quinone methide formation should then lead to the mononuclear cross-link (17.4).

The fact, that the rate of vulcanization is greatly increased by addition of acidic catalysts is interpreted by Giller[5] in favour of an ionic chain reaction mechanism. Metal halides are formed, even if in small quantities, by the interaction of basic

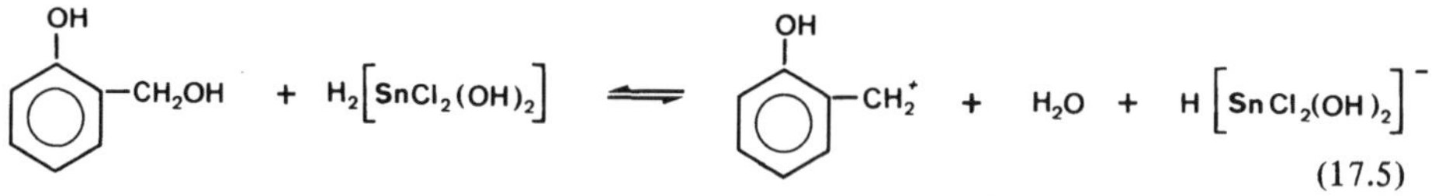

$$C_6H_4(OH)-CH_2OH + H_2[SnCl_2(OH)_2] \rightleftharpoons C_6H_4(OH)-CH_2^+ + H_2O + H[SnCl_2(OH)_2]^- \qquad (17.5)$$

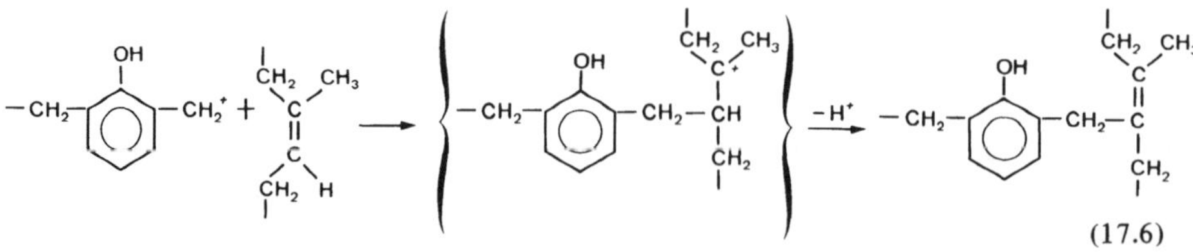

(17.6)

oxides and halogenated polymers, which then act as Lewis acid-type catalysts, leading to carbonium ion formation (17.5). The dibenzyl ether bridge is also split by the action of acidic catalysts. The phenolic vulcanization is best explained by the ionic chain mechanism as shown in equation (17.6). Additional cross-linking reactions may occur with synthetic rubbers according to their functional groups.

17.2. Thermosetting Alloy Adhesives

In structural adhesives the phenolic resins for metal/metal and rubber bonding are combined with thermoplastic polymers or elastomers. By these combinations, called alloys, the characteristic properties of both can be utilized and specific requirements met by the ratio of the resins. As soft components, polyvinylacetals, NBR, polyamides and polyacrylates are preferred. The elongation, elasticity and resiliency of the phenolic resin is improved considerably, especially at low temperatures. The toughening action of such polymer blends is attributed to chemical reaction, but even more to the morphology of the cured system. Because of the limited solubility of the thermoplastic or elastomeric component a discrete, well dispersed, discontinuous phase develops in the cured phenolic matrix. The crack propagation is considerably minimized by the two-phase system and the impact strength therefore improved.

17.2.1. Vinyl-Phenolic Structural Adhesives

Polyvinylacetal-phenolic resin blends are most frequently used for structural metal bonding. They are used in the airplane industry, as copper adhesives for printed circuits, to bind brake linings to brake shoes, for various honeycomb constructions, and in the ski manufacturing industry.

High temperature shear strength is higher for polyvinylformal (PVF)/phenol blends than for the polyvinylbutyral (PVB)/phenol combination, but the latter system yields higher peel strength[6)]. For the curing reaction a temperature between 140 and 180 °C and pressure of 0.3 to 1.0 N/mm^2 are necessary. By addition of accelerators, for instance resorcinol-formaldehyde resins, the temperature can be reduced to approximately 115 °C; however, some loss in strength must be taken into consideration.

The ratio of phenol/acetal resin can range from 0.3 : 1 to 2 : 1, depending upon the desired elastic modulus, tensile strength, creep and temperature resistance. High molecular weight is preferred for the thermoplastic component; however, sufficient fusion and wetting during the curing reaction must be guaranteed to exhibit high adhesive and cohesive strength.

For phenol/PVB combinations alcohols may be used as solvents, while PVF can be dissolved only in special solvent combinations, such as toluene/ethanol, ethylene dichloride/MEK/ethanol or ethylene dichloride/dioxane/ethanol. Solutions with about 12–15% solids are applied by brushing as copper adhesives in the manufacture of phenolic-paper laminates for printed circuits. High peel strength and solvent- and blister resistance when in contact with tin solder at 260 °C are the specific requirements for this application. The laminates according to NEMA FR-2 and

XXX-PC are cured and at the same time bonded with a 35 μm copper foil at 160 °C and 10 N/mm^2 pressure for approximately one hour. The adhesive coat amounts to between 25 and 40 g/m^2 (see Section 12.1).

The high viscosity of polyacetal/phenol resin solutions and the necessity of draining-off the high quantity of solvents is very inconvenient in some cases. If so, either the metal substrate is coated first with the liquid phenol resol and then strewn with powdered acetal resin after airing, or prefabricated adhesive foils or tapes are used. The use of adhesive foils offers a series of advantages. Taking a low weight carrier (25–65 g/m^2) of fabric or non-woven, an even joint thickness is obtained. Here too, the carrier is first impregnated with the resol solution and then the tacky prepreg is powdered with pulverized acetal resin.

The phenolic resins applied in this field are resols catalyzed by sodium hydroxide. The hydroxymethyl group can react with the hydroxyl group of the polyvinylacetal resin. Consequently, the hydroxyl and acetal content is an important factor in the selection of the polyvinylacetal component. Furthermore, their grain size distribution is important with regard to the manufacturing and application of the adhesive.

17.2.2. Nitrile-Phenolic Structural Adhesives

NBR, despite of its higher compatibility, reacts more easily with phenolic resins than polymers which are only unsaturated in the polymer chain (17.7).

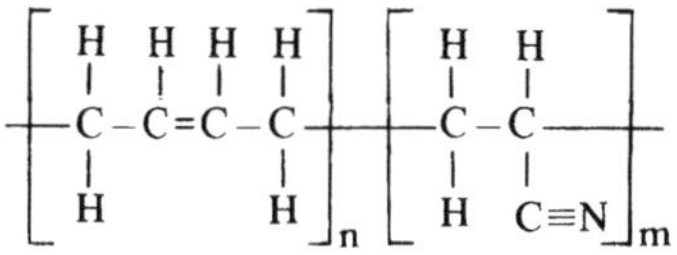

$$\left[-CH_2-CH=CH-CH_2- \right]_n \left[-CH_2-CH(C\equiv N)- \right]_m \qquad (17.7)$$

Customary trade names for such elastomers are: Perbunan (Bayer AG), Hycar (Goodrich Chemical Co.) and Breon (British Geon Ltd.). Some types contain methacrylic acid as copolymer.

The stronger reinforcement of NBR or even more, of carboxylated NBR-terpolymers suggests that further reactions between the nitrile and carboxyl group and the methylol group may occur (17.8 and 17.9). NBR-phenolic cements are used for metal-to-metal and rubber-to-metal bonding. For some applications priming of the substrate with chlorinated rubber or polyurethane is necessary. Zinc oxide, carbon black, iron oxide, and sulfur are common additives. Magnesium oxide is a more active curing agent for carboxylated NBR than zinc oxide. Compared to vinyl-

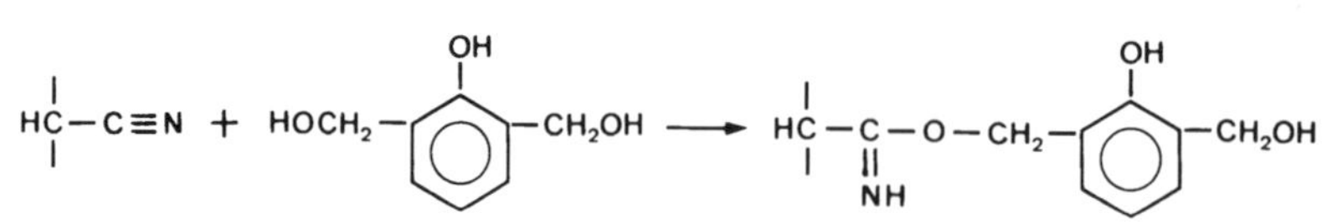

$$>HC-C\equiv N + HOCH_2-C_6H_3(OH)-CH_2OH \longrightarrow >HC-C(=NH)-O-CH_2-C_6H_3(OH)-CH_2OH \qquad (17.8)$$

$$>HC-C(=O)OH + HOCH_2-C_6H_3(OH)-CH_2OH \longrightarrow >HC-C(=O)-O-CH_2-C_6H_3(OH)-CH_2OH \qquad (17.9)$$

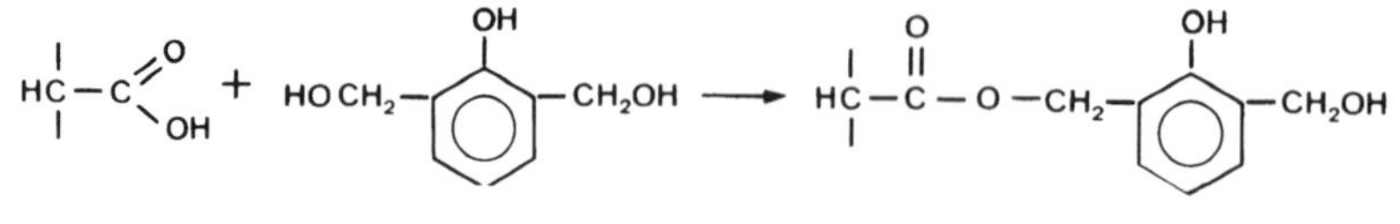

phenolic adhesives, NBR-phenolic blends have higher peel strength and impact resistance. Humidity and salt spray resistance is very good, the poor fileting characteristic[6] is disadvantageous.

17.3. Phenolic Resins in Contact Adhesives

Adhesives based on polychloroprene (neoprene) and NBR exhibit high green strength and excellent adhesion to various substrates. The strength, temperature and creep resistance can be improved and adhesive cost reduced by the addition of phenolic resins. They are used in shoe manufacturing (to bind leather, cloth, plastic and rubber), in the automotive- (interior upholstery), furniture- and construction industries. Chloroprene adhesives yield high peel strength and have outstanding green strength properties. Nitrile adhesives exhibit excellent oil and grease resistance. Heat and metal oxide reactive phenolic resins according to formula (17.1) are used (see section 3.4.3). *p*-tert.-Butyl phenol resins offer the best hot cohesive strength and form a single phase system with CR.

17.3.1. Chloroprene-Phenolic Contact Adhesives

The polymerization of the 2-chlorobutadiene may occur as 1,4- as well as 1,2-addition (Formula 17.10). The chlorine bound in the allyl position in the case of 1,2-addition, is far more reactive and can easily be split-off. Known tradenames for CR are Neoprene (Du Pont) and Baypren (Bayer). Some CR-types show a marked tendency towards crystallization because of stereochemical reasons depending also on the ratio of *cis*- and *trans*-configuration. Setting by crystallization is a reversible process, which can be reversed by increased temperature or dynamic stress.

$$-CH_2-\overset{\displaystyle Cl}{\overset{|}{C}}=CH-CH_2- \qquad\qquad -CH_2-\overset{\displaystyle Cl}{\overset{|}{\underset{\displaystyle CH=CH_2}{\underset{|}{C}}}}- \qquad (17.10)$$

1,4-polychlorobutadiene 1,2-polychlorobutadiene

Thin films, prepared from such elastomer solutions, show a high contact adhesiveness and a high cohesive strength due to rapid crystallization. This property suits them for application as contact adhesive[7]. The crystallization, however, is cancelled at temperatures above 60–70 °C. The temperature and solvent resistance can be considerably improved by additional cross-linking with phenol resins, which also improve tack and adhesion. The thermal resistance is enhanced further by the addition of bivalent oxides of the second main group and sub group metals such as calcium-, zinc-and cadmium oxide. Experience has shown that magnesium oxide is the most suited compound. In addition, they prevent "burning" during milling, contribute favorably to the stability of the formulation and act as hydrochloric acid acceptors[8]. Heat aging resistance is enhanced through addition of common antioxidants.

By increasing the amount of resin the heat resistance is raised; however, the elongation will be reduced and therefore the brittleness of the adhesive layer enhanced. A 40–45% resin level seems to be the best compromise.

CR-types with a medium or strong crystallization tendency are preferred. CR is soluble in aromatic chlorinated hydrocarbons, in some esters, ketones, and also in mixtures of certain non-solvents. The metal oxide-prereacted phenolic resin is also soluble in these solvents. An important feature of a contact adhesive is the relationship between open time and green strength. For an ideal system a rapid build-up of green strength and long tack retention is desired. Normally, the green strength increases with open time to a maximum value and then begins to decrease. The dried bond strength is greater at shorter open times. The nature of the phenolic resin used influences the open time and peel strength to a great extent. The content of methylol groups and dimethylene ether bridges are the key factors which have to be coordinated with the crystallization tendency of the CR. The higher the proportion of hydroxyl- and dimethylene ether groups, the shorter the open time; peel strength and elevated temperature resistance will be increased considerably. This holds if the rubber shows a medium crystallization rate. If the rate is high, non- or less base reactive phenolic resins are of advantage[9)].

These room temperature curing adhesives (Table 17.1) are formulated in two parts. Neoprene AC is masticated on a two-roll mill to improve solubility; the aging inhibitor is worked in, and at last zinc oxide is added. The temperature of the rubber should not exceed 60–80 °C. The homogenized rolled sheet is then cut to pieces and dissolved in the solvent mixture. For the second component the alkyl phenol resin is prereacted with magnesium oxide in toluene by addition of minor quantities of water (2%). Maintainig certain temperature (25–30 °C) and reaction time (5–15 hours) is important for the adhesive quality. The solubility of the rubber can be improved by adding a part of the solid resin during the mastication. Sometimes flocculation or settling of the metal oxides and filler particles and phase separation may occur depending on the type of neoprene, resin and solvent system. The MWD of the phenolic resins has a certain influence on the stability; low molecular weight portions are detrimental[11)]. A suitable MWD of a *p*-tert.-butylphenol resol resin is shown in the Figure 17.1. The adhesive application method for bonding materials used for shoe manufacturing shall be described briefly. The strips to be adhered are first buffed with emery paper. The coat is applied with a brush to both sides, and the solvents driven out. The bonding can be affected by the pres-

Table 17.1. Guide formulation for a neoprene-phenolic contact adhesive[10)]

Polychloroprene (Neoprene AC)	100
Phenolic resin (Bakelite KA 773)	40
Magnesium oxide	5
Zinc oxide	4
Antioxidant (Neozone A)	2
Toluene	150
Ethyl acetate	150
White spirit	150

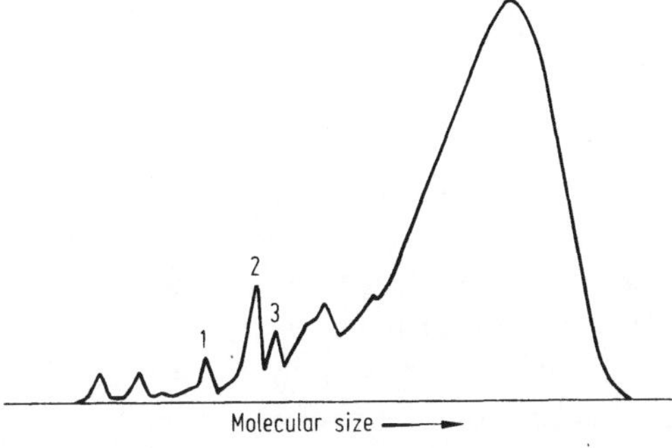

Fig. 17.1. Molecular weight distribution (GPC) of a *p*-tert.-butylphenol resin for contact adhesives.
1 = *p*-cresol; *2* = hydroxymethyl-*p*-cresol; *3* = bis(hydroxymethyl)-*p*-cresol

ence of small amounts of residual solvents or by heat activation with hot air or by IR radiation. The activation temperature is approximately 90 °C. The strips are then put together under pressure of 0.3–0.7 N/mm^2 for 30 seconds. Several days are required for maximum bond strength to be obtained.

Water-borne neoprene adhesives are receiving considerable attention owing to their ecological advantages. Phenolic resins based on tert.-butylphenol can be emulsified and blended with neoprene latex[12, 13]. Neoprene latex has a higher molecular weight than bulk polymers which must be handled in solution; therefore, better strength retention at elevated temperatures can be expected, and the use of metal oxides can be omitted in some cases. MgO cannot be used in aqueous dispersions because it leads to coagulation.

The limits of CR-adhesives in the shoe industry[14] are clearly marked, although this industry still represents the largest market. The bonding of greasy uppers and especially of PVC materials is very problematic. Further disadvantages are the poor resistance against plasticisers (PVC), poor light resistance and the very complicated adhesive formulation. They are being replaced by polyurethanes to a certain extent and therefore, the market for alkyl phenol resins in adhesives is stagnant or even retrograde.

17.3.2. Nitrile-Phenolic Contact Adhesives

NBR adhesives do not have some of the disadvantages which limit the application of CR adhesives so that they can be considered as alternatives to urethanes. They offer good adhesion to PVC and good resistance against plasticisers. The oil and white spirit resistance is clearly higher according to the nitrile content. The open time and the adhesion to sole rubber is, however, not always adequate. Raw material costs are higher than for neoprene/phenolic cements. The phenolic resin addition improves tack, solvent release and hot strength.

Table 17.2. Guide formulation for a NBR-phenolic contact adhesive[10)]

NBR	(Perbunan 3810)	100
Phenol resol	(Bakelite KP 785)	40
Magnesium oxide		5
Methylethyl ketone		450

The adhesive formulation is clearly simpler (Table 17.2) compared to a neoprene type adhesive.

17.4. Phenolic Resins in Pressure-Sensitive Adhesives

Pressure-sensitive adhesives are permanent tacky materials which upon light pressure develop adhesive functions in contact with a second substrate. In contrast, the tack of contact adhesives is limited in time. Generally, an adhesive tape consists of a backing, tie coat, pressure-sensitive adhesive and in some cases a release liner. Backings of polyester or polyethylene foils possess poor adhesive properties so that the application of a coupling layer is indispensable. For soft PVC foils the priming has the additional function as a barrier to prevent migration of the plasticiser, because natural and neoprene rubber are not resistant against them. In general, the adhesive consists of a base resin, tackifiers, plasticisers, fillers and stabilisers. It can be applied as an emulsion or latex, solution or hot melt. Even if adhesives containing solvents are prevalent, hot melts offer favorable economic advantages, such as fast processing rate, no solvent costs, and a small working area because drying tunnels are not required[15].

In most cases, the base resin is rubber (SBR, BR, IR or EPDM), but different thermoplastic polymers, for example, vinyl polymers, acrylates or urethanes, can also be used. For optimum adhesive function, the key properties adhesion and tack, must be balanced out. Incremental addition of a tackifier to the base resin increases tack to a maximum and then with further addition the tack decreases. In this first stage a continuous phase of the base resin saturated with the tackifying resin is formed. By further addition, a second disperse phase with appropriate distribution of both polymers among each other is formed. Finally, a coat of tackifying resin forms at the surface.

Heat setting and thermoplastic phenolic resins may be used to provide building tack among hydrocarbon resins, rosin and cumaron resins[16]. The phenolic hydroxyl group has an important function, it is easily observed that the effectiveness of the resin decreases if the hydroxyl group is etherified. Resins with high or medium molecular weight are needed in minor quantities to obtain optimum tackiness rather than those with lower molecular weights, however, a wide distribution of molecular weight is necessary. As substitutent, the octyl group is more effective than the tert.-butyl group; the methyl group shows only a weak action.

Terpene-phenolic resins are reaction products of phenols with terpenes, e. g. limonene[17]. The alkylation is performed in acidic medium whereby sulfuric acid, *p*-toluene sulfonic acid, boron trifluoride or acidic ion exchangers are preferred as catalysts. These thermoplastic resins have also found application in coatings. Also phenol-acetylene resins, made from *p*-tert.-butylphenol and acetylene using mercury salts or organic bases as catalysts, are appropriate tackifiers for rubber (Koresin, BASF).

17.5. Rubber-Reinforcing Resins

Phenol novolaks in combination with HMTA in fine particle form can easily be incorporated into rubber blends[18]. During processing they function as a processing aid and reduce compound viscosity. At vulcanization, the cross-linking reaction of the novolak resin with HMTA leads to a second solid phase. No chemical reaction with rubber occurs in general, they act only as fillers. PF-resins with a P/F molar ratio of approximately 1 : (0.75–0.85) are used. The compatibility with rubber can be improved by resin modification with cashew nut shell liquid or tall oil. Reinforcing resins are recommended for SBR and NBR; the resin addition can amount to up to 50% or even more for NBR. Homogeneous (up to 100% addition) and transparent nitrile rubber compounds with high hardness and strength as well as oil resistance are obtained[16]. Phenolic reinforcing resins find application in shoe soles and heel top lift compounds and solid and pneumatic tires. In general, they impart increased hardness and stiffness, which is maintained even at elevated temperature, increased tear resistance and decreased elongation[19]. Growing use is expected in tire construction, particularly in the non dynamic areas of radial tire applications[20].

17.6. Resorcinol-Formaldehyde Latex Systems

Resorcinol-formaldehyde prepolymers[21,22] have been used in combination with vinyl pyridine copolymer latex for tire cord adhesion even since rayon replaced cotton in the early 1940s. They find application in bonding almost all currently employed tire reinforcing materials, including rayon, nylon, polyester, glass, aramide fibers and wire. Common bonding systems include silica/resorcinol prepolymer/HMTA, resorcinol prepolymer/HMTA and resorcinol resin/ methoxymethyl melamine resin combinations. A blocked, latent source of formaldehyde leads to single component resins which are stable at ambient working conditions. The curing rate is influenced by temperature and pH as shown in Figure 9.22, Chapter 9. The alkalinity of most rubber compounds is sufficient to affect complete resin cure at normal vulcanization times and temperatures of at least 145 °C[20].

References

1. Coleman, M. M., Shelton, J. R., Koening, J. L.: Ind. Engng. Chem. Prod. Res. Develop., Vol. *13,* 154 (1974)
2. Wildschut, A. J.: Rec. Trav. Chim. *61,* 898 (1942)
3. Hultzsch, K.: Chemie der Phenolharze: Berlin, Göttingen, Heidelberg: Springer 1950
4. van der Meer, S.: Rec. Trav. Chim. *63,* 147 (1944)
5. Giller, A.: Kautschuk u. Gummi-Kunststoffe *19,* 188 (1966)
6. Eby, L. T. Brown, H. P.: Thermosetting Adhesives. In: R. L. Patrick (ed.): Treatise on Adhesion and Adhesives, Vol. II. New York: Marcel Dekker 1969
7. Bayer AG: Baypren Chloroprenkautschuk. Technical Bulletin
8. Sender, H.: Kunstharz-Nachrichten (Hoechst AG) *6,* 40, (1970)

9. Schunck, E.: Kunstharz – Nachrichten (Hoechst AG) *4,* 33, (1973)
10. Bakelite GmbH: Alkylphenolharze für Polychloropren-Klebstoffe, Technical Bulletin
11. Garret, R. R., Lawrence, R. D.: Adhesives Age, October 1966, 22
12. Azrak, R. G., Barth, B. P.: Adhesives Age, June 1975, 23
13. Union Carbide Corp.: DE-AS 2 420 684
14. Meuser, H.: Adhäsion *12,* 332 (1975)
15. Mooncai, W. W.: Adhesives Age, October 1968, 28
16. Giller, A.: Gummi, Asbest, Kunststoffe *29,* 766 (1976)
17. Schmelzer, H.: Adhäsion *5,* 174 (1973)
18. Hoechst AG: DE-AS 1 769 456
19. Yurcick, P. A.: Rubber World, October 1976, 47
20. Hooker Chemicals: Durez Resins for the Rubber Industry, Technical Bulletin
21. Gils, G. E.: I & EC Product Research and Development, Vol. 7/2 151 (1968)
22. The General Tire & Rubber Co.: DE-AS 24 22 769 (1974)

18. Phenolic Antioxidants

Most plastics will be subjected to thermal or oxidative stresses during processing or lifetime which restricts their use temporarily and qualitatively. Oxygen and UV-radiation in combination have an even greater destructive effect. The oxidation of polymers by molecular oxygen is in most instances a free radical chain reaction, yielding a hydroperoxide as the primary product. Stabilizers can be effective as peroxide decomposers, metal deactivators, chain terminators and UV absorbers.

In contrast to aminic AO's, phenolic compounds do not cause any or only minor discoloration of rubber or thermoplastics[1,2,3].

Two peroxy radicals can be destroyed by one phenolic molecule yielding a 2,4-peroxycyclohexadienone which can be isolated easily at low temperatures[4].

OH O˙ O O

2 R–O–O˙ + ... → [... ⇌ ...] → ... + ROOH

OOR (18.1)

Evidence has been given that cyclohexanediones can be transformed back to the phenolic structure by dilaurylthiodipropionates, which are often used as efficient co-stabilizers in polyethylene and polypropylene.

The phenoxy radicals formed in the first stage are resonance stabilized by delocalization of the unpaired electron over the aromatic ring. They have a considerably longer lifetime than alkyl radicals and do not remove hydrogen from C–H bonds[5]. The lifetime of the non-substituted phenoxy radical was estimated to be about 10^{-3} seconds[6]. Further stabilization through steric hindrance is obtained by alkyl substitution. Some of these substituted aryloxy radicals are stable over periods of hours or even days. They are paramagnetic and can be highly colored in solution. In the solid state they can exist as colorless dimeric quinol ethers[5]. The relative rate of hydrogen abstraction from mononuclear phenols by peroxy radicals indicates a good correlation with Hammet's σ-values, but steric effects also have considerable influence on AO-efficiency.

O˙ Ph Ph 2 Ph ⇌ O Ph Ph Ph Ph O Ph Ph (18.2)

2,4,6-triphenylphenoxy radical
colored in solution, paramagnetic

dimeric quinol ether
solid, colourless and diamagnetic

For alkyl substituted phenols (18.3) maximum efficiency was found for:

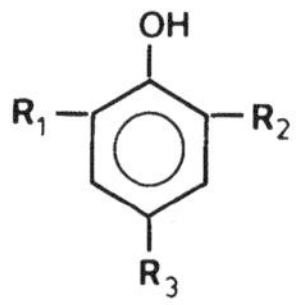

R_1 = methyl or tert.-butyl,
R_2 = tert.-butyl and
R_3 = methyl (18.3)

Bulky group in the *para* position decrease AO-activity.

This indicates that although steric hindrance of the phenolic hydroxyl group is required, too much hindrace renders the compound less effective[8]. Electron releasing groups (methyl, tert.-butyl, methoxy) in general increase the AO activity. Electron attracting substituents, such as nitro, carboxyl and halogen, have the opposite effect.

It has been shown that mononuclear phenolic compounds represent effective antioxidants. They have, however, certain disadvantages. Their vapour pressure is relatively high so that during the extrusion or molding operation at temperatures between 180–220 °C considerable quantities may be lost by vaporization. Many plastics are used as sheets and fibers and further loss may arise due to migration and volatilization of low molecular weight AO's at even lower temperatures. Two- and poly-nuclear phenols, however, have in addition to lower vapour pressure further advantages. They are able to bind metals under complex formation. Many metal ions, for example copper and manganese, catalyze the decomposition of hydroperoxides and accelerate the oxidative degradation. This catalytic activity of metals with variable valency is attributed to the formation of a coordinative complex with hydroperoxide followed by electron transfer between the hydroperoxide and the metal ion. Through competitive interaction with stronger chelating agents, the catalytic activity of the metal can be suppressed. Instead of formaldehyde, diolefins can be used for alkylation of phenols, for example dicyclopentadiene or limonene, in the presence of Lewis acids as catalysts.

Of the mononuclear phenols the 2,6-di-tert.-butyl-4-methylphenol, also called BHT, Ionol or Topanol O, is the best known compound[7]. It is obtained by reaction of isobutylene and *p*-cresol as a colorless crystalline powder with a MP of 70 °C. BHT is approved to be used in the food sector. Styrenated phenols, valuable rubber AO's, are obtained by arylation of phenol or *p*-cresol with styrene (18.4).

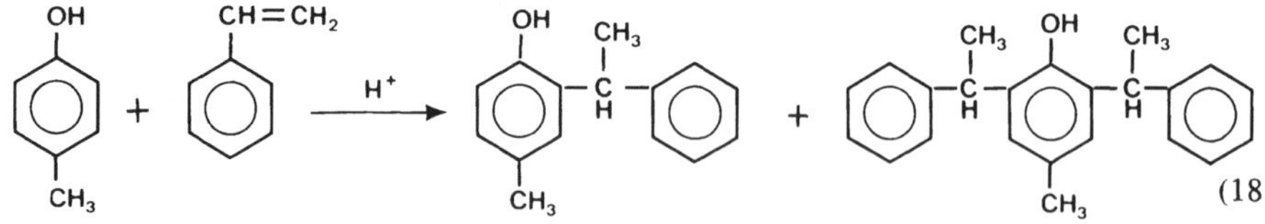

(18.4)

Styrenated phenols are liquid, this applies also to alkylated or terpene substituted phenols. Bisphenols or higher molecular weight compounds, on the other hand, are crystalline or grindable solids, have lower vapor pressure and higher migration stability.

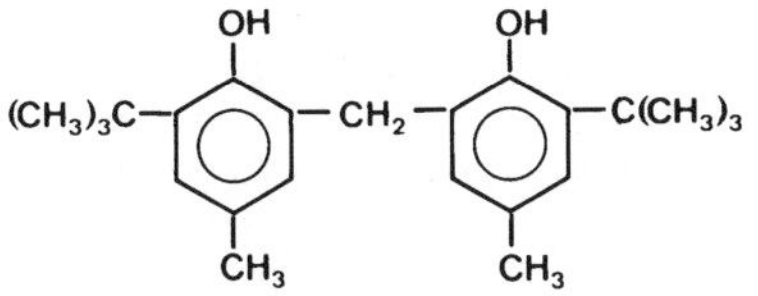

2,2'-methylene-bis-(4-methyl-6-tert.-butylphenol)
Orthophene OMB (Bakelite)
Anti-ageing agent BKF (Bayer)
Anti-oxidant 2246 (Cyanamide) (18.5)

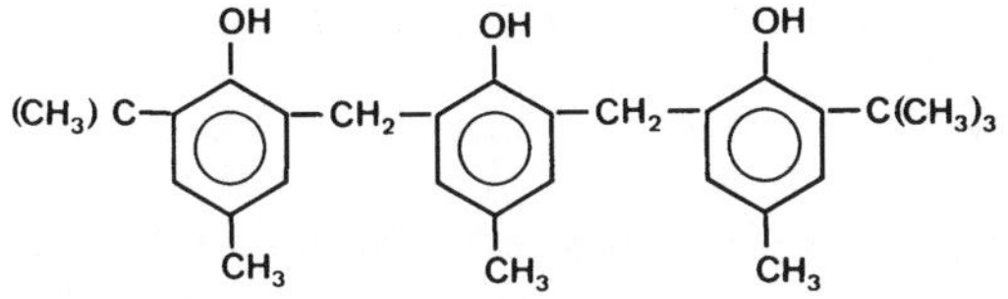

trisphenol (18.6)

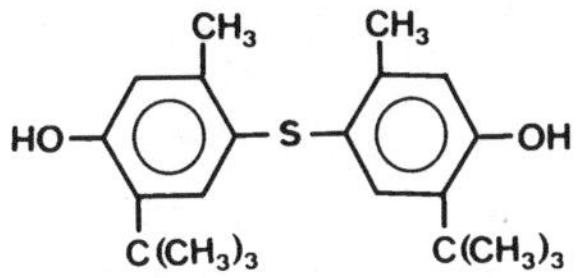

4,4'-thio-bis-(3-methyl-6-tert. butylphenol)
Rütenol (Chem. Fabrik Weyl) (18.7)

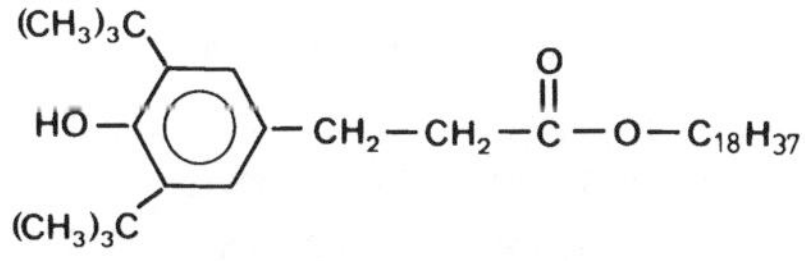

β-(3,5-di-tert.-butyl-4-hydroxyphenyl)-propionic acid-n-octadecylester, Irganox 1076 (Ciba-Geigy) (18.8)

Amines, e. g. N,N'-dialkylphenylenediamines, are effective stabilizers in rubber compounds containing carbon black. Conversely, hindered phenols are more effective than aromatic amines in rubber formulations without carbon black reinforcement. As a rule, phenolic AO's are used in polyolefins. Bisphenols and multivalent phenols are preferred because of their increased effectiveness. The use of synergistic mixes of phenols with organic sulfides especially β,β'-thiodipropionic acid esters, is wide spread.

References

1. Lloyd, D. G.: Kautschuk u. Gummi-Kunststoffe *27*, 477 (1974)
2. D'Ianni, J. D., Widmer, H.: Gummi, Asbest, Kunststoffe 9, 718 (1973)
3. Voigt, J.: Stabilisierung der Kunststoffe gegen Licht und Wärme. Berlin, Heidelberg, New York: Springer 1966
4. Rampley, D. N., Hasnip, J. A.: J. Oil Col. Chem. Assoc. *59*, 356 (1976)
5. Mihajlovič, M. L., Cekovič, Z.: Oxidation and Reduction of Phenols. In: S. Patai, (ed.): The Chemistry of the Hydroxyl Group. New York: Interscience 1971
6. Stone, T. J., Waters, W. A.: J. Chem. Soc. *213* (1964)
7. Kurze, W.: Antioxidantien. In: Ullmanns Encyclopädie d. techn. Chem. 4. Ed. Weinheim: Verlag Chemie 1976
8. Ohkatsu, Y., Haruna, T., Osa, T.: J. Macromol. Sci.-Chem., A11, 1975 (1977)

19. Other Applications

19.1. Carbon and Graphite Materials

Elemental carbon does not melt at normal pressure, is temperature resistant up to approximately 3,000 °C (sublimation temperature 3,650 °C) in inert atmosphere and has excellent resistance to high-corrosive materials. Carbon is resistant to phosphoric acid, hydrochloric acid, sulfuric acid, organic acids and also to aggressive gases like hydrogen chloride and sulfur dioxide. Graphite is corroded only by strongly oxidizing chemicals like nitric acid and chromic acid or by fluorine and sulfur vapor at high temperatures[1]. This explains the manifold fields of application in engineering[2], especially in the construction of chemical equipment (heat exchangers, reactors, columns and pipes), as lining or cladding material for blast furnaces, melting furnaces, foundry molds and high temperature insulation material. Further fields of application are moderators, reflectors, cladding for fuel rods in nuclear power equipment and matrices for carbon fiber composites.

There are several reasons for the application of phenolic resins in the "artificial" carbon and graphite technology. Phenolic resins are used as impregnating resins in order to increase the density, i.e. impermeability for gases[3], and the strength of the molded part, or as carbon precursor for the production of glassy carbon and carbon foams[4].

For special applications, phenol resins are used as temporary binders for the production of shaped parts instead of coal tar pitches. The shaped parts are either only cured or also carbonized afterwards. In comparison to coal tar pitch, phenolic bonded structures have higher strength and lower gas permeability[1, 2].

Ground and sized petroleum coke or pitch coke is coated with coal tar pitch (at 100–170 °C) or phenolic resins and the plastic material is extruded or molded. The preform is fired in a furnace protected from oxygen by a coke layer. The bonding agent is carbonized and adheres the carbon particles together to a solid material of micro crystalline structure and high porosity, due to the separation of gases during the pyrolysis of the binder, which is within the range of between 15–25%. Homogeneity and impermeability are obtained by repeated impregnation with either coal-tar pitch followed by a firing cycle or with phenol resins (Table 19.1). The cycle lasts approximately 20 days.

Carbon materials which are baked up to approximately 1,300 °C are called "artificial" carbons, treated up to 3,000 °C "artificial" graphites. Graphitized structures exhibit considerably higher thermal and electrical conductivity.

Glassy carbon which is obtained by carbonization of phenol resin shows an excellent impermeability to gases and resistance to chemicals. It has found numerous applications, particularly in laboratory equipment (tubes, crucibles). The phenol resin is hardened in appropriate molds and then carbonized in accordance with a

Fig. 19.1. Phenol resin bonded carbons (coke). (Photo: Sigri Elektrographit GmbH, D-8901 Meitingen)
dark = phenol resin; light = coke

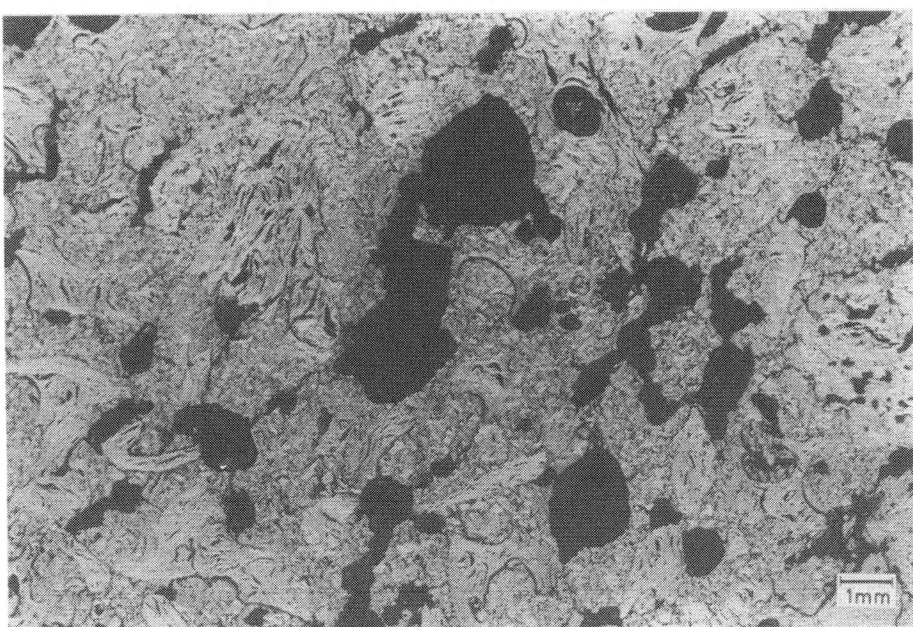

Fig. 19.2. Phenol resin impregnated graphite materials. (Photo: Sigri Elektrographit GmbH, D-8901 Meitingen)
dark = phenol resin; light = graphite

Table 19.1. Physical properties of phenol resin treated "artificial" carbon and graphite[1, 4)]

	Unit	PF-resin treated graphite	PF-resin bonded carbon
Density	g/cm^3	1.75	1.60
Flexural strength	N/mm^2	50	70
Compressive strength	N/mm^2	100	180
Modulus of elasticity	N/mm^2	$2.7 \cdot 10^4$	$3 \cdot 10^4$
Coefficient of thermal expansion K^{-1}		$2 \cdot 10^{-6}$	$3 \cdot 10^{-6}$
Permeability coefficient	cm^2/s	$1 \cdot 10^{-5}$	$5 \cdot 10^{-5}$
Thermal conductivity	W/m K	70	4.6

defined treating schedule at 800–3,000 °C. The carbon obtained is very hard, resembling to glass and stronger and stiffer than conventional carbon and graphites[4)].

Carbon foams are obtained by carbonizing foamed phenol resins. Such foams show a thermal resistance up to 3,000 °C in an inert gas or in vacuum, which is superior to that of refractory bricks. The most important applications of carbon foam are high temperature insulations, filters for corrosive agents and support materials for catalysts. Carbon-carbon composites[4, 5)], i.e. carbon fiber reinforced artificial carbons, have been developed for use in aerospace technology. Such composites do not show any reduction of strength up to 2,500 °C (in inert atmosphere). Apart from their use for rocket propulsion and reentry systems they are employed for airplane disc pads. Carbon brake linings show a very uniform braking effect and do not bind. Another field of application is machine construction. The carbon-carbon composites are produced from carbon fibers and a resinous matrix. As matrix and carbon precurser, phenol resins or coal tar pitch are used.

Liquid resols are preferred for these technologies (dry resin content of 75–85%); novolaks can also be used with HMTA. The impregnation with resols is performed by the application of vacuum and pressure.

19.2. Phenolics for Chemical Equipment

In the twenties, the Säureschutz Gesellschaft, Berlin, developed acid resistant molding materials for the manufacture of chemical equipment based on asbestos and phenol resins. This material was soon known under the trade name Haveg. Those materials, later also made in an alkali resistant form, are used nowadays in the chemical industry and comparable fields where they are subjected to high chemical stress and elevated temperatures[6, 7)]. The chemical properties of asbestos are described in Section 10.2.2. Anthophyllite and crocidolite[8)] are especially suited for these applications because they have a smaller percentage of acid soluble components and adequate fiber length. For special acid resistant performance, the asbestos is pretreated with hydrochloric acid in order to remove the basic parts, or carbon or graphite are used as fillers.

The alkali resistance of phenol resols can be improved by etherification of the phenolic hydroxyl group[9]. In addition to primary alcohols dichlorohydrin was especially recommended. At the most, one third of the phenolic hydroxyl groups are etherified so that the curing rate is not too much reduced. An addition of approximately 2% of MgO accelerates the curing rate and extent and improves the properties. Cresol resols with a high part of *m*-cresol are especially suitable. Further improvement of the chemical resistance of etherified resols is possible by a combination with furfuryl alcohol or furfuryl alcohol resins[10]. Excellent alkali resistance can be obtained with furane resins which are produced from furfuryl alcohol and formaldehyde[7].

Treated asbestos with a fiber length of 5–15 mm is impregnated with liquid phenolic resin and completely coated. A plastic, dough-like material is obtained which can easily be molded. According to the shape of the molded part, the material can be applied to levelled or vertical surfaces in a thin coat or can be stamped into wooden or metal molds. The optimum wall thickness is between 15–35 mm. The material can be applied in several layers, always after intermediate curing, in order to obtain the required wall thickness. In pipe manufacturing the material is wrapped around a steel plug, pressed with rolls and smoothed. Manufacturing of pipes by extrusion is not possible because it causes fiber orientation and thereby anisotropic strength. Smaller molded parts in larger numbers of pieces are manufactured in steel molds by compression molding. Larger shaped parts are cured in an autoclave at a pressure of 6–8 bar and at slowly rising temperature up to about 130–140 °C. The curing cycle lasts approximately 12 hours. The cooling must also be performed slowly.

The cured parts can be mechanically treated (for instance by grinding, drilling, sawing, turning and milling) so that close tolerances can be maintained. Individual parts are glued together with acid-hardening phenolic resin putties to make larger equipment. Putties based on resols harden at 20 °C within two hours to such an extent that the part can safely be handled. The surface of the molded part is, in most cases, painted with a phenolic resin varnish. Examples for this application are vessels, vessel covers, pipes, pumps and column trays.

Molding materials produced with phenol- and cresol resins are completely resistant to all nonoxidizing inorganic and organic acids of all concentrations, to salt solutions as well as to most organic solvents at temperatures up to 140 °C. They have no or limited resistance to strong alkalis and all oxidizing compounds like nitric acid, chromic acid, sulfuric acid with a concentration of more than 70%, chlorine in aqueous

Table 19.2. Physical properties of asbestos molding materials for chemical equipment

Specific gravity	g/cm^3	1.6–1.8
Compression strength	N/mm^2	70–80
Tensile strength	N/mm^2	15–20
Flexural strength, at 20 °C	N/mm^2	40–60
Flexural strength, at 150 °C	N/mm^2	28–40
Modulus of elasticity	N/mm^2	$8–10 \cdot 10^3$
Coefficient of thermal expansion	K^{-1}	$11–16 \cdot 10^{-6}$
Heat resistance (Martens)	°C	170–200

Fig. 19.3. Activated carbon exhaust air treating and solvent recovering unit, manufactured of corrosion resistant phenolic resin/asbestos molding material. (Photo: L. Siegrist GmbH, D-7500 Karlsruhe)

solution, alkaline hypochlorite solutions, sodium chlorite, sodium chlorate, sodium perchlorate and potassium permanganate. Minor resistance can also be observed towards strongly polar solvents like acetone and methylene chloride[1].

19.3. Phenolic Resin/Fiber Composites

High temperature resistance, high strength and low smoke generation in case of fire are the main reasons for the use of phenolic resin composites in the aircraft industry and for hypersonic missile application. Glass fabrics, Kevlar fabric and carbon fibers are used as reinforcement for the fabrication of complex shapes in the 250 °C structural range[13]. Room temperature curing laminating resins are formulated from high-solids aqueous resols and strong inorganic or preferably organic acids which may be integrated into the macromolecular backbone, thus preventing corrosion problems. However, considerably higher strength is obtained with heat cured resols. MIL-R-9299 C covers two grades of phenolic resins which may be supplied in liquid form or as prepreg[14]. Grade A is specified as having normal properties and grade B as having improved properties. Similar specifications are published[15] in West Germany by Normenstelle Luftfahrt (WL 8.4361.6).

Prepregs made of thermoplastic resin modified phenolics exhibiting adhesive properties are used to some extent for interior aircraft secondary structures because of their low smoke generation. They are cured at 130 °C.

Table 19.3. Minimum strength values of phenolic resin laminates according to MIL-R-9299 C. Glass fabric 181 style, 37% resin content

		Phenolic MIL-R-9299 C Grade A	Grade B	Epoxy MIL-9300 B Structural
Flexural strength	N/mm^2	350	510	525
Tensile strength	N/mm^2	280	320	330
Compression strength	N/mm^2	245	400	350
Modulus of elasticity	N/mm^2	21,000	24,500	22,500
Interlaminar shear strength	N/mm^2	20[1]	20[1]	28[2]

[1]WL 8.4361.6.
[2]WL 8.4321.6.

Asbestos fibers are superior in high temperature performance and rigidity in comparison to E-glass fibers (Table 10.8). Asbestos composites are used in high-performance missile systems operating at hypersonic speeds and heat shields in reentry vehicles (see Chapter 7). Their ablative performance is superior to carbon fiber composites.

19.4. Phenolic Resin Fibers

The phenolic fiber production process was developed by Carborundum, USA[16)]. The technical production was started in Japan by the Nippon Kynol Inc. It appears that present production (Mitsubishi Chemical) does not exceed 300 tons per year. The most important advantage of phenolic resin fibers is their flame resistance. The non-flexibilized fiber does not burn, it is carbonized in the flame retaining its form and a certain strength. The smoke density is very low. For these reasons, phenolic fibers have been recommended for the manufacturing of fire fighter suits and curtains for auditoriums, hotels and meeting rooms. Their temperature resistance also is remarkable, but still inferior in comparison to aromatic polyamides. No reduction in weight occurs at temperatures below or around 150 °C; however the tensile strength and elasticity are considerably reduced[17)].

The production of phenol fibers of novolaks with a MW of between 800–1,000 and very low free phenol content (0.1%), is performed in a melt spinning process. The continuous yarns are cured in acidic, aqueous formaldehyde solutions at temperatures between 85–100 °C for several hours. The fiber forming ability of novolaks is poor; they are therefore modified with polyamides, polyesters or other thermoplasts[18, 19)]. The flame retardancy, however, is decreased by the modification. The essential disadvantage of phenolic fibers is their poor abrasion resistance and brittleness; they are hardly suitable for processing on high-efficiency weaving machines. The natural yellow-brown color and limited light fastness restricts the dyeing of the fiber to dark shades like dark red, brown, blue and black. Light fastness can be improved by acetylating of the phenolic hydroxyl group.

19.5. Blast Furnace Taphole Mixes

Up to now, mainly tar bonded mixes were used for blast furnace tapholes. These materials consist of finely granulated Al_2O_3, SiO_2, SiC and coke and the binder coal tar or pitch[20]. Since blast furnaces have become larger and are operated under high top pressure, reliable taphole mix performance has become increasingly important. Conventional tar bonded mixes are relatively weak in resistance to heat and erosion, give off black smoke, and take a long time to harden. In order to avoid these disadvantages, phenolic resin bonded mixes were developed[21, 22]. However, the new materials are two to three times more expensive than tar bonded mixes. The early strength in the lower and middle temperature range is guaranteed by the resin bond. A relatively strong, abrasion resistant vitreous carbon bond is formed at higher temperatures due to the carbonization reaction. Since with these materials a high strength is obtained in a short period, the taphole can be stuffed up without problems, and the stuffing machine moved back after about 10 minutes.

The material possesses high hot compression strength and erosion resistance and low smoke density. A phenolic resin bonded material[22] contains 15–25% clay, 25–35% chamotte, 15–30% silicon carbide, 15–25% coke and approximately 12–18% bonding agent which consists of a phenol novolak (MW 400–800), HMTA and a plasticizing solvent containing hydroxyl groups (diethylene glycol, glycerin).

19.6. Photo-Resists

Positive resists used in microelectronics and manufacturing of printing plates for relief printing[23] consist of a cresol novolak[24] and a sensitiser of the series 5-substituted diazonaphthoquinones and appropriate solvents. The unexposed positive

SO_3H — σ-Diazoquinone $\xrightarrow{h\nu}$ [SO_3H] + N_2 ⟶ (19.1)

Ketene (SO_3H) $\xrightarrow{+H_2O}$ Indene-1-carboxylic acid (SO_3H) (19.2)

lacquer is not soluble in buffered aqueous alkali solutions. When exposed to UV-rays, the *o*-diazonaphthoquinone is converted to an indene carboxylic acid over a ketene as intermediate[25].

Due to the carboxyl and phenolic hydroxyl groups now present, the exposed coating is soluble in aqueous NaOH. The insolubility of the unexposed lacquer is attributed to an azo coupling between phenol polymer and sensitiser[25] of perhaps to a strong hydrogen bond between the phenolic hydroxyl and the carbonyl group. Cresol resins with a high portion of *m*-cresol and very exactly specified MWD are used; low as well as very high molecular portions reduce the resolution of the system.

A further application for PF-resins in the printing field is manufacturing of printing plates, called flong. Sheets of pure sulfate cellulose are impregnated with a novolak/HMTA solution. One side is coated with a suspension of a pulverized phenolic/wood flour molding compound. Sheet and print picture forming is performed in a heated mold. This process seems to be of limited importance today.

19.7. Socket Putties

Phenol resins are used as socket putties for light bulbs, radio valves, and the like because of their high temperature resistance and strong adhesion. The putties consist of a powdery mixture of fast curing phenol novolak/HMTA resin and mineral fillers. The part of phenol resin is 12–15%. They are mixed to a paste with 10 pbw of a solvent like ethanol or isopropanol. This paste is filled into the hopper of a socket machine and sprayed into the metal jacket by a metering device. The glass bulb is then placed on the jacket under spring pressure and the putty hardened in a tunnel oven or carousel at temperatures of 180–200 °C or more.

19.8. Brush Putties

The brush manufacturing industry changed over to phenolic resins to fasten the hairs and bristles because of their water and solvent resistance some time ago. High-solids (80%) phenol resols, differing in viscosity and reactivity, are used in combination with inorganic acids as hardening agents. The acids are diluted with alcohols or glycol. The formulation of the putty almost always includes silica flour. Cure is effected either at 90–95 °C in an oven within 6–8 hours or at room temperature within 2 days.

19.9. Tannins

Synthetic organic tannins (syntanes) are used for tanning either alone or more frequently in combination with synthetic inorganic tannins, plant tannins or those from sulfite cellulose waste lye. They are divided into pre-, post-, exchange-, alum-, bleach-, shrink- and combination tannins according to their application. Depending upon their chemical constitution, they have a more or less strong (synthetic replacement tannins) or no self-tanning effect (auxiliary tannins)[27].

Synthetic aromatic tannins are condensation products of aryl-, mono- or polynuclear phenol-sulfonic acids or carboxylic acids with formaldehyde or other carbonyl compounds, whereby further aromatic compounds can be incorporated (for instance amines, carbonamides, polynuclear aromatic compounds, lignin-sulfonic acids).

The tanning effect of phenols and PF-resins is due to the ability of the phenolic hydroxyl group to form strong hydrogen bonds with the polypeptide groups and thereby have a cross-linking effect upon the collagen micellae[28]. Reactive polymers polymerize in the cutaneous structure and, after the displacement of water, lead to the isolation of the collagen fibrils and thereby to the irreversible change of the skin[29]. Condensation products of sulfonated phenols and formaldehyde were recommended for the first time as tanning substances by Stiasny[30]. The sulfonation (and neutralization) is necessary in order to obtain the water solubility required for technical tanning processes. The sulfomethylation of phenols with sodium sulfite and formaldehyde (19.3) or later sulfonation of novolaks (19.4) are frequently applied reactions[31, 32]. Naphthols and resorcinol are also recommended[33].

$$C_6H_5OH + CH_2O + Na_2SO_3 \longrightarrow HOC_6H_4{-}CH_2{-}SO_3Na + NaOH \xrightarrow{CH_2O} {-}CH_2{-}C_6H_2(OH)(CH_2{-})(CH_2{-}SO_3Na) \quad (19.3)$$

$$\text{Sulfonated novolak: } {-}C_6H_2(OH)(SO_3H){-}CH_2{-}C_6H_2(OH)(SO_3H){-}CH_2{-}C_6H_2(OH)(SO_3H){-}OH \quad (19.4)$$

Resins similar in structure are also used for ion-exchange resins and concrete additives.

19.10. Ion-Exchange-Resins

By incorporation of different active groups, such as $-SO_3H$, $-COOH$, in the phenolic resin backbone additional ion-exchange abilities are obtained[33, 34]. The chelating ability of non-modified phenolic polymers is indicated in Section 3.4.3. The high resistance to water, solvents and usual regenerating agents (diluted acids and bases) is the advantage of cured phenolic resins as matrix for functional groups with exchanging abilities. The desired grain shape is obtained by grinding and screening (first generation resins) or by a suspension condensation process in a reversed-phase system[34]. However, polycondensation polymers have been replaced by polymerization polymers to a great extent in this field of application.

19.11. Casting Resins

Casting-type resins which are used for the production of decorative articles, bowling balls and the like are prepared using a high formaldehyde excess (P/F ratio 1 : 2–3) with sodium hydroxyde or alkaline earth hydroxides as catalysts[3]. After completed reaction the water is removed as far as possible and diluents, e.g. ethylene glycol, polyethylene glycols or glycerol may be added to reduce the viscosity and enhance the flexibility of the casting. The pH is generally adjusted to about 5–7 with weak organic acids. Fillers and pigments may be added. The resin is poured or cast into open molds and maintained below the boiling point of water. Usually a period of 70–200 hrs is required at temperatures between 75–95 °C.

The curing temperature can be reduced to ambient temperature by the addition of organic acids, e.g. lactic acid, phenoxyacetic or sulfamic acid. However, post curing at temperatures between 40 and 50 °C is necessary. The light transparency depends upon the choice of the acidifying agent[3]. Lactic acid for example gives opaque castings while phenoxyacetic acid gives transparent products.

References

1. Fitzer, E., Heym, M.: Chem. Ind. *29,* 527 (1977)
2. Winnacker, K., Küchler, L.: Chemische Technologie, Vol. 1, München: Carl Hanser 1960
3. Brandmeier F.: Werkstoffe u. Korrosion, 17, 10 (1966)
4. Vohler, O., Reiser, P. L., Overhoff, D., Martina, R.: Angew. Chem., Int. Ed. *9,* 414 (1970); Sigri Elektrographit GmbH: Diabon/Durabon R, Technical Bulletin
5. Fitzer, E., Terwiesch, B.: Werkstofftechnik/J. Materials Technology *5/2,* 53 (1974)
6. Dr. C. Otto & Co. GmbH: Haveg, Technical Bulletin
7. Kessler, G., Schacht, E.: Rohre und Apparaturen aus Phenol-Formaldehyd-Harzmassen. In: Haus der Technik-Vortragsveranstaltungen *287,* 31 (1972)
8. Barton, H. D.: Ind. Eng. Chem. *56,* 16 (1964)
9. Campbell, N. R.: Werkstoffe-Korrosion *22,* 219 (1971)
10. Röbe-Oltmanns, G.: DE-OS 1965165 (1969)
11. Hefele, J.: DE-OS 1494099 (1962)
12. Bakelite GmbH: Harze für den Apparate- und Behälterbau, Technical Bulletin
13. Lubin, G. (ed.): Handbook of Fiberglass and Advanced Composite Materials. New York: Van Nostrand Reinhold Co. 1969
14. Edwards, V.: Composites, May 1974, 122
15. Werkstoffleistungsblatt WL 8.4361.6 (Jan. 1978), Köln: Beuth-Verlag
16. Economy, J., Wohren, L., Frechette, F.: Kynol-A New Flame Resistant Fiber, 39th Annual Meeting Textile Research Institute, New York, April 1969
17. Fasern, einzelne organische. In: Ullmanns Encyclopädie d. Techn. Chem., Vol. 11, P. 339, 4. Ed., Verlag Chemie: Weinheim (1976)
18. Carborundum Co.: DE-OS 1910419
19. Nippon Kynol Inc.: DE-OS 2328313
20. Rütgerswerke AG: DE-PS 2021469, DE-OS 1960461, DE-OS 1960462
21. Bové, F.: DE-OS 2424936
22. Nippon Steel Corp.: DE-OS 2624288
23. Kosar, J.: Light Sensitive Systems, New York: Wiley 1965
24. Howson-Algraphy Ltd.: DE-OS 1809248
25. Maas, K. (ed.): Themen zur Chemie der Reproduktionsverfahren, Heidelberg: Hüthig 1964
26. Rosshaupter, E., Hundt, D.: Chemie unserer Zeit, 1971, P. 147
27. Scheer, W.: Zur Nomenklatur der Textilhilfsmittel, Leder und Pelzhilfsmittel, Papierhilfsmittel, Gerbstoffe und Waschrohstoffe. Verband der Gerbstoffindustrie e.V., Frankfurt (1970), Limburger Vereinsdruckerei GmbH

28. Faber, K., Komarek, E.: Leder und Gerbung. In: Ullmanns Encyclopädie d. techn. Chem., Vol. 11, 3. Ed., P. 595, München: Urban und Schwarzenberg
29. Stather, F.: Gerbereichemie und Gerbereitechnologie. Berlin: Akademie Verlag 1967
30. Stiasny, E.: DE-PS 262558 (1911)
31. Küntzel, A., Plapper, J.: Leder *6*, 176 (1955); *7*, 60 (1956)
32. Noerr, H., Mauthe, G.: DE-PS 693923
33. Wegler, R., Herlinger H.: Polyaddition u. -kondensation v. Carbonyl- u. Thiocarbonylverbindungen. In Houben-Weyl: Methoden d. Org. Chem., Vol. XIV/2, Thieme: Stuttgart 1963
34. Unitaka Ltd.: DE-OS 2403158 (1974)
35. Helfferich, F., et al.: Ionenaustauscher. In: Ullmanns Encyclopädie d. techn. Chem., Vol. 13, 4. Ed., P. 299, Weinheim: Verlag Chemie 1977
36. Sandler S. R., Karo, W : Polymer Syntheses, Vol. II, P. 57, New York: Academic Press 1977
37. Harris, T. G., Neville, H. A.: J; Polymer Sci. *4*, 673 (1953)

The life teaches, not books.
They help to understand.

Subject Index

Zeitfracht Medien GmbH
Ferdinand-Jühlke-Straße 7
99095 Erfurt, Deutschland
produktsicherheit@kolibri360.de